D0535886

DESIGN AND ANALYSIS OF EXPERIMENTS
in the Animal and Medical Sciences

VOLUME 2

DESIGN AND ANALYSIS
in the Animal

THE IOWA STATE UNIVERSITY

OF EXPERIMENTS
and Medical Sciences

JOHN L. GILL

VOLUME 2

PRESS / *Ames, Iowa, U.S.A.*

John L. Gill is professor of biometry in the Department of Dairy Science at Michigan State University. He received the M.S. and Ph.D. degrees in animal genetics and statistics at Iowa State University. Besides this book, he is author of many articles in the journals of his fields, and he has had long experience as a teacher of applied statistics and statistical consultant in the biological sciences.

Printed by
The Iowa State University Press
Ames, Iowa 50010

First edition, 1978

Library of Congress Cataloging in Publication Data

Gill, John L.
Design and analysis of experiments in the animal and medical sciences.

Bibliography: p.
Includes index.
1. Zoology, Experimental—Statistical methods. 2. Medicine, Experimental—Statistical methods. I. Title. [DNLM: 1. Research design. 2. Models, Theoretical. W20.5 G475d]
QH323.5.G549 591'.01'82 78-8516
ISBN 0-8138-0060-9

CONTENTS

CONTENTS FOR VOLUME 2

CONTENTS FOR VOLUME 1

CONTENTS FOR VOLUME 3

PREFACE

The general plan and motivation for this text are sketched in the preface to Vol. 1. Readers may profit from reviewing the philosophy of Sec. 1.2 on orderly planning of an experiment. In Vol. 2, emphasis is turned from analysis of data to design of experiments. However, after one absorbs some details about the merits, limitations, and structure of a particular design, it is necessary to return to analysis of data accrued from use of the design. In analysis one usually encounters only minor extensions of the various procedures of analysis of variance discussed in Vol. 1. I have chosen analytical descriptions suitable for desk or hand calculators for two reasons: (1) many small sets of data may be manipulated just as easily by hand calculation as by programmed computer (all necessary efforts considered), and (2) one is less likely to miss erratic behavior of data and underlying biological phenomena when results are handled personally. In many practical situations efficient computation is served best by using the various library programs for large computers. The purposes for which such programs were intended, their limitations, and the format for input of data and output of results should be investigated thoroughly, lest one's efforts be reduced to the level of "garbage in-garbage out." The details of proper analysis of results are as important as the details of the laboratory procedures that produced or refined them. Too often, scientists neglect this matter, sometimes with regard to personal assurance of correctness but even more often with respect to reporting the procedures in sufficient detail that the reader could duplicate the analysis.

A major hurdle will arise for the student in the study of designs. It will become increasingly necessary to enlarge one's horizons of thought to encompass a variety of designs that may be "competitors" in reducing error for a specific experiment. One should cultivate the habit of weighing the relative merits of different designs against the particular economic and physical limitations imposed by the experimental situation. Indeed, some sophisticated designs are required only when practical restrictions dictate that other, less complex designs are not feasible.

The volume begins (Chap. 5) with the examination of the simplest class of designs, randomized complete blocks, which often are nearly ideal for the purpose of reducing error arising largely from innate differences among subjects *if* the number of treatments or treatment combinations is not large.

If the number of treatments is large; if the experimentation is very costly in subjects, facilities, or time; or if physical restrictions are severe, one may find that the necessary blocking (grouping) of like subjects can be accomplished best (or only) by the use of *incomplete* blocks that contain fewer subjects each than the number of treatments in the experiment (see Chap. 6).

Experiments that may be made more efficient by two-way elimination of heterogeneity (e.g., natural differences among litters of animals and differences in weight among animals of the same litter) may be performed in Latin squares or designs of similar structure (Chap. 7).

A class of experiments being used more frequently involves nonrandom repeated measurement of subjects. The experiments are of two general forms: (1) exposure of each subject to a sequence of treatments and (2) analysis of trend in response over time to a single treatment on a given subject. Although the structural relationships of factors affecting the results appear to be the same as in other designs, rather special analytical problems arise from intercorrelations among the data. These problems are addressed from univariate and multivariate points of view in Chap. 8.

The analytical and design problems associated with response surfaces resulting from application of two or more quantitative treatment factors are discussed in limited detail in Chap. 9, and a classification of experimental designs and a key for selecting a design are included at the end of this volume.

Necessary statistical tables and charts, a glossary of symbols, and answers to odd-numbered exercises are bound separately as Vol. 3.

I am indebted to the literary executor of the late Sir Ronald A. Fisher, F.R.S.; to Dr. Frank Yates, F.R.S.; and to Oliver & Boyd, Edinburgh, for permission to reprint as Table A.7.1 part of Table 23 from their book, *Statistical Tables for Biological, Agricultural, and Medical Research.* I also thank the management of the Rand Corporation (Santa Monica, California); CIBA-GEIGY Ltd. (Basel); Aerospace Research Laboratories of the U.S. Air Force (Dayton, Ohio); Prentice-Hall (Englewood Cliffs, N.J.); Cambridge University Press; Iowa State University Press (Ames); the editors of the *Annals of Mathematical Statistics, Biometrics, Biometrika, Journal of the American Statistical Association, Journal of Dairy Science,* and *Technometrics*; and Glenn Beck, A. H. Bowker, G. E. P. Box, I. D. J. Bross, H. A. David, C. M. Dayton, A. L. Dudycha, C. W. Dunnett, F. E. Grubbs, H. L. Harter, N. A. Hartmann, R. B. Howe, J. S. Hunter, D. R. Jensen, G. J. Lieberman, R. K. Lohrding, Bernard Ostle, C. N. Park, E. S. Pearson, W. D. Schafer, S. S. Shapiro, F. J. Wall, and M. B. Wilk for permission to reproduce various tables or charts published or produced by them. Specific acknowledgments are given as footnotes to the appropriate tables and charts in the text and in App. A of Vol. 3.

The patience and precision of Mrs. Linda Wilson, who typed several drafts of the manuscript, are much appreciated. Many others have given help in various ways. I am grateful for their contributions.

Inevitably, errors of various kinds will be found. For those I accept sole responsibility, and I invite correspondence concerning them.

DESIGN AND ANALYSIS OF EXPERIMENTS
in the Animal and Medical Sciences

VOLUME 2

CHAPTER 5

Randomized Complete Block Designs

5.1. NUISANCE VARIABLES; ASSUMPTIONS

An experiment in a randomized block design requires prior knowledge of the existence of one or more nuisance variables that influence the trait of primary interest but cannot be physically controlled or stabilized for the experiment.

The distinction between blocking factors and treatment factors, noted in Sec. 1.2.3, in some cases depends on the stage of experimentation or "state of the art" in a given field of inquiry. For example, some classification factors, such as groups of animals based on body weight, may initially have been considered "treatment" factors when their influence on a new trait was of primary interest; later, after their influence was clearly established, they became nuisance factors to be recognized and accounted for when experimenting with the influence of new treatments on the same trait.

The principle of maintaining a broad inferential base is discussed in Sec. 1.2.4. That is, one may apply the blocking principle of design to circumvent the effects of nuisance variables (sex, age, weight, strain, etc.) while maintaining sufficient diversity of experimental units or subjects, to represent properly the population of interest.

In Sec. 1.2.5, the stratification of subjects into homogeneous groups (blocking) is discussed as one means of reducing experimental error. As seen in Chap. 3, block designs and the analysis of covariance often are alternatives open as ways devised to minimize error. Blocking in the animal and medical sciences commonly is based on the stratification of subjects into groups prior to the application of treatments, in such a manner that individuals within one group are relatively homogeneous with respect to one or more nuisance variables characteristic of the subjects. Blocking also may be based on nuisance variables external to the subjects, such as the time or place in which data are obtained. In such cases, a *complete* block (exactly one subject for each treatment) often is called a *replicate*.

The use of blocking to reduce experimental error is especially important when precise estimates of treatment means cannot be obtained by using large numbers of subjects because of cost or limited facilities. When the number of treatments or treatment combinations is large, it may not be possible to find enough homogeneous subjects to form complete blocks. In such cases, various *incomplete* block designs may be valuable (Chap. 6). Double blocking, based on stratification of subjects by two nuisance criteria (two-way elimination of heterogeneity) also may be more effective in some instances. Latin squares and other designs that utilize double blocking are discussed in Chap. 7 and Sec. 8.1. A distinctive form of blocking, often very effective in reducing experimental error, is the use of a subject as a block--several nonrandom observations on each subject represent classes of a factor of interest, such as time. Special requirements for analyzing this important class of designs are discussed in Chap. 8.

The main goal in all blocking activity is the reduction of experimental error, accomplished by partitioning variation caused by differences between blocks from the total sum of squares so that such variation does not fall into the usual residual or error term.

In all designs that utilize blocking, a critical assumption involved in the analysis of resulting data is that effects of the nuisance variable(s) do not interact with effects of treatments. For example, if the nuisance variable is body weight, an interaction exists if the array of true mean differences among treatments is not the same for animals or subjects in different weight classes. Violation of the assumption results in an inflated estimate of error and therefore inaccurate interval estimation and testing of treatment effects. When an experimenter is unsure about assuming that a blocking factor is independent of the treatments proposed, the blocking factor should be designed into the experiment as an additional "treatment" factor (as in a completely randomized design with more than one subject per combination), even though the interest lies in the potential interaction with treatments rather than in the main effects of the blocking factor. John (1963) developed designs in which it is not necessary to replicate every combination. If it becomes reasonably clear that a particular nuisance variable does not interact more than trivially with a class of treatments of interest, then in future experiments one may safely utilize the nuisance variable as a blocking factor (provide only one subject per block-treatment combination). When for physical reasons the number of subjects or units per block is necessarily larger than the number of treatments, one may use *extended* complete block designs generated from the balanced incomplete block designs of Chap. 6 (Trail and Weeks 1973).

5.2. RESTRICTED RANDOMIZATION; RANDOMIZATION ERROR

Blocking constitutes a restriction on the randomization of assignment of subjects or units to treatment groups. In experiments with complete blocks, the usual (and most effective) procedure is to assign to each block as many subjects (similar with respect to one or more nuisance characteristics) as there are treatments or treatment combinations. Then *within* each block the subjects are assigned randomly one to each treatment. The importance of actually carrying out this randomization should not be overlooked; blocking and randomization are not maneuvers that take place in the statistician's office after the data have been collected.

Because of the separate randomization of subjects to treatments within each block, Anderson (1970) has argued for a recognition in the model of randomization or restriction error. The suggestion is analogous to the situation in factorial experiments with only one observation per treatment combination, in which the highest ordered interaction is completely confounded with experimental error but both terms are written into the model [e.g., (2.123)]. To include inestimable terms in the model means that variance components for such effects are represented in expected mean squares derived from the model (Sec. 2.4.2), forcing the experimenter to recognize and account for a more realistic structure of the sources of variation in constructing proper tests of hypotheses or estimators of components of variance. At the very least, one becomes aware of potential biases, even if they cannot be avoided.

Little danger exists in ignoring the restriction error for simple randomized complete block designs (RCBD) in which one is not interested in comparing block means. However, occasionally experimenters wish to randomize the levels of one factor within the levels of another and still be able to test both factors. Such an experiment is not a completely randomized design (CRD) as commonly supposed, but the structure of randomization and analysis are the same as for a complete block design with several subjects per block-treatment combination. For example, suppose the subjects to be treated may be classified a priori according to some group status of interest, such as strain, breed, race, etc. If subjects of each group are randomly assigned to the treatments and one wishes to compare group means as well as treatment means, the randomization error is important in a realistic appraisal of the validity of the procedure for testing group means.

5.3. ESTIMATION AND TESTING

5.3.1. Analysis of Variance

Consider a model for a randomized complete block design with r blocks of t experimental units or subjects each, i.e., one for each of the t treatments:

$$Y_{ij} = \mu + \tau_i + \delta_j + J_j + (\tau\delta)_{ij} + E_{(ij)} \quad (i = 1, 2, \ldots, t;\ j = 1, 2, \ldots, r) \tag{5.1}$$

where τ_i is the fixed effect of the ith treatment, δ_j is the fixed effect of the jth block (D_j if random), J_j is the random effect of the randomization or restriction error in the jth block (distributed normally with mean zero and component of variance σ_J^2), $(\tau\delta)_{ij}$ is the interaction of the effects of treatments and blocks, and $E_{(ij)}$ is the random experimental error or sum of unnamed effects peculiar to each experimental unit (distributed normally with mean zero and component of variance σ^2). At this point it should be obvious that δ_j and J_j are completely confounded (not separable) as are $(\tau\delta)_{ij}$ and $E_{(ij)}$.

The rules of Sec. 2.4.2 for obtaining sums of squares and expected mean squares are also applicable to this model, which has the basic properties of a factorial model with only one observation per treatment combination. The total sum of squares (ss_y) may be partitioned into sums of squares for treatments (ss_T), blocks (ss_D), and error (ss_E) according to

$$ss_T = \left(\sum_{i=1}^{t} y_{i.}^2/r\right) - (y_{..}^2/n) \quad (n = tr) \tag{5.2}$$

$$ss_D = \left(\sum_{j=1}^{r} y_{.j}^2/t\right) - (y_{..}^2/n) \tag{5.3}$$

$$ss_E = ss_y - ss_T - ss_D \tag{5.4}$$

A summary of the analysis of variance, including expected mean squares, is shown in Table 5.1.

Table 5.1. Analysis of Variance for RCBD

Source of Variation	df	ss	ms	$E[MS]$
Treatments	$t-1$	ss_T	ms_T	$\sigma^2+r\sum_{i=1}^{t}\tau_i^2/(t-1)$
Blocks	$r-1$	ss_D	ms_D	$\sigma^2+t\sigma_J^2+t\sum_{j=1}^{r}\delta_j^2/(r-1)$
Error	$(t-1)(r-1)$	ss_E	ms_E	$\sigma^2+\sum_{i=1}^{t}\sum_{j=1}^{r}(\tau\delta)_{ij}^2/(t-1)(r-1)$

Perusal of the expected mean squares should clarify why it is useful to include J_j and $(\tau\delta)_{ij}$ in the model (although they are customarily ignored). First, ms_E clearly is an appropriate measure of experimental error for use in making interval estimates of and testing hypotheses about treatment means only if no interaction exists between the effects of treatments and blocks, i.e., $(\tau\delta)_{ij}$ is negligible. Second, because of σ_J^2 no unbiased test of the equality of block means exists. Granted, such a test may be of little interest as the significance of block effects usually is assumed a priori in designing the experiment.

If the $(\tau\delta)_{ij}$ are negligible, $\sqrt{ms_E/r}$ represents an accurate appraisal of the standard error of estimated treatment means $\bar{y}_{i.}$ and may be used in confidence intervals (CI) for the parameters $(\mu + \tau_i)$:

$$(1 - \alpha)100\%(\text{CI})(\mu + \tau_i) = \bar{y}_{i.} \pm t_{\alpha/2,\nu_E}\sqrt{ms_E/r} \tag{5.5}$$

where $\nu_E = (t - 1)(r - 1)$.

Also, for fixed block effects, the test of $H{:}\tau_i = 0$ (for all i) is unbiased only if all $(\tau\delta)_{ij} = 0$:

$$f = ms_T/ms_E \tag{5.6}$$

versus $f_{\alpha,t-1,(t-1)(r-1)}$. However, if block effects may be considered random (as for repetitions over time of a basic experiment with one unit or subject per treatment) the component of variance for interaction, σ_{TD}^2, appears in $E[MS_T]$ as well as in $E[MS_E]$. Then the test is unbiased regardless of the assumption about interaction, but the inferences drawn may be too broadly sketched if strong interaction exists (e.g., treatment differences varying in magnitude with weight of animals, the blocking factor). Orthogonal contrasts and other procedures for making specific comparisons among treatment means may be applied.

If one's interest in the block means extends only to suggesting whether blocking has been effective in reducing the experimental error used to test treatment effects, one may test the composite hypothesis, $H{:}(\sigma_J^2 + \delta_j) = 0$ (for all j), by comparing

$$f = ms_D/ms_E \tag{5.7}$$

with $f_{\alpha,r-1,(t-1)(r-1)}$. However, if one is interested in the block means per se, no exact test of $H{:}\delta_j = 0$ (for all j) exists. The statistic of (5.7) would be biased upward to the degree that $\sigma_J^2 > 0$; i.e., the hypothesis $H{:}\delta_j = 0$ would be falsely rejected more often than specified by the nominal level of Type I error α.

5.3.2. Test for Nonadditivity

A partial check (a posteriori) on the validity of assuming additivity of τ_i and δ_j [negligible $(\tau\delta)_{ij}$] was proposed by Tukey (1949). The procedure tests not $H{:}(\tau\delta)_{ij} = 0$, but $H{:}\varepsilon = 0$, where $\varepsilon(\tau_i)(\delta_j)$ is one parametric form of nonadditivity or interaction between the effects of treatments and blocks. The sum of squares for nonadditivity (with 1 degree of freedom) is

$$ss_{\text{NA}} = n[\sum_{i=1}^{t}\sum_{j=1}^{r} y_{ij}(\bar{y}_{i.} - \bar{y}_{..})(\bar{y}_{.j} - \bar{y}_{..})]^2/[(ss_T)(ss_D)] \tag{5.8}$$

and the test statistic,

$$f = ss_{\text{NA}}/[(ss_E - ss_{\text{NA}})/(\nu_E - 1)] \tag{5.9}$$

may be compared with $f_{\alpha,1,\nu_E-1}$. For this test (as for many preliminary tests) one should not select a small value of α because then the test is not very sensitive, and one must be able to detect nonadditivity to avoid making conclusions (from the main test of $H{:}\tau_i = 0$) that would be too general if interaction exists. Perhaps $\alpha = 0.25$ will be satisfactory in most cases.

Rojas (1973) derived an equivalent test that is generalized for ease of application to other designs. The squares of fitted values [e.g., $\hat{y}_{ij}^2 = (\hat{\mu} + \hat{\tau}_i + \hat{\delta}_j)^2$ for the complete block model] are computed for use as values of a covariate, say x_{ij}. Then a covariance analysis is performed with y_{ij} and x_{ij}, and the test of nonadditivity is equivalent to the test for significant slope of regression [see (3.18), (3.19)]. For computational convenience one may replace $\hat{y}_{ij}^2$ by $a(\hat{y}_{ij} - b)^2$, where a and b are arbitrary coding constants.

If nonadditivity is significant, there may be no satisfactory solution short of redesigning the experiment as a completely randomized factorial with two or more observations per combination of the classes of former treatments and blocks, using them as factors *A* and *B*. Some writers (e.g., Kirton 1971) have advocated transforming the data at hand to eliminate nonadditivity and allow proper testing for the effects of treatments averaged over diverse conditions of blocks, which may be sufficient in some cases. However, when interaction is apparent on the original scale, more often it is the block-treatment combinations pertaining to specific inferences that are meaningful. For example, if the blocking factor is age of the subject and interaction between age classes and treatments is discovered, the experimenter will probably want specific insights into the relative responses to different treatments by subjects of each age group.

For further partitioning of the residual into other terms measuring nonadditivity, the reader is referred to Mandel (1961) and Kirton (1966, 1971). Johnson and Graybill (1972) have discussed a more general likelihood ratio test of interaction for cases with one observation per cell. Cornell (1974) has discussed use of extended complete block designs that permit easy separation of nonadditivity and experimental error.

5.3.3. Heterogeneous Variance

Among the usual assumptions for the analysis of variance, perhaps the one most often seriously violated is the assumption of homogeneous variance. One common case of heterogeneous variance in biology involves the dependence of variability on level of response. Anscombe and Tukey (1963) developed a procedure to test for the presence of such dependence. First, define the fitted or predicted values of the variable Y (derived by the theory of least squares) as

$$\hat{Y}_{ij} = \hat{\mu} + \hat{\tau}_i + \hat{\delta}_j = \bar{y} + (\bar{y}_{i.} - \bar{y}) + (\bar{y}_{.j} - \bar{y}) = (\bar{y}_{i.} + \bar{y}_{.j} - \bar{y}) \tag{5.10}$$

Then, the residuals or estimates of errors of model (5.1), are

$$e_{ij} = y_{ij} - \hat{Y}_{ij} \tag{5.11}$$

To test the hypothesis that the linear regression of the squared residuals on the fitted values is zero, compare the test statistic,

$$f = n[\sum_{i=1}^{t} \sum_{j=1}^{r} e_{ij}^2(\hat{Y}_{ij} - \bar{y})]^2/\{[2\nu_E ms_E^2/(\nu_E + 2)][(t-2)(r-1)ss_T + (t-1)(r-2)ss_D]\} \tag{5.12}$$

with critical value $f_{\alpha,1,\nu_E}$, where ν_E represents the degrees of freedom associated with experimental error.

Han (1969) presented two entirely different and more general approaches for testing variance heterogeneity among treatments when the variances for each block are equal. One method requires computation of the squared coefficient of multiple correlation among t defined variables and is computationally unwieldy unless programmed for analysis by computer. Unfortunately, the correlation depends on the magnitude of the block effects, which if large drastically reduce the power of the test. The other method is a shortcut procedure similar to the f_{max} procedure described in Sec. 2.1.5; it requires modification of percentage points from the distribution of the range (Harter 1969).

Shukla (1972) introduced a test of variance heterogeneity among treatments that is invariant to changes in the nuisance parameters.

Several alternatives of analysis are available if variance heterogeneity is found. Anscombe and Tukey discussed power transformations, but transformations in general are difficult to apply and interpret unless the simple log transformation is appropriate. They also indicated how to obtain unbiased estimates of variance for each of t treatments ($t \geq 3$):

$$s_i^2 = [t \sum_{j=1}^{r} e_{ij}^2 - (r - 1)ms_E]/[(t - 2)(r - 1)] \tag{5.13}$$

Note that negative estimates should be taken as zero for purposes of further analysis. One may utilize the s_i^2 to evaluate the significance of the test statistic [(5.6)] for $H{:}\tau_i = 0$ (for all i) by applying Box's (1954b) method of approximate degrees of freedom. The critical value is $f_{\alpha,\hat{\nu}_T,\hat{\nu}_E}$, where

$$\hat{\nu}_T = (t - 1)/g \tag{5.14}$$

$$\hat{\nu}_E = (t - 1)(r - 1)/g \tag{5.15}$$

$$g = 1 + \sum_{i=1}^{t} (s_i^2 - \overline{s^2})^2/[(t - 1)(\overline{s^2})^2] \tag{5.16}$$

Note that $\overline{s^2}$ is the average of the s_i^2 for the t different treatments. The approximation will be considerably better if the number of blocks is moderately large (say $r \geq 10$). The approximate critical value also may be used in conjunction with the various s_i^2 in Scheffé's interval for contrasts among means (Sec. 2.2.3.2).

A unique advantage of the RCBD is that a separate error variance with $r - 1$ df may be calculated for each of several orthogonal contrasts. Suppose $t - 1$ contrasts are indexed $k = 1, 2, \ldots, t - 1$. Define the kth contrast within the jth block as

$$q_{jk} = \sum_{i=1}^{t} c_{ik}y_{ij} \qquad \left(\sum_{i=1}^{t} c_{ik} = 0\right) \tag{5.17}$$

Then the error sum of squares for the kth contrast is

$$(ss_E)_k = \left[\sum_{j=1}^{r} q_{jk}^2 - r\left(\sum_{i=1}^{t} c_{ik}\bar{y}_{i.}\right)^2\right] / \sum_{i=1}^{t} c_{ik}^2 \tag{5.18}$$

and has $r - 1$ df. The test statistic for the kth contrast in the jth block,

$$f = q_{jk}^2/\left[\left(\sum_{i=1}^{t} c_{ik}^2\right)(ss_E)_k/(r - 1)\right] \tag{5.19}$$

may be compared with $f_{\alpha,1,r-1}$. These tests are not very sensitive if the number of blocks is small.

The problem of detecting outliers has been discussed by Stefansky (1972).

5.4. MISSING CELLS

In experimental practice generally and especially in experiments with animals, one or more observations may be missing from the final analysis. An animal may be sick or die or an error may be made in the laboratory analysis of specimens from the animal. Such missing data do not create difficult problems in most completely randomized designs, because techniques are available for analyzing data with unequal replication per treatment or treatment combination and usually are easily applied without a computer. However in other designs, such as RCBDs, any given observation may be the only representative of a block-treatment combination or cell, and a missing observation implies a missing cell with consequences for analysis beyond mere unequal replication.

Several methods of estimating values for missing cells are available [e.g., covariance technique (Sec. 3.1) or iteration (Kirk 1968)]; the method of minimum mean square error is useful for relatively simple designs, such as RCBDs. Haseman and Gaylor (1973) have shown how to extend the method to multiple classifications and more complex designs.

To illustrate the procedure, consider an artificial sample of data shown in Table 5.2. The simplest expression, computationally, for the sum of squares of error is

Table 5.2. RCBD with Missing Cells

	Block				
Treatment	1	2	3	4	$y_{i.}$
1	3	4	3	y_{14}	$10+y_{14}$
2	4	y_{22}	6	5	$15+y_{22}$
3	5	2	4	y_{34}	$11+y_{34}$
$y_{.j}$	12	$6+y_{22}$	13	$5+y_{14}+y_{34}$	$36+y_{14}+y_{22}+y_{34}=y_{..}$

$$ss_E = \sum_{i=1}^{t}\sum_{j=1}^{r} y_{ij}^2 - \sum_{i=1}^{t} y_{i.}^2/r - \sum_{j=1}^{r} y_{.j}^2/t + y_{..}^2/n \tag{5.20}$$

$$= (156 + y_{14}^2 + y_{22}^2 + y_{34}^2) - [(10 + y_{14})^2 + (15 + y_{22})^2 + (11 + y_{34})^2]/4$$

$$- [313 + (6 + y_{22})^2 + (5 + y_{14} + y_{34})^2]/3 + (36 + y_{14} + y_{22} + y_{34})^2/12$$

The goal is to estimate values for y_{14}, y_{22}, and y_{34} such that ss_E is a minimum; it may be accomplished by differentiating the expression in (5.20) with respect to each of the missing values, setting each resulting equation equal to zero, and solving simultaneously. Obviously, if several values (say more than 5) are missing in a large experiment, the simultaneous solutions will be much easier to obtain by using a computer. So one may as well use a general computer routine for analysis of the data.

For this example, the equations resulting from differentiating ss_E with respect to y_{22}, y_{14}, and y_{34} and equating results to zero are

$$6\hat{y}_{22} + \hat{y}_{14} + \hat{y}_{34} = 33 \tag{5.21}$$

$$\hat{y}_{22} + 6\hat{y}_{14} - 3\hat{y}_{34} = 14 \tag{5.22}$$

$$\hat{y}_{22} - 3\hat{y}_{14} + 6\hat{y}_{34} = 17 \tag{5.23}$$

where circumflexes over symbols for the missing values indicate specific values that fulfill the condition of minimizing ss_E. Solutions to (5.21)-(5.23) are obtained by the usual algebraic methods or by applying matrix manipulations (Chap. 4). The matrix of coefficients of the $\hat{y}_{ij}$ and the inverse of the matrix, respectively, are

$$\begin{bmatrix} 6 & 1 & 1 \\ 1 & 6 & -3 \\ 1 & -3 & 6 \end{bmatrix} \quad \text{and} \quad \begin{bmatrix} 0.188 & -0.062 & -0.062 \\ -0.062 & 0.243 & 0.132 \\ -0.062 & 0.132 & 0.243 \end{bmatrix}$$

Multiplying each row of the inverse matrix in turn times the vector of right-hand sides of (5.21)-(5.23), i.e., [33, 14, 17], provides the proper solutions:

$$\hat{y}_{22} = 0.188(33) - 0.062(14) - 0.062(17) = 4.28 \tag{5.24}$$

$$\hat{y}_{14} = -0.062(33) + 0.243(14) + 0.132(17) = 3.60 \tag{5.25}$$

$$\hat{y}_{34} = -0.062(33) + 0.132(14) + 0.243(17) = 3.93 \tag{5.26}$$

These values may now be placed with the observed values to compute sums of squares in the usual manner for randomized complete block experiments. The only alteration needed in the analysis of variance is subtraction of 1 df from the usual degrees of freedom for error, $(t - 1)(r - 1)$, for each estimated missing cell.

If the test of $H{:}\tau_i = 0$ (for all i) is marginally significant, one should adjust the treatment sum of squares for bias. It is biased upward relative to the error sum of squares, which has been minimized in the process of estimating the missing values. Minimization of the error sum of squares is equivalent to setting the residuals $e_{ij} = 0$ for the missing values (Sclove 1972). The adjusted sum of squares for treatments is

$$ss_{T(\text{adj})} = ss_T - [(ss_y - ss'_y) - (ss_D - ss'_D)] \tag{5.27}$$

where ss'_y and ss'_D are sums of squares of observed data only, computed for total data and blocks in the manner of sums of squares for the one-way analysis of variance with unbalanced replication. That is, for an experiment with m missing values, compute

$$ss'_y = \sum_{ij}\sum y^2_{ij} - (\sum_{ij}\sum y_{ij})^2/(n - m) \tag{5.28}$$

$$ss'_D = \sum_{j=1}^{r} y^2_{.j}/t_j - (\sum_{ij}\sum y_{ij})^2/(n - m) \tag{5.29}$$

where t_j treatments are represented in the observed data of the jth block. For the data of Table 5.2, $ss'_y = 12$ and $ss'_D = 3.3$. Also, when estimated values are included, $ss_y = 12.24$, $ss_T = 4.41$, and $ss_D = 1.41$. Therefore, the adjusted sum of squares for treatments is

$$ss_{T(\text{adj})} = 4.41 - [(12.24 - 12) - (1.41 - 3.3)] = 2.28$$

The drastic reduction from ss_T to $ss_{T(\text{adj})}$ by removing bias is not typical but is a consequence of using an artificial example with 25% of the data missing.

For a single missing cell, the formula for $\hat{y}_{ij}$ is

$$\hat{y}_{ij} = (ty'_{i.} + ry'_{.j} - y'_{..})/[(t - 1)(r - 1)] \tag{5.30}$$

where $y'_{i.}$, $y'_{.j}$, and $y'_{..}$ are totals of observed data for the pertinent treatment, pertinent block, and entire experiment, respectively. Also, the adjusted sum of squares for treatments may be computed from

$$ss_{T(\text{adj})} = ss_T - [y'_{.j} - (t - 1)\hat{y}_{ij}]^2/[t(t - 1)] \tag{5.31}$$

Taylor (1948) derived the variance of the difference between 2 treatment means that have m_1 and m_2 estimated missing values, respectively:

$$\hat{V}[\bar{y}_{1.} - \bar{y}_{2.}] = ms_E[1/(r - m_1 - m_2/\nu_T) + 1/(r - m_2 - m_1/\nu_T)] \tag{5.32}$$

where $\nu_T = t - 1$ df for treatments. For the case with $m_1 = m_2 = 1$,

$$\hat{V}[\bar{y}_{1.} - \bar{y}_{2.}] = 2ms_E/[r - t/(t - 1)] \tag{5.33}$$

or for $m_1 = 1$, $m_2 = 0$, it becomes

$$\hat{V}[\bar{y}_{1.} - \bar{y}_{2.}] = ms_E[1/(r - 1) + (t - 1)/(n - r - 1)] \tag{5.34}$$

5.5. ESTIMATION OF SAMPLE SIZE REQUIRED

One may estimate the amount of replication per treatment (r = number of blocks) required to have power $1 - \beta$ to detect a specified array of treatment effects $\{\tau_{di\cdot}\}$ at a given level of significance (α) by utilizing the principles discussed for the analysis of variance of one-way classifications (see Sec. 2.1.3). Briefly, one uses App. Figs. A.15.17-A.15.21 (for $\nu_1 = t - 1 \leq 4$ and various α), noting if the β on the vertical scale, which corresponds to the intersection of $\phi = \sqrt{(r/t)\Sigma(\tau_{di\cdot}/\hat{\sigma})^2}$ on the horizontal scale and the line for ν_2 for a given trial value of r, is small enough. If not, increase r and try again. It is necessary to express the array of treatment effects so that $\sum_{i=1}^{t} \tau_{di\cdot} = 0$. The main differences between the procedures for one-way classifications and block experiments are that $\nu_2 = (t - 1)(r - 1)$ for block analysis and prior estimate of σ must reflect only the variation within blocks and treatments. For example, if blocking is based on weights of animals, use a prior estimate of σ from a prior similar RCBD study (perhaps from recent scientific literature) for the trait of interest Y for animals of similar weights treated alike. Alternatively, one may have some theoretical or practical notions about the potential range of responses for animals of the same weight treated alike; then the crude estimate of the range may be converted into an estimate of σ by referring to App. A.14. In practice one may wish to divide the range by 5, equivalent to assuming that the estimated range was based on a sample of 100 hypothetical observations, to obtain an estimate of σ. Or a more uncertain estimate of the range may be divided by 4 or 3, which is approximately correct for hypothetical samples of 30 and 10, respectively. Obviously, estimation of sample size is a crude business at best, but the point is to avoid doing hopelessly insensitive or needlessly expensive experiments.

As for one-way analysis of variance, if the specification of an entire array of treatment effects $\{\tau_{di\cdot}\}$ that one wishes to detect seems difficult and arbitrary, orthogonal contrasts (Sec. 2.2.1) may offer an easier way of specifying important differences. The important magnitude of mean difference Δ between the two sides of a contrast ($\nu_1 = 1$) is relatively easy to specify; the principles are identical to those in Sec. 2.2.1, where $\phi = (\Delta/\sigma)\cdot \sqrt{(r/8)[(\Sigma|c_{i\cdot}|)^2/\Sigma c_{i\cdot}^2]}$ for a contrast with coefficients $c_1, c_2, \ldots, c_t$ ($\Sigma c_{i\cdot} = 0$), except that $\nu_2 = (t - 1)(r - 1)$ and (as above) σ must be estimated within blocks and treatments. Kastenbaum et al. (1970b) have provided an alternative way of specifying differences to be detected.

5.6. EFFICIENCY OF BLOCKING

The relative efficiency of one estimator to another is based on Fisher's (1925b) concept of *information* contained in a sample. The Cramér-Rao inequality (Kendall and Stewart 1967) expresses the minimum variance bound for an unbiased estimator of a parameter as the inverse of information, which is derived from the likelihood of occurrence of a sample, given a parametric distribution (Sec. 1.5.3). Most estimators used are asymptotically normally distributed because of the central limit theorem; they depend only on the parameters of mean and variance. Also, most estimators are consistent and asymptotically unbiased (Sec. 1.5.2), leaving variance as the chief criterion of efficiency of estimators. Generally, the relative efficiency of one estimator to another is the ratio of their relative amounts of information (or the inverse ratio of variances of the estimators).

For a normal distribution, with unknown value of σ^2 estimated by $\hat{\sigma}^2$ with ν df, the information for estimating the mean from a sample of n observations (Fisher 1935) is

$$[(\nu + 1)/(\nu + 3)](n/\hat{\sigma}^2)$$

Therefore the efficiency of design 1 relative to design 2 (with means in each case estimated from the same number of observations) is

$$E_{1:2} = [(\nu_1 + 1)/(\nu_1 + 3)][(\nu_2 + 3)/(\nu_2 + 1)](\hat{\sigma}_2^2/\hat{\sigma}_1^2)100\% \qquad (5.35)$$

For a comparison of a randomized complete block design (RCBD) with a completely randomized design (CRD), both involving $n = tr$ observations, consider comparing the relative variances expected with and without blocking in the absence of treatment effects [Kempthorne (1952); Peng (1967)]. With blocking,

$$\hat{\sigma}^2_{\text{RCBD}} = [(t - 1) + (t - 1)(r - 1)]ms_E/[r(t - 1)] = ms_E \qquad (5.36)$$

with $\nu_{\text{RCBD}} = r(t - 1) = n - r$. Without blocking,

$$\hat{\sigma}^2_{\text{CRD}} = \{(r - 1)ms_D + [(t - 1) + (t - 1)(r - 1)]ms_E\}/(n - 1) \qquad (5.37)$$

with $\nu_{\text{CRD}} = n - 1$. Then the relative efficiency of blocking is

$$E_{\text{RCBD:CRD}} = [(n - r + 1)/(n - r + 3)][(n + 2)/n](\hat{\sigma}^2_{\text{CRD}}/\hat{\sigma}^2_{\text{RCBD}})100\% \qquad (5.38)$$

For example, with $t = 3$ treatments, $r = 6$ blocks ($n = 18$), and $ms_D = 0.02$, $ms_E = 0.01$,

$$\hat{\sigma}^2_{\text{CRD}} = [5(0.02) + 12(0.01)]/17 = 0.013$$

and the relative efficiency is

$$E_{\text{RCBD:CRD}} = (13/15)(20/18)(0.013/0.01)100\% = 125\%$$

The adjustment in (5.38) for difference in degrees of freedom is very near unity (0.97) and may be ignored in practice except for cases with only 2 treatments or with 3 treatments and 8 or fewer blocks.

In some situations with a rather large number of treatments, formation of complete blocks is not efficient because the many units required for each block are not homogeneous with respect to the nuisance variable used in blocking. Then further reduction of experimental error may be accomplished by using incomplete block designs (every treatment does not occur in every block).

5.7. NUMERICAL EXAMPLE

A study was made of the gains in weight (kg) of pigs in a comparative feeding trial over a fixed period. Nearly half of the normal variation in this trait was known to be genetically determined, so some natural variation could be excluded from experimental error by forming blocks of related pigs (choosing pigs from the same litter to form a block). Data from the experiment are shown in Table 5.3. A summary of the analysis of variance, with effects of litters assumed to be random, is shown in Table 5.4.

Table 5.3. Gains in Weight (kg) of Pigs in RCBD

	Treatment						
Litter	1	2	3	4	5	$y_{.j}$	$\hat{D}_j=\bar{y}_{.j}-\bar{y}$
1	79.5	80.9	79.1	88.6	95.9	424.0	+7.74
2	70.9	81.8	70.9	88.6	85.9	398.1	+2.56
3	76.8	86.4	90.5	89.1	83.2	426.0	+8.14
4	75.9	75.5	62.7	91.4	87.7	393.2	+1.58
5	77.3	77.3	69.5	75.0	74.5	373.6	-2.34
6	66.4	73.2	86.4	79.5	72.7	378.2	-1.42
7	59.1	77.7	72.7	85.0	90.9	385.4	+0.02
8	64.1	72.3	73.6	75.9	60.0	345.9	-7.98
9	74.5	81.4	64.5	75.5	83.6	379.5	-1.16
10	67.3	82.3	65.9	70.5	63.2	349.2	-7.22
$y_{i.}$	711.8	788.8	735.8	819.1	797.6	$y_{..}=3853.1$	
$\hat{\tau}_i=\bar{y}_{i.}-\bar{y}$	-5.88	+1.82	-3.48	+4.85	+2.70	$\hat{\mu}=\bar{y}=77.06$	

Table 5.4. Analysis of Gains in Weight of Pigs from RCBD

Source of Variation	df	*ss*	*ms*	$E[MS]$
Treatments	4	808.1	202.0	$\sigma^2+[\sigma^2_{TL}]+10\sum_{i=1}^{5}\tau_i^2/4$
Litters	9	1291.4	143.5	$\sigma^2+[5\sigma^2_J]^*+5\sigma^2_L$
Error	36	1813.7	50.4	$\sigma^2+[\sigma^2_{TL}]$
Total	49	3913.2		

$^*\sigma^2_J$ is the component of variance caused by restricted randomization.

One can see from the expected mean squares that if effects of litters are truly random (or at least representative of a large class of litter effects), interaction between treatments and litters (σ_{TL}^2) will not bias the f test for general inference about treatments. However, more specific inferences are needed if interaction exists. To illustrate the procedure of checking for interaction, consider Tukey's test for nonadditivity. The sum of squares for nonadditivity [(5.8)] is

$$ss_{\text{NA}} = 50(1085.85^2)/[(808.1)(1291.4)] = 56.5$$

and the test statistic is

$$f = 56.5/[(1813.7 - 56.5)/35] = 1.125$$

which is less than $f_{0.25,1,35} = 1.37$. Therefore, one may conclude that little evidence exists for nonadditivity between treatments and litters, although a complete test of interaction can be obtained only from designs that have replication in the cells (block-treatment combinations).

The dependence of variance on response level may be tested after computing fitted values ($\hat{Y}_{ij}$) and residuals (e_{ij}) according to (5.10) and (5.11). The results shown in Table 5.5 are $\hat{Y}_{ij} - \bar{y} = \hat{\tau}_i + \hat{D}_j$ and $e_{ij} = y_{ij} - \hat{\mu} - (\hat{\tau}_i + \hat{D}_j)$. Note that values of $\{\hat{\tau}_i\}$ and $\{\hat{D}_j\}$ are shown in Table 5.3. The test statistic, (5.12), is

$$f = 50(2854.18^2)/\{[2(36)(50.4^2)/38][3(9)(808.1) + 4(8)(1291.4)]\} = 1.34$$

which is slightly less than $f_{0.25,1,36} = 1.37$ and considerably less than $f_{0.10,1,36} = 2.86$. Therefore, one may accept the hypothesis that variance does not depend on level of response.

The general hypothesis that the effects of all treatments are zero, $H{:}\tau_i = 0$ (for all i), may be tested with the statistic

$$f = ms_T/ms_E = 202/50.4 = 4.01$$

which exceeds $f_{0.01,4,36} = 3.94$. Therefore, one may conclude with high confidence that the treatments differ in mean effect on gain in weight. More specific comparisons may be of interest. Suppose it had been decided a priori that the following orthogonal contrasts should be performed: treatment 1 (control) versus others, treatments 2 and 3 (jointly) versus 4 and 5, treatment 2 versus 3, and treatment 4 versus 5. The estimated mean differences for the 4 contrasts [(2.50)] are $\hat{\Delta}_1 = 7.35$, $\hat{\Delta}_2 = 4.60$, $\hat{\Delta}_3 = 5.22$, and $\hat{\Delta}_4 = 2.15$, respectively. The corresponding values of the contrast coefficients c_{ik} provide values of $\sum_{i=1}^{5} c_{ik}^2 = 20, 4, 2,$ and 2, and values of $\sum_{i=1}^{5} |c_{ik}| = 8, 4, 2,$ and 2, respectively. Confidence intervals of 95% for the Δ_k [see (2.52)] may be computed from

Table 5.5. Deviations of Fitted Values from the Mean and Residuals from RCBD

	1		2		3		4		5	
Litter	$\hat{Y}_{ij}-\bar{y}$	e_{ij}	$\hat{Y}_{ij}-\bar{y}$	e_{ij}	$\hat{Y}_{ij}-\bar{y}$	e_{ij}	$\hat{Y}_{ij}-\bar{y}$	e_{ij}	$\hat{Y}_{ij}-\bar{y}$	e_{ij}
1	+1.86	+0.58	+9.56	-5.72	+4.26	-2.22	+12.59	-1.05	+10.44	+8.40
2	-3.32	-2.84	+4.38	+0.36	-0.92	-5.24	+7.41	+4.13	+5.26	+3.58
3	+2.26	-2.52	+9.96	-0.62	+4.66	+8.78	+12.99	-0.95	+10.84	-4.70
4	-4.30	+3.14	+3.40	-4.96	-1.90	-12.46	+6.43	+7.91	+4.28	+6.36
5	-8.22	+8.46	-0.52	+0.76	-5.82	-1.74	+2.51	-4.57	+0.36	-2.92
6	-7.30	-3.36	+0.40	-4.26	-4.90	+14.24	+3.43	-0.99	+1.28	-5.64
7	-5.86	-12.10	+1.84	-1.20	-3.46	-0.90	+4.87	+3.07	+2.72	+11.12
8	-13.86	+0.90	-6.16	+1.40	-11.46	+8.00	-3.13	+1.97	-5.28	-11.78
9	-7.04	+4.48	+0.66	+3.68	-4.64	-7.92	+3.69	-5.25	+1.54	+5.00
10	-13.10	+3.34	-5.40	+10.64	-10.70	-0.46	-2.37	-4.19	-4.52	-9.34

$$\hat{\Delta}_k \pm t_{0.025,36} 2\sqrt{ms_E/r}\,[\sqrt{\Sigma c_{ik}^2}/\Sigma|c_{ik}|]$$

The results are 7.35 ± 5.10, 4.60 ± 4.56, 5.22 ± 6.44, and 2.15 ± 6.44, respectively. Since zero is not included in the first two intervals but is included in the last two, one may conclude that the experimental treatments exceed the control and treatments 4 and 5 collectively are better than 2 and 3. However, evidence is not clear that treatment 2 is better than 3 or that 4 is better than 5, as the sample means suggest.

Suppose that in a future experiment of the same type one would like to have at least 0.8 probability (power) to detect a true mean difference of 7 kg between treatments 2 and 3 at a 95% significance level. One may apply App. Fig. A.15.17, with $\alpha = 0.05$ and $\Phi = \sqrt{r/4}(\Delta/\sigma)$ as in (2.58). For $\Delta = 7$ and $\hat{\sigma} = \sqrt{ms_E} = 7.1$, Φ becomes $0.986\sqrt{r/4}$. With $\nu_2 = (t - 1)(r - 1)$ and $\beta \leq 0.2$, the figure shows that $2 < \Phi < 2.5$ for $t = 5$. Therefore, try $r = 18$, which provides $\Phi = 2.09$ and $\nu_2 = 68$. For that combination, $\beta \simeq 0.17$, which may be satisfactory. Therefore, 18 litters (or more) should be used.

Finally, one may wish to estimate the efficiency of blocking. Estimate σ^2_{RCBD} by $ms_E = 50.4$, with $\nu_{\text{RCBD}} = n - r = 40$. Also, σ^2_{CRD} may be estimated [(5.37)]:

$$\sigma^2_{\text{CRD}} = [9(143.5) + 40(50.4)]/49 = 67.5$$

with $\nu_{\text{CRD}} = n - 1 = 49$. The efficiency of the RCBD relative to a CRD of the same size is [see (5.38)]

$$E_{\text{RCBD:CRD}} = (67.5/50.4)100\% = 134\%$$

if one ignores the trivial correction for difference in degrees of freedom. Thus, it is estimated that blocking based on genetic (and other) differences between litters is 34% more efficient than ignoring litter differences when studying gains in weight.

5.8. FACTORIAL EXPERIMENTS WITH COMPLETE BLOCKS

In many experimental situations that involve a nuisance variable suitable for blocking, the treatments of interest are factorial in nature, i.e., treatment combinations are to be randomly assigned to the units in each block. If the number of treatment combinations is small enough that an equivalent number of relatively homogeneous units can be collected for each block, the experiment may be performed with complete blocks. However, for some factorial experiments the number of treatment combinations is so large that the equivalent number of units needed to form each block are not homogeneous with respect to the nuisance variable. Then further reduction of experimental error may be accomplished by using incomplete block designs in which every treatment combination does not appear in every block (Secs. 6.3-6.5).

As an example of a 2-factor experiment performed with complete blocks, consider an experiment designed to study gains in weight of dairy heifers assigned to combinations of 4 diets (A) and 3 styles of housing (B), with initial weight used in blocking. Since there are 12 treatment combinations, each complete block must consist of 12 heifers of similar initial weights. A general model for experiments of such nature is

$$Y_{ijk} = \mu + \alpha_i + \beta_j + (\alpha\beta)_{ij} + \delta_k + J_k + (\tau\delta)_{ijk} + E_{(ijk)} \qquad (5.39)$$

where $i = 1, 2, \ldots, a$, $j = 1, 2, \ldots, b$, $k = 1, 2, \ldots, r$; α_i, β_j, and $(\alpha\beta)_{ij}$ represent the main effects and interaction of factors A and B; and the other terms are as defined for (5.1), with $(\tau\delta)_{ijk}$ interpreted as the interaction of block effects with treatment combinations. As before, restriction effect (J_k) is completely confounded with block effect (δ_k) and interaction effect $(\tau\delta)_{ijk}$ is completely confounded with error $(E_{(ijk)})$, but the terms are included in the model so that the expected mean squares will serve to warn the investigator about assumptions involved.

Sums of squares for blocks (ss_D) and for main effects and interaction of factors A and B (ss_A, ss_B, ss_{AB}) may be computed as for a CRD with 3 factors (the proper design if one were uncertain whether block effects would interact with treatment effects). Refer to Sec. 2.4.4. As usual, the sum of squares for error may be obtained by difference:

$$ss_E = ss_y - (ss_A + ss_B + ss_{AB} + ss_D) \qquad (5.40)$$

where ss_y is the total sum of squares.

A summary of the analysis of variance is shown in Table 5.6, where it is

Table 5.6. Analysis of Variance for Two-Factor Experiment with Complete Blocks

Source of Variation	df	*ss*	*ms*	*E[MS]* (fixed effects)
A	$a-1$	ss_A	ms_A	$\sigma^2 + rb\sum_{i=1}^{a}\alpha_i^2/\nu_A$
B	$b-1$	ss_B	ms_B	$\sigma^2 + ra\sum_{j=1}^{b}\beta_j^2/\nu_B$
AB	$(a-1)(b-1)$	ss_{AB}	ms_{AB}	$\sigma^2 + r\sum_{i=1}^{a}\sum_{j=1}^{b}(\alpha\beta)_{ij}^2/\nu_{AB}$
Blocks	$r-1$	ss_D	ms_D	$\sigma^2 + ab\sigma_J^2 + ab\sum_{k=1}^{r}\delta_k^2/\nu_D$
Error	$(ab-1)(r-1)$	ss_E	ms_E	$\sigma^2 + \sum_{i=1}^{a}\sum_{j=1}^{b}\sum_{k=1}^{r}(\tau\delta)_{ijk}^2/\nu_E$

assumed that the effects of both factors and blocks are fixed. If one or both factors are considered random or if blocks are considered random, examine expected mean squares (see Sec. 2.4.2, e.g., Table 2.18) to determine appropriate procedures for testing hypotheses and estimating components of variance.

Of course the factorial scheme may be extended to as many factors and combinations as necessary. The rules of Sec. 2.4.2 for models, sums of squares, and expected mean squares and the inferential techniques discussed in Secs. 2.4.4 and 2.4.5 apply equally well.

5.9. NONRANDOM HANDLING OF TREATMENT GROUPS

In some experimental situations it is inconvenient (or impossible) to handle treatment groups in a random manner to minimize confounding of effects of latent nuisance variables with effects of treatments. For example, in animal group feeding trials, each treatment (diet) is commonly applied to an entire treatment group that is physically separated from other groups although the response (gain in weight) is to be measured on each animal separately. In such cases it is difficult to avoid confounding the effects of location, housing, order of handling, feeding, etc., with the effects of treatments. Two basic faults commonly found in carrying out trials of this nature are: (1) instead of blocking on a nuisance variable (such as initial weight of each animal), the experimenter assigns animals to treatment groups so as to equalize ("balance") the mean values of the nuisance variable; and (2) the experimenter assigns all animals in the same treatment group to the same pen, cage, or facility.

Failure to block on the nuisance variable does not allow reduction of experimental error by removal of a sum of squares for blocks; and the process of "balancing" groups tends to create negatively correlated errors within each group, causing the estimate of experimental error to be possibly larger than if the nuisance variable were ignored completely. Of course, the intent of balancing is to avoid bias in the resulting treatment means, but similar insurance as well as reduced error is obtained in a properly randomized block design.

Assigning all animals in the same treatment group to the same pen may cause positively correlated errors (because of common environment, order of handling, etc.) and confounding of pen effects with treatment effects so that comparisons of treatments are biased. This problem should be avoided by assigning the animals on a given treatment to 2 or more pens if enough are available (*space* replication) or by considering repeating the entire experiment one or more times if the number of pens is limited (*time* replication).

First, consider space replication. One should randomly assign p pens ($p \geq 2$) for each of t treatments and allocate tp homogeneous animals to each of r blocks; $n = tpr$ animals in the entire experiment. Within each block, p animals should be randomly assigned to each treatment, one to each pen receiving that treatment. Suppose there are $t = 6$ treatments (rations for steers) and 12 available pens, or $p = 2$ per treatment. If one forms $r = 8$ blocks on the basis of weight with $tp = 12$ animals per block, one would randomly assign 2 pens per treatment and within each block randomly assign $p = 2$ animals to each treatment, 1 to each pen. This procedure leads to 8 animals per pen, 1 from each of 8 blocks, or 16 steers per treatment (96 animals in all).

A model for space replication may be written

$$Y_{ijk} = \mu + \tau_i + \delta_j + J_j + P_{(i)k} + (\tau\delta)_{ij} + (\delta P)_{(i)jk} + E_{(ijk)} \qquad (5.41)$$

where $i = 1, 2, \ldots, t$, $j = 1, 2, \ldots, r$, and $k = 1, 2, \ldots, p$ for each i; $P_{(i)k}$ represents the random effects associated with the environment and handling in the kth pen assigned to the ith treatment; and $(\delta P)_{(i)jk}$ represents the interaction of effects of blocks and pens (inseparable from $E_{(ijk)}$ and usually thought to be trivial). All other effects are as defined for (5.1). Sums of squares and expected mean squares may be obtained as in Sec. 2.4.2. A summary of the analysis of variance is shown in Table 5.7.

Table 5.7. Analysis of Variance for RCBD with Treatments Replicated in Pens

Source of Variation	df	*ss*	*ms*	*E*[*MS*] (pen effects random)
Treatments (*T*)	$t-1$	ss_T	ms_T	$\sigma^2+r\sigma_P^2+rp\sum_{i=1}^{t}\tau_i^2/\nu_T$
Blocks (*D*)	$r-1$	ss_D	ms_D	$\sigma^2+\sigma_{DP}^2+tp\sigma_J^2+tp\sum_{j=1}^{r}\delta_j^2/\nu_D$
Pens/*T* (*P*)	$t(p-1)$	ss_P	ms_P	$\sigma^2+r\sigma_P^2$
TD	$(t-1)(r-1)$	ss_{TD}	ms_{TD}	$\sigma^2+\sigma_{DP}^2+p\sum_{i=1}^{t}\sum_{j=1}^{r}(\tau\delta)_{ij}^2/\nu_{TD}$
Error (*DP*)	$t(p-1)(r-1)$	ss_E	ms_E	$\sigma^2+\sigma_{DP}^2$

From the expected mean squares one can see that the appropriate procedure for testing treatment effects $H:\tau_i = 0$ (for all i) is to compare the test statistic,

$$f = ms_T/ms_P \tag{5.42}$$

with critical value $f_{\alpha,t-1,t(p-1)}$. This test is likely to be insensitive to treatment differences unless there are many treatments or several pens assigned to each treatment. However, this design permits one to test for the existence of interaction between treatments and blocks by comparing

$$f = ms_{TD}/ms_E \tag{5.43}$$

with $f_{\alpha,(t-1)(r-1),t(p-1)(r-1)}$. Also, if one is willing to assume that the component of variance for interaction of blocks and pens (σ_{DP}^2) is zero or trivial, pen effects (i.e., the hypothesis $H:\sigma_P^2 = 0$) may be tested by comparing

$$f = ms_P/ms_E \tag{5.44}$$

with $f_{\alpha,t(p-1),t(p-1)(r-1)}$. If that hypothesis is clearly acceptable (say at some relatively large value of α such as 0.25) or if $f < 2f_{0.50,t(p-1),t(p-1)(r-1)}$, one may consider pooling the sums of squares and degrees of freedom for pens and error (and possibly treatment-block interaction as well, if clearly trivial) to obtain more power in the test of treatment effects. For example, if $H{:}(\tau\delta)_{ij} = 0$ (for all i and j) and $H{:}\sigma_P^2 = 0$ both are clearly acceptable, a pooled error mean square may be calculated from

$$ms'_E = (ss_P + ss_{TD} + ss_E)/(n - t - r + 2) \tag{5.45}$$

and be used to test $H{:}\tau_i = 0$ (for all i) by comparing

$$f = ms_T/ms'_E \tag{5.46}$$

with $f_{\alpha,t-1,n-t-r+2}$. Of course, if pen effects are found to be trivial, future experiments of similar type in the same facility may be designed in ordinary complete blocks, ignoring pens.

The problem of prior estimation of sample size may be approached by the principles discussed in Sec. 5.5, except that the abscissa scale is

$$\phi = \sqrt{(pr/t)\sum_{i=1}^{t}\tau_{di}^2/(\hat{\sigma}^2 + r\hat{\sigma}_P^2)} \tag{5.47}$$

(which requires an estimate of pen variation as well as error variation) and $\nu_2 = t(p - 1)$. Alternatively, it is easier to specify a mean contrast difference Δ to be detected than to specify an entire array of treatment effects $\{\tau_{di}\}$. In that case, $\nu_1 = 1$ and

$$\phi = \sqrt{(pr/8)\Delta^2[(\sum_{i=1}^{t}|c_i|)^2/\sum_{i=1}^{t}c_i^2]/(\hat{\sigma}^2 + r\hat{\sigma}_P^2)} \tag{5.48}$$

where the c_i are contrast coefficients of treatment totals such that $\sum_{i=1}^{t} c_i = 0$.

In some situations, the facilities available limit the size of the experiment. If the total number of pens (tp) is fixed and the size of pen permits no more than r animals (r blocks), one may wish to determine the minimum mean difference detectable at a given power by an experiment of such size. For simple orthogonal contrasts between two means, $(\Sigma|c_i|)^2/\Sigma c_i^2 = 2$; and one can solve (5.48) for Δ to obtain

$$\Delta_{\min} = \phi\sqrt{4[(\hat{\sigma}^2/pr) + (\hat{\sigma}_P^2/p)]} \tag{5.49}$$

where ϕ is determined by reading the power chart for the intersection of Type II error allowable (β) and degrees of freedom $\nu_2 = t(p - 1)$.

For example, consider the feeding trial of steers mentioned earlier with $t = 6$, $p = 2$, and $r = 8$. Suppose only 12 pens that will hold a maximum of 8 animals each are available. Suppose prior estimates of variation are $\hat{\sigma} = 0.05$ kg/day (among animals of same weight treated and handled alike) and $\hat{\sigma}_P = 0.02$ kg/day (among pens). For $\alpha = 0.05$, $\beta = 0.2$ (for example), and $\nu_1 = 1$, $\nu_2 = 6(1) = 6$, App. Fig. A.15.17 shows $\phi \simeq 2.4$. One may estimate the minimum mean difference between two treatments, detectable with 80% chance by an orthogonal contrast:

$$\Delta_{\min} = 2.4\sqrt{4(0.05^2/16 + 0.02^2/2)} = 0.090 \text{ kg/day}$$

Time replication may be necessary if facilities are so restricted that replication of pens for each treatment is not possible at any given time. Time replication is less desirable than space replication because it is impossible to separate the effects of treatments from the temporary effects of pens --peculiarities of microenvironment or handling during one particular repetition of the experiment. Orthogonal separation of the effects of treatments from the permanent effects of pens--physical effects, location, order in routine handling procedures, etc.--is possible only if the experiment is designed with pens as balanced incomplete blocks unless the experiment is performed t times for t treatments. Then the experiment may be designed as a Latin square (Chap. 7), with repetitions as rows and pens as columns and blocks (e.g., weight) formed within each repetition. However, treatments will not be assigned randomly within each block but in a fixed order across pens as indicated by a particular Latin square design. For example, suppose an experiment with $t = 4$ treatments is performed 4 times using the same facilities (4 pens), with $r = 6$ blocks each time so that $n = t^2r = 96$ animals. Ignoring interactions involving blocks (not estimable), one may write the model

$$Y_{ijkl} = \mu + \tau_i + R_j + \delta_{(j)k} + P_l + (\tau R)_{ij} + (\tau P)_{il} + (RP)_{jl} + (\tau RP)_{ijl} + E_{(ijkl)} \quad (5.50)$$

where $i, j, l = 1, 2, 3, 4$, $k = 1, 2, \ldots, 6$ for each j; R_j is the random effect of the jth repetition, $\delta_{(j)k}$ is the fixed effect of the kth block in the jth repetition, etc. A summary of the analysis of variance is shown in Table 5.8.

From the expected mean squares one can see that there is no exact test of $H{:}\tau_i = 0$ (for all i); but one may resort to Satterthwaite's procedure (as in the analysis of experiments with 3 factors when 2 have random effects) [see (2.174)].

Alternatively, one may test for the existence of the various interactions and pool an interaction sum of squares with error sum of squares for any case in which $f < 2f_{0.50}$, as in (5.45).

Pimentel-Gomes (1970) has described a method of jointly analyzing several randomized block experiments having some treatments (but not all) in common and varying amounts of replication from experiment to experiment.

Table 5.8. Analysis of Variance for Block Experiment with t=4 Treatments, Repeated 4 Times, in Latin Square Design

Source of Variation	df	ss	ms	$E[MS]$ (treatments and blocks fixed)
Treatments (T)	$t-1=3$	ss_T	ms_T	$\sigma^2+r\sigma^2_{TRP}+tr\sigma^2_{TR}+tr\sigma^2_{TP}+t^2r\sum_{i=1}^{t}\tau_i^2/\nu_T$
Repetitions (R)	$t-1=3$	ss_R	ms_R	$\sigma^2+tr\sigma^2_{RP}+t^2r\sigma^2_R$
Blocks/R (D)	$t(r-1)=20$	ss_D	ms_D	$\sigma^2+t^2\sigma^2_J+t^2\sum_{j=1}^{t}\sum_{k=1}^{r}\delta^2_{(j)k}/\nu_D$
Pens (P)	$t-1=3$	ss_P	ms_P	$\sigma^2+tr\sigma^2_{RP}+t^2r\sigma^2_P$
TR	$(t-1)^2=9$	ss_{TR}	ms_{TR}	$\sigma^2+r\sigma^2_{TRP}+tr\sigma^2_{TR}$
TP	$(t-1)^2=9$	ss_{TP}	ms_{TP}	$\sigma^2+r\sigma^2_{TRP}+tr\sigma^2_{TP}$
RP	$(t-1)^2=9$	ss_{RP}	ms_{RP}	$\sigma^2+tr\sigma^2_{RP}$
TRP	$(t-1)^3=27$	ss_{TRP}	ms_{TRP}	$\sigma^2+r\sigma^2_{TRP}$
Error	$n-t(t^2+r-1)=12$	ss_E	ms_E	σ^2
Total	$n-1=95$			

EXERCISES

5.1. Hamsters were blocked by weight into groups of 3 for the comparison of 3 diets varying in fat content. Final weights are given in grams.

	Block					
Treatment	1	2	3	4	5	6
1	96	96	94	99	99	102
2	103	101	103	105	101	107
3	103	104	106	108	109	110

(a) Test for treatment differences.
(b) Suppose the fat content was equally spaced so that a polynomial interpretation is easy to compute. Use orthogonal polynomial contrasts to test linear and quadratic effects.
(c) Estimate the relative efficiency of blocking.

5.2. Twelve populations of flies bred for resistance to an insecticide were blocked by strain and randomly assigned to 3 doses of the insecticide. Percent mortality was recorded for each population (experimental unit).

	Block			
Treatment	1	2	3	4
1	66	55	43	32
2	71	57	44	37
5	79	63	51	44

(a) Test for overall differences in mean mortality for the 3 doses.
(b) Use orthogonal polynomials for unequal spacings to test linear and quadratic response to dose.
(c) Estimate the relative efficiency of blocking.

5.3. Mean lymphocyte counts ($1000/mm^3$) were compared for mice given one of 2 experimental drugs or a placebo (control). Littermate mice of the same sex were used to form homogeneous blocks of 3 mice each; within each block, the 3 treatments were assigned at random. It seemed reasonable to assume that the treatment effects would be relatively constant for various genotypes of mice represented by different litters (Schefler 1969).

Treatment	Block 1	2	3	4	5	6	7
Drug 1	6.0	4.8	6.9	6.4	5.5	9.0	6.8
Drug 2	5.1	3.9	6.5	5.6	3.9	7.0	5.4
Placebo	5.4	4.0	7.0	5.8	3.5	7.6	5.5

(a) Test for nonadditivity and for dependence of variance on level of response.
(b) Test for differences among treatment means, including Dunnett's procedure to compare each drug with the control.
(c) If in future experiments of this type one wished to have 90% chance

of detecting a mean contrast difference of 280 lymphocytes/mm^3 between 2 experimental drugs at a 95% significance level, how many blocks should be used?

(d) Estimate the relative efficiency of blocking.

(e) Suppose that the value for drug 2 in litter 4 (y_{24}) was unobtainable because the mouse died. Estimate a value for the missing cell and compute the bias in the sum of squares for treatments that would result.

5.4. An experiment was designed to test the effectiveness of feeding chemicals (isobutyrate, isovalerate, 2-methylbutyrate, phenylacetate, indoleacetate, and imadozoleacetate) as a supplement to urea to stimulate the growth of rumen cellulolytic organisms in lactating Holstein cows. Cows were blocked in groups of 3 on the basis of quantity of milk produced during the first 5 weeks of lactation prior to treatment. Corn silage rations were supplemented with (1) 80% soy protein, (2) urea, or (3) urea plus chemicals. Average daily milk production (kg) for 60 days was recorded.

	Block							
Ration	1	2	3	4	5	6	7	8
1	17.3	22.6	22.1	21.7	24.7	28.8	25.9	29.9
2	14.5	17.2	19.0	18.1	19.2	20.8	24.2	27.5
3	18.0	16.7	19.7	21.7	21.6	24.5	26.0	26.2

(a) Test for nonadditivity and for dependence of variance on level of response.

(b) Test for differences among treatment means, including Tukey's test of all possible pairs. Draw conclusions.

(c) Suppose a future experiment involves only rations 2 and 3. How many pairs of cows will be needed to have 90% chance of detecting a mean difference of 2.8 kg at a 95% significance level?

(d) Estimate the relative efficiency of blocking.

(e) Suppose the value for ration 1, block 4 (y_{14}), was missing because that cow became ill during the trial. Estimate a value for the missing cell and compute the bias in the sum of squares for treatments.

5.5. Sixteen random litters of rats were used as complete blocks in a design to study the effects of 4 diets on the weights of rats measured 3 months after weaning. Suppose the total sum of squares was 150 and the sums of squares for diets and litters each were 30.

(a) Estimate the relative efficiency of blocking. What inference should be made about designs for future experiments of this type?

(b) If a true interaction between treatments (diets) and blocks (litters) exists ($\sigma^2_{TD} = 1$) but is unknown to the experimenter, how much bias exists in the f ratio used to test dietary effects?

(c) Given $\sigma^2_{TD} = 1$, reestimate the relative efficiency of blocking and compare inference with that in (a).

(d) Given knowledge that $\sigma^2_{TD} \simeq 1$, estimate the number of litters that should be used in a future similar experiment with 5 diets to have 0.9 power to detect a mean difference of 1.1 g between two diets in an orthogonal contrast, at a 99% significance level.

(e) Estimate the amount of correlation between weaning weight and weight 3 months later that would have to exist for a covariance analysis (x = weaning weight) to be as efficient as the randomized block design (with litters = blocks) was estimated to be in (c).

5.6. A factorial experiment was planned to study the effects of early feeding of colostrum on serum levels of immunoglobulin in dairy calves. Factor A was quantity fed (0.5 or 1.5 kg) and factor B was time of first feeding (1, 2, 6, or 12 hours after birth). At a later time the experiment was repeated. Consider the 2 experiments as blocks with random effects. Values recorded are turbidimetric units relative to a barium sulfate standard of 20 when blood was sampled 48 hours after birth. Colostrum was pooled to eliminate variation among dams.

		Time of First Feeding			
Block	Amount	1	2	6	12
1	0.5	7.9	10.2	6.1	2.3
	1.5	11.7	10.7	9.9	5.4
2	0.5	9.5	6.0	7.8	7.1
	1.5	15.0	11.7	9.4	7.2

(a) Test for interaction and for main effects of A and B.

(b) Compute the standard errors for A means, B means, and combination means.

(c) Estimate the relative efficiency of blocking.

5.7. The pyridoxine content of 10 random lots (blocks) of evaporated milk was determined by microbial assays with A = 2 different organisms (*Saccharomyces carlsbergensis* and *Neurospora sitophilus*) from aliquots taken B = before and after sterilizing (Hodson 1956). The response is recorded as y = 0.6 + log(mg/L). Note that one value is missing.

	S.c.		*N.s.*	
Block	Before	After	Before	After
1	0.131	0.079	0.214	0.111
2	.206	.207	.167	.129
3	.168	.156	.160	.134
4	.124	.133	.161	.104
5	.159	.201	.180	.135
6	.087	...	.197	.126
7	.109	.161	.189	.076
8	.175	.070	.179	.073
9	.142	.112	.161	.145
10	0.197	0.174	0.156	0.090

(a) Estimate the missing value (y_{126}) and compute the bias in the sum of squares for treatment combinations.
(b) Test for main effects of organisms (A) and sterilization (B) and for interaction. What additional conclusions should be made?

5.8. An experiment is to be planned to compare the rate of gain of lambs fed different diets (standard = moderate rate of substitution of dehydrated, sterilized poultry feces; high rate of the same). Lambs will not be fed individually but by pens; 72 lambs and 12 pens are available. The lambs should be blocked by initial weight.
(a) Describe the randomization and blocking procedure.
(b) Complete a table for the analysis of variance, including sources of variation, degrees of freedom, and expected mean squares. How large must an f statistic be for 95% confidence that treatments differ?
(c) A prior estimate of the standard deviation among lambs of similar weight treated alike is 0.08 kg/day. A prior estimate of the component of variance for pen effects is $\hat{\sigma}_P^2 = 4 \times 10^{-4}$. Estimate the minimum mean difference between the two rates of substitution that one could detect with 80% chance for an orthogonal contrast at a 95% significance level.

CHAPTER 6

Incomplete Block Designs and Fractional Replication

In designing experiments with a relatively large anticipated number of treatments or treatment combinations (t), it may be difficult to select replicates (complete blocks) of t experimental units that are relatively homogeneous with respect to one or more nuisance variables (e.g., age and weight of animals). Also, block size sometimes is restricted by practical considerations, such as unavailability of enough subjects at the same time or inability to perform the mechanics of the experiment on many subjects at the same time.

It is desirable to form small blocks of homogeneous subjects to control experimental error variance and thus to obtain powerful tests of hypotheses and precise estimates of parameters. Therefore, designs are needed that permit small groupings of subjects (fewer subjects per block than the number of treatments) for experiments with many treatments.

For some experiments with several treatment factors, resources may be so restricted (as to availability or cost) that not even one replicate of the desired treatment combinations is feasible. If the experimenter is unwilling to eliminate any combinations, perhaps because of potential interactions among the factors involved, designs are needed that permit estimation of the more important factorial effects with only fractional replication.

In all the incomplete block and fractionally replicated designs some *confounding* occurs; i.e., certain effects are inseparable or only partially separable by analysis of data from the designed experiment. In factorial experiments, incomplete block effects usually are confounded with effects of one or more of the higher ordered interactions thought to be negligible. In fractional designs the lower ordered factorial treatment effects (main effects and 2-factor interactions) are deliberately confounded with higher ordered effects thought to be negligible.

Whereas incomplete block designs sometimes are quite valuable, their use requires a commitment on the part of the experimenter to observe the "rules of the game" with utmost scrupulousness. As the consultant W. E. Deming once observed, "Many a study, launched on the ways of elegant statistical design, later boggled in execution, ends up with results to which the theory of probability can contribute little."

6.1. CRITERIA FOR CLASSIFYING DESIGNS

Incomplete block designs may be divided into 2 classes: (1) designs with only one blocking restriction (usually based on a single nuisance variable) and (2) designs with 2 or more blocking restrictions (usually derivatives of Latin squares). The first group is divided into nonfactorial experiments and factorial experiments for discussion in this chapter. Designs with 2 or more restrictions are discussed in Chap. 7 (see Classification of Experimental Designs following Chap. 9).

Designs for nonfactorial experiments may be further classified according to *dimensionality* and degree of *balance*. The incomplete blocks in one-dimensional designs cannot be grouped to form a complete block (replicate) in which each treatment appears exactly once, whereas that is possible in multidimensional designs because the number of treatments involved is an integral multiple or power (usually the square) of the block size. Multidimensional designs often are easier to construct and many permit very small block size, but most are quite restrictive in that only a limited set of numbers of treatments can be used. Designs in which each treatment occurs an equal number of times in the same block with each of the other treatments are said to be completely balanced. Complete balance permits simplified analysis and equal precision for all estimates of treatment differences. In partially balanced designs all treatments are not paired in the same block equally often (some treatments may never be paired in the same block). Although the analysis of data produced from such designs is complex, the designs are useful when complete balance requires too many expensive replications or is impossible or unwarranted because of natural or practical restrictions on block size. (For example, a litter of rats is a natural block of restricted size.)

Incomplete block designs for factorial experiments are *symmetrical* if all factors involved have the same number of levels or *asymmetrical* if they do not. Commonly, effects of one or more higher ordered interactions thought to be negligible are deliberately confounded with differences between incomplete blocks as a method of constructing a design with the desired reduction in block size. Split-plot designs involving subjects within blocks may be symmetrical or asymmetrical, but they involve confounding of one or more main effects of factors instead of interactions between factors. That situation often arises out of practical restrictions on the ability to randomize all factors simultaneously, or less commonly when the main effect of one factor is well known a priori but interactions involving that factor are of interest. Because of the similarity of such split-plot designs to some designs that involve repeated measurement, they are considered in Chap. 8.

Designs for factorial experiments also may be classified by amount of replication and degree of confounding. Some replicated designs (those having more than one subject per treatment combination) are constructed to involve partial confounding, i.e., different interactions are confounded with blocks in each replicate. In such cases all factorial effects are estimable but not with equal precision. Unreplicated designs (those having exactly one subject per treatment combination) always involve complete confounding of one or more interactions with blocks, so the confounded interactions are not estimable. Fractionally replicated designs may involve regular fractions (e.g., 1/2, 1/3, 1/4) or irregular fractions (e.g., 2/3, 3/4) and also may be classified by degree of *resolution* and by orthogonality (versus nonorthogonality) of estimated effects. In fractional designs, main effects and 2-factor interactions are confounded with higher ordered interactions, so proper interpretation of results depends on the validity of assuming that higher ordered interactions have negligible effects. The degree of resolution (ϕ) is defined for designs that permit estimation of factorial effects up to order $m < \phi/2$ when all interactions of order $\phi - m$ and higher are negligible ($m = 1$ for main effects, $m = 2$ for 2-factor interactions, etc.). Most-used plans are of resolution ϕ = III, IV, or V. Resolution III implies a main-effect plan with all interactions assumed to be negligible. Only 3-factor and higher ordered interactions are assumed to be negligible in designs of resolution IV. Resolution V permits estimation of 2-factor interactions as well as main effects if all higher ordered interactions are negligible.

In this text designs for factorial experiments are divided into symmetrical 2-level designs, symmetrical 3-level designs, and other designs. The classification of incomplete block designs may be placed in context by examining the broader outline of designs in the classification following Chap. 9.

6.2. NONFACTORIAL EXPERIMENTS

Incomplete block designs were developed, for experiments concerned with a relatively large number of treatments, to obtain improved precision by reducing block size without sacrificing completely the information on any treatment effect. The goal is achieved by partial confounding of the effects of all treatments with differences among incomplete blocks, leading to a more complex experimental plan and analysis. In some fields of research, use of these designs is dictated by physical limitations. For example, in food-tasting experiments the preparations scored by one judge form an incomplete block if the number of items to be compared exceeds capacity to distinguish differences in taste at one sitting. In some nutritional studies with large animals (e.g., energy balance trials), a shortage of complex equipment may preclude testing of more than 3 or 4 animals at one time or there may be limitations because specially adapted animals are required (e.g., fistulated ruminants or ovariectomized females). In animal behavior research, the intensive conditioning and training of animals often required may severely limit the number of animals available for any one trial (block). Some studies performed with multiparous species, such as rats or pigs, involve more treatments than can be assigned to a single litter (block) of animals. In clinical trials, cohorts of patients (i.e., groups of the same age) available over a short period of time may be rather small.

6.2.1. One-Dimensional Balanced Designs

6.2.1.1. *General Structure of Balanced Designs*.

Consider a proposed experiment with t treatments. The experimental units or subjects to be used will be grouped into b blocks of k each ($k < t$), the subjects within each block being homogeneous with respect to one or more nuisance variables such as age or weight. If block size (k) is fixed by choice or practical limitations, the number of blocks necessary for complete balance remains to be determined. Note that the total number of subjects in the experiment will be $n = bk = rt$, where r is the number of replications of each treatment. For complete balance, each treatment must occur exactly λ times in the same block with each of the other treatments ($\lambda = 1, 2, 3, \ldots$). A particular treatment (say treatment i) can be paired in the same block somewhere in the design with $t - 1$ other treatments, i.e., $\lambda(t - 1)$ pairs in the experiment involve treatment i. The same treatment also occurs in r blocks and in each is paired with $k - 1$ other treatments, i.e., $r(k - 1)$ pairs in the experiment involve treatment i. Thus $\lambda(t - 1) = r(k - 1)$, or $\lambda = r(k - 1)/(t - 1)$. Because λ must be a positive integer, the possibilities for complete balance are somewhat limited. An additional practical restriction is the cost of replication required for complete balance. The minimum number of replications required is $r = \lambda(t - 1)/(k - 1)$. Suppose, for example, that an experiment is to involve $t = 25$ treatments and block size is restricted to $k \leq 8$ (e.g., blocks = litters). Then $r = 24\lambda/(k - 1)$ and it is clear, for small λ, that possible values of k are 2, 3, 4, 5, and 7. The corresponding minimal amounts of replication required (with $\lambda = 1$) are $r = 24$, 12, 8, 6, and 4 animals per treatment, or $n = 25r$ = 600, 300, 200, 150, and 100 total animals. The last case ($k = 7$) must be eliminated because the number of blocks required is $b = n/k = 100/7$, not an integer. Actually, the only design plan readily available at this time is for $k = 4$ (Plan 11.36, Cochran and Cox 1957). However, this example permits more choices for k than are available in most cases. An experiment with $t = 20$ treatments, where $\lambda = r(k - 1)/19$, does not permit complete balance for $n <$ 380. One would have to use a partially balanced design or reduce the treatments to a number that permits complete balance (e.g., $t = 19$). Table 11.3 of Cochran and Cox provides a listing of some published design plans.

As a simple example of a design plan, consider the case of $t = 7$ treatments with block size of $k = 3$. Then $\lambda = 2r/6$, so $r = 3$ replicates per treatment for the smallest balanced experiment with $n = tr = 21$ subjects. Since $n = bk$, then $b = 7$ incomplete blocks. For $\lambda = 1$, each treatment is to occur exactly once in the same block with each of the other 6 treatments. The following assignment of treatments to blocks fulfills the requirements: block 1 = 1, 2, 4; block 2 = 1, 3, 7; block 3 = 1, 5, 6; block 4 = 2, 3, 5; block 5 = 2, 6, 7; block 6 = 3, 4, 6; block 7 = 4, 5, 7. Designs for which the number of treatments t is not a multiple of the block size k are necessarily one dimensional, i.e., incomplete blocks cannot be grouped such that a group of blocks forms exactly one complete block or replicate. No completely general method of constructing these designs exists. Some plans are obtainable from prime power factorial systems (Sec. 6.5.1.2), completely orthogonal squares (Chap. 7), or cyclic enumeration. Bose (1939) described the use of finite geometries and symmetrically repeated differences to construct plans. Systematic trial and error, using the fewest replicates possible, is often fairly simple. Fortunately, many practical plans for assigning treatments to blocks have been provided by Cochran and Cox.

6.2.1.2. *Intrablock Analysis.* A model for intrablock analysis may be written in exactly the same way as for complete blocks [(5.1)]. For simplicity, the randomization error discussed in Sec. 5.3 will be omitted without loss of perspective if the block effects are of no interest per se (the usual case in incomplete block experiments). Letting τ_i represent the fixed effect of the ith treatment and δ_j represent the fixed effect of the jth block, one may write

$$Y_{ij} = \mu + \tau_i + \delta_j + E_{(ij)} \qquad (i = 1, 2, \ldots, t;\ j = 1, 2, \ldots, b) \tag{6.1}$$

where $E_{(ij)}$ is the random experimental error, assumed to be normally distributed with mean zero and component of variance σ^2. Note again that any interaction between the effects of treatments and blocks is completely confounded with error. For simplicity in the development of the analysis, the interaction effect is deleted from the model because it is assumed to be negligible.

Although (6.1) is a simplified version of (5.1) used for complete block designs, there is a distinction. The effects of treatments and blocks are orthogonal for complete block designs but are not orthogonal for incomplete block designs because all treatments do not occur in every block.

The basic problem of analysis is to disentangle the effects of treatments and blocks, i.e., to estimate the treatment means ($\mu + \tau_i$) unbiasedly. To illustrate, consider a particular treatment assigned to r subjects somewhere in the incomplete block design and having effect $\tau_{i'}$. Restricting summation to the r blocks in which treatment i' occurs, one may write a parametric expression for the total of responses to that treatment:

$$\sum_{j(i')}^{r} Y_{i'j} = r\mu + r\tau_{i'} + \sum_{j(i')}^{r} \delta_j + \sum_{j(i')}^{r} E_{(i'j)} = r\bar{Y}_{i'.} \tag{6.2}$$

Similarly, restricting summation to the k treatments that occur in a particular block j', one may write a parametric expression for that block total:

$$\sum_{i(j')}^{k} Y_{ij'} = k\mu + \sum_{i(j')}^{k} \tau_i + k\delta_{j'} + \sum_{i(j')}^{k} E_{(ij')} \tag{6.3}$$

Next, consider the total of all blocks in which treatment i' occurs (rk subjects):

$$\sum_{j(i')}^{r} \left(\sum_{i(j)}^{k} Y_{ij} \right) = rk\mu + \sum_{j(i')}^{r} \left(\sum_{i(j)}^{k} \tau_i \right) + k \sum_{j(i')}^{r} \delta_j + \sum_{j(i')}^{r} \left(\sum_{i(j)}^{k} E_{(ij)} \right)$$

Note that summation of treatment responses on index i is restricted to blocks in which treatment i' occurs. The problem of dependency between the equations for estimating μ and the τ_i (discussed in Chap. 2) arises here also. As before, an arbitrary but reasonable solution is to define the treatment effects as deviations from μ such that $\sum_{i=1}^{t} \tau_i = 0$. Therefore, $\tau_{i'} = -\sum_i \tau_i$ $(i \neq i')$ and

$$\sum_{j(i')}^{r} \left(\sum_{i(j)}^{k} \tau_i \right) = r\tau_{i'} + \lambda \sum_{i \neq i'} \tau_i = (r - \lambda)\tau_{i'}$$

where λ is the number of times treatment i' appears in the same block with each of the other treatments. Now, the total of all blocks in which treatment i' occurs may be written:

$$\sum_{j(i')}^{r} \left(\sum_{i(j)}^{k} Y_{ij} \right) = rk\mu + (r - \lambda)\tau_{i'} + k \sum_{j(i')}^{r} \delta_j + \sum_{j(i')}^{r} \left(\sum_{i(j)}^{k} E_{(ij)} \right) = \sum_{j(i')}^{r} k\bar{Y}_{.j} \tag{6.4}$$

Equations (6.2) and (6.4) involve no treatment effects except $\tau_{i'}$ and involve the same set of block effects $\left(\sum_{j(i')}^{r} \delta_j \right)$ that in (6.4) are multiplied by k. Since (6.2) involves r subjects and (6.4) involves rk subjects, k(6.2) - (6.4) is a proper contrast and will eliminate μ and block effects completely, i.e., the difference becomes

$$kr\bar{Y}_{i'.} - \sum_{j(i')}^{r} k\bar{Y}_{.j} = (kr - r + \lambda)\tau_{i'} + k \sum_{j(i')}^{r} E_{(i'j)} - \sum_{j(i')}^{r} \left(\sum_{i(j)}^{k} E_{(ij)} \right) \tag{6.5}$$

Solving (6.5) for $\tau_{i'}$,

$$\tau_{i'} = [(kr\overline{Y}_{i'.} - \sum_{j(i')}^{r} k\overline{Y}_{.j}) - k \sum_{j(i')}^{r} E_{(i'j)} + \sum_{j(i')}^{r} (\sum_{i(j)}^{k} E_{(ij)})]/(kr - r + \lambda) \tag{6.6}$$

If the experimental errors are truly random and have zero expectation as assumed, the expected value of the right-hand side of (6.6) with Es deleted equals $\tau_{i'}$, so an unbiased estimator of $\tau_{i'}$ is

$$\hat{\tau}_{i'} = (kr\overline{Y}_{i'.} - \sum_{j(i')}^{r} k\overline{Y}_{.j})/(kr - r + \lambda) \tag{6.7}$$

This result also satisfies the conditions of the theory of least squares estimation (Secs. 1.5.3, 2.1.2). Note that

$$kr - r + \lambda = r(k - 1) + \lambda = \lambda(t - 1) + \lambda = t\lambda$$

Therefore, in practice one may compute an unbiased estimate of treatment effect $\tau_{i'}$ from

$$\hat{\tau}_{i'} = (k/t\lambda)(r\overline{y}_{i'.} - \sum_{j(i')}^{r} \overline{y}_{.j}) \tag{6.8}$$

where $\overline{y}_{i'.}$ is the sample treatment mean and the $\overline{y}_{.j}$ are the sample means of the blocks that contain treatment i'. As usual, μ may be estimated unbiasedly by $\hat{\mu} = \overline{Y}$, the overall sample mean, so an unbiased estimator of $\mu + \tau_{i'}$ may be obtained from $\hat{\mu} + \hat{\tau}_{i'}$. The variance of this estimator is

$$V[\overline{Y}] + (k^2/t^2\lambda^2)\{r^2V[\overline{Y}_{i'.}] + \sum_{j(i')}^{r} V[\overline{Y}_{.j}] - 2r \sum_{j(i')}^{r} \text{Cov}[\overline{Y}_{i'.}, \overline{Y}_{.j}]\}$$

$$= (\sigma^2/rt) + (k^2/t^2\lambda^2)\{r^2(\sigma^2/r) + r(\sigma^2/k) - 2r \sum_{j(i')}^{r} \text{Cov}[(Y_{i'.}/r), (Y_{.j}/k)]\}$$

where the treatment total $Y_{i'.}$ and each relevant block total $Y_{.j}$ have exactly one subject response in common. Therefore, each covariance term involves the variance σ^2 of the one common response. Then the variance becomes

$$(\sigma^2/rt) + (k^2/t^2\lambda)[r\sigma^2 + (r/k)\sigma^2 - 2r \sum_{j(i')}^{r} (\sigma^2/rk)]$$

$$= (\sigma^2/rt) + (k^2/t^2\lambda^2)[r(k-1)\sigma^2/k] = (\sigma^2/rt)\{1 + (k^2/t^2\lambda^2)[(rt)\lambda(t-1)/k]\}$$

$$= (\sigma^2/rt)[1 + kr(t - 1)/t\lambda]$$

The standard error of a sample estimate of $\mu + \tau_{i'}$, obtained from $\bar{y}$ plus (6.8), is

$$s_{\hat{\mu}+\hat{\tau}_{i'}} = \sqrt{(ms_E/rt)[1 + kr(t - 1)/t\lambda]} \quad (6.9)$$

where ms_E is the mean square error from the analysis of variance to be described.

The difference in effects of 2 treatments may be estimated by $\hat{\tau}_1 - \hat{\tau}_2$ from (6.8). The variance of the estimator is

$$V[\hat{\tau}_1] + V[\hat{\tau}_2] - 2\mathrm{Cov}[\hat{\tau}_1, \hat{\tau}_2]$$

From the derivation of (6.9), the sum of the variances is $(2\sigma^2/rt)[kr(t - 1)/t\lambda]$; and it can be shown (Graybill 1961) that the covariance between estimators of two different treatment effects is $-(k/t^2\lambda)\sigma^2$. Combining the two results, one obtains

$$V[\hat{\tau}_1 - \hat{\tau}_2] = (2\sigma^2/r)(kr/t\lambda) \quad (6.10)$$

The standard error of a sample estimate of the difference is

$$s_{\hat{\tau}_1-\hat{\tau}_2} = \sqrt{(2ms_E)(k/t\lambda)} \quad (6.11)$$

Recall that the standard error of the difference between two treatment means in completely randomized or complete block designs is $\sqrt{2ms_E/r}$. If the true experimental error variance pertaining to the incomplete block design is σ_1^2 and that pertaining to the complete block design with the same numbers of treatments and replications is σ_2^2, the relative efficiency of the incomplete block design with respect to precise estimation of treatment differences is

$$E_{1:2} = (2\sigma_2^2/r)/[(2\sigma_1^2/r)(kr/t\lambda)] = (t\lambda/kr)(\sigma_2^2/\sigma_1^2) \quad (6.12)$$

If $\sigma_1^2 = \sigma_2^2$, the relative efficiency is $(t\lambda/kr) = E$. This ratio, which is usually reported for specific design plans (Cochran and Cox 1957), is really the minimal efficiency of the incomplete block design, because often the purpose (and usual outcome) of forming small blocks is to obtain $\sigma_1^2 < \sigma_2^2$. Consider a design with $t = 16$ treatments, $k = 6$ subjects per block, $r = 6$ subjects per treatment, and $\lambda = 2$. Then $E = [(16)(2)/(6)(6)] = 0.89$. If the use of incomplete blocks is expected to be effective in reducing σ_1^2 to the degree that $\sigma_1^2/$

$\sigma_2^2 < E$, it is expected that $E_{1:2} > 1$, i.e., the incomplete block design is likely to be more efficient than the corresponding complete block design. For some incomplete block plans, E is rather low. For example, if $t = 11$ treatments are to be compared using block size of $k = 2$ (e.g., two brothers, twins, or other pairs), complete balance with $\lambda = 1$ requires $r = 10$ subjects per treatment and $E = [(11)(1)/(2)(10)] = 0.55$. In that case reduction of σ_1^2 may not be sufficient to reach 100% efficiency. Another way to improve efficiency, if the effects of blocks are random instead of fixed, is by recovery of interblock information (Sec. 6.2.1.3). Unfortunately, such recovery rarely has been found to be worthwhile in terms of precision unless $k \geq 5$ subjects per block. Since most practical plans having low E also have $k < 5$ (Cochran and Cox), the use of such plans should be restricted to cases with good opportunity to improve the homogeneity of subjects by very small groupings (such as $k = 2$ identical twins) or with natural limitations on block size that preclude using any other design.

The partition of variation without recovery of interblock information is the same as for complete block designs with the exception of the sum of squares for treatments. The total sum of squares may be computed from

$$ss_y = \sum_{i=1}^{t} \sum_{j=1}^{r} y_{ij}^2 - (\sum_{i=1}^{t} \sum_{j=1}^{r} y_{ij})^2/n = \sum_{i=1}^{t} \sum_{j=1}^{r} y_{ij}^2 - n\bar{y}^2 \tag{6.13}$$

as usual, except that j is used here to denote replications of treatments without regard to their blocks. The sum of squares for blocks may be computed from

$$ss_D = \sum_{j=1}^{b} y_{\cdot j}^2/k - (\sum_{i=1}^{t} \sum_{j=1}^{r} y_{ij})^2/n = k \sum_{j=1}^{b} \bar{y}_{\cdot j}^2 - n\bar{y}^2 \tag{6.14}$$

where $y_{\cdot j}$ is the total of k responses in a block. The sum of squares of treatments adjusted for block effects may be computed from

$$ss_{T(\text{adj})} = (k/t\lambda) \sum_{i=1}^{t} (r\bar{y}_{i\cdot} - \sum_{j(i)}^{r} \bar{y}_{\cdot j})^2 \tag{6.15}$$

where $\bar{y}_{i\cdot}$ and $\bar{y}_{\cdot j}$ are treatment and block means, respectively, as defined in the development of (6.8). The sum of squares for experimental error (residual) may be obtained by difference,

$$ss_E = ss_y - ss_D - ss_{T(\text{adj})} \tag{6.16}$$

and has $\nu_E = n - t - b + 1$ degrees of freedom (df). Treatments may be tested by the usual f ratio of mean squares for treatments and error, with $t - 1$ and ν_E df. Orthogonal contrasts among the $\hat{\tau}_i$ of (6.8) (i.e., $\bar{q}_k = \sum_{i=1}^{t} c_{ik}\hat{\tau}_i$ with

$\sum_{i=1}^{t} c_{ik} = 0$) have sample variance, $\hat{V}[\bar{q}_k] = \sum_{i=1}^{t} c_{ik}^2 ms_E/(rE)$, and thus are significant if $t = \bar{q}_k/\sqrt{\hat{V}[\bar{q}_k]}$ is outside the limits of $\pm t_{\alpha/2,\nu_E}$.

6.2.1.3. *Recovery of Interblock Information.* Yates (1940) showed that not all the information about treatments is contained in the intrablock analysis. Equation (6.4) expresses the total of responses for all blocks in which treatment i' occurs as $\sum_{j(i)}^{r} k\bar{Y}_{\cdot j}$. The difference between such totals for 2 treatments having λ blocks in common is

$$k\left(\sum_{j(1)}^{r} \bar{Y}_{\cdot j} - \sum_{j(2)}^{r} \bar{Y}_{\cdot j}\right) = (r - \lambda)(\tau_1 - \tau_2) + k\left(\sum_{j(1,\bar{2})}^{r-\lambda} \delta_j - \sum_{j(2,\bar{1})}^{r-\lambda} \delta_j\right)$$

$$+ \left[\sum_{j(1,\bar{2})}^{r-\lambda} \left(\sum_{i(j)}^{k} E_{(ij)}\right) - \sum_{j(2,\bar{1})}^{k-\lambda} \left(\sum_{i(j)}^{k} E_{(ij)}\right)\right] \qquad (6.17)$$

where the summation index $j(1,\bar{2})$ indicates blocks containing treatment 1 but not treatment 2. If one divides both sides of the equation by $r - \lambda$, it is apparent that the expectation of the resulting left-hand side (assuming fixed effects of treatments and random errors) involves

$$(\tau_1 - \tau_2) + [k/(r - \lambda)]\left(\sum_{j(1,\bar{2})}^{r-\lambda} \delta_j - \sum_{j(2,\bar{1})}^{r-\lambda} \delta_j\right)$$

If the effects of blocks are considered fixed (blocks based on a scaled variable, such as weight or age of animals), the second term represents the bias of an interblock estimator of $\tau_1 - \tau_2$, i.e., the bias is k times the average difference between effects of two groups of blocks. In each group only one of the two treatments in question appears. Because of this bias, recovery of interblock information is not recommended when the effects of blocks are fixed. However if the effects of blocks are considered random (e.g., blocks based on litters of rats), the term involving the δ_j (or D_j for consistent notation for random effects) has zero expectation so that an unbiased interblock estimator of the difference between two treatments is

$$\hat{\tau}_1 - \hat{\tau}_2 = [k/(r - \lambda)]\left(\sum_{j(1)}^{r} \bar{Y}_{\cdot j} - \sum_{j(2)}^{r} \bar{Y}_{\cdot j}\right) \qquad (6.18)$$

where the $\bar{Y}_{\cdot j}$ are means of blocks in which treatments 1, 2, or both are involved. The variance of the interblock estimator is

$$V[\hat{\tau}_1 - \hat{\tau}_2] = [k^2/(r - \lambda)^2]\left(\sum_{j(1,\bar{2})}^{r-\lambda} V[D_j] + \sum_{j(2,\bar{1})}^{r-\lambda} V[D_j]\right) \qquad (6.19)$$

$$+ [1/(r - \lambda)^2]\{\sum_{j(1,\bar{2})}^{r-\lambda} \sum_{i(j)}^{k} V[E_{(ij)}] + \sum_{j(2,\bar{1})}^{r-\lambda} \sum_{i(j)}^{k} V[E_{(ij)}]\}$$

$$= [k^2/(r - \lambda)^2][2(r - \lambda)\sigma_D^2] + [1/(r - \lambda)^2][2k(r - \lambda)\sigma^2]$$

$$= 2(\sigma^2 + k\sigma_D^2)/[(r - \lambda)/k]$$

where σ_D^2 is the component of variance among the random effects of blocks. Recall [(6.10)] that the corresponding variance of the intrablock estimator of $\tau_1 - \tau_2$ is $(2\sigma^2/rE)$, where $E = (t\lambda/kr)$.

One may combine the intrablock and interblock estimators of $\tau_1 - \tau_2$. Graybill (1961) has shown that the best (minimum variance) linearly combined estimator of any parameter is obtained by weighting the individual estimators by the inverse variances. When the true variances are not known one must obtain unbiased estimates. Letting the intrablock and interblock estimators be represented by $\hat{\Delta}_1$ and $\hat{\Delta}_2$ and the respective estimated variances by $\hat{\sigma}_1^2$ and $\hat{\sigma}_2^2$ [see (6.10), (6.19)], one may write the combined estimator as

$$(\hat{\tau}_1 - \hat{\tau}_2)_C = (\hat{\sigma}_2^2\hat{\Delta}_1 + \hat{\sigma}_1^2\hat{\Delta}_2)/(\hat{\sigma}_1^2 + \hat{\sigma}_2^2) = \hat{\phi}\hat{\Delta}_1 + (1 - \hat{\phi})\hat{\Delta}_2 \tag{6.20}$$

where $\hat{\phi} = \hat{\sigma}_2^2/(\hat{\sigma}_1^2 + \hat{\sigma}_2^2)$. Ignoring the errors of estimating the variances (assuming $\hat{\phi}$ to be ϕ), one may describe the approximate variance of the combined estimator as

$$V[\hat{\tau}_1 - \hat{\tau}_2]_C \simeq \hat{\phi}^2 V[\hat{\Delta}_1] + (1 - \hat{\phi})^2 V[\hat{\Delta}_2] \simeq \hat{\phi}^2\hat{\sigma}_1^2 + (1 - \hat{\phi})^2\hat{\sigma}_2^2 \tag{6.21}$$

Graybill shows that the intrablock and interblock estimators are independent and therefore have zero covariance.

Graybill also shows that an unbiased combined estimator of a single treatment effect is

$$(\hat{\tau}_i)_C = (\bar{Y}_{i.} - \bar{Y}) + \{\sigma_D^2/[(t - 1)\sigma^2 + t(k - 1)\sigma_D^2]\}\{(t - k)(\bar{Y}_{i.} - \bar{Y})$$

$$- [k(t - 1)/r][\sum_{j(i)}^{r} \bar{Y}_{.j} - r\bar{Y}]\} \tag{6.22}$$

Note that the first term, $\bar{Y}_{i.} - \bar{Y}$, is the appropriate estimator of τ_i in a complete block analysis. To be able to compute estimates of τ_i or $\tau_1 - \tau_2$ using the combined estimators of (6.22) and (6.20), one must obtain estimates of σ^2 and σ_D^2. The computations for sums of squares described by (6.13)-(6.16) are required. One must also compute the sum of squares for blocks adjusted for treatments, $ss_{D(adj)}$. Since $ss_D + ss_{T(adj)} = ss_{D(adj)} + ss_T$, it follows that one may compute

$$ss_{D(adj)} = ss_D + ss_{T(adj)} - ss_T \tag{6.23}$$

where the first two terms come from (6.14) and (6.15), and the unadjusted sum of squares for treatments is

$$ss_T = (\sum_{i=1}^{t} y_{i.}^2/r) - (\sum_{i=1}^{t} \sum_{j=1}^{r} y_{ij})^2/n = r\sum_{i=1}^{t} \bar{y}_{i.}^2 - n\bar{y}^2 \tag{6.24}$$

as in complete block analysis. Graybill shows that expected value of the adjusted block sum of squares is

$$E[ss_{D(adj)}] = (b - 1)\sigma^2 + (n - t)\sigma_D^2$$

The analysis of variance is summarized in Table 6.1. From the expected mean squares it is clear that the components of variance for error and blocks may

Table 6.1. Analysis of Variance for Random Effects of Incomplete Blocks

Source of Variation	df	ss	Equation	ms	E[MS]
Treatments	$t-1$	ss_T	(6.24)	...	
Treatments (adj.)	...	$[ss_{T(adj)}]$	(6.15)	...	
Blocks	...	$[ss_D]$	(6.14)	...	
Blocks (adj.)	$b-1$	$ss_{D(adj)}$	(6.23)	ms_D	$\sigma^2+[(n-t)/(b-1)]\sigma_D^2$
Intrablock error	$n-t-b+1$	ss_E	(6.16)	ms_E	σ^2.
Total	$n-1$	ss_Y	(6.13)		

be estimated unbiasedly by

$$\hat{\sigma}^2 = ms_E \tag{6.25}$$

$$\hat{\sigma}_D^2 = (ms_D - ms_E)/[(n - t)/(b - 1)] \tag{6.26}$$

If a negative estimate of σ_D^2 is obtained, σ_D^2 is taken to be zero and the recovery of interblock information is abandoned.

Zelen (1957) has shown how to make a combined test of treatment effects, using both intrablock and interblock information, if the number of blocks (b)

exceeds the number of treatments (t). It requires partitioning ss_D [see (6.14)] into an interblock treatment sum of squares,

$$ss_{T'} = k \sum_{i=1}^{t} \left(\sum_{j(i)}^{r} \bar{y}_{\cdot j} - r\bar{y} \right)^2 / (r - \lambda) \tag{6.27}$$

with $t - 1$ df, and the remainder,

$$ss_{D'} = ss_D - ss_{T'} \tag{6.28}$$

with $b - t$ df. Then $ms_{T'} = ss_{T'}/(t - 1)$, $ms_{D'} = ss_{D'}/(b - t)$, and the interblock test statistic for treatment effects is

$$f' = ms_{T'}/ms_{D'} \tag{6.29}$$

Using App. A.5, one should determine by interpolation the value of α (say α_2) at which $f' \simeq f_{\alpha,t-1,b-t}$. Also, for the intrablock test statistic,

$$f = ms_{T(\text{adj})}/ms_E \tag{6.30}$$

one should determine the value of α (say α_1) at which $f \simeq f_{\alpha,t-1,n-t-b+1}$. Then using the principles of Sec. 1.6.8, one obtains a combined test statistic,

$$q = -2(\log_e \alpha_1 + \log_e \alpha_2) \tag{6.31}$$

which indicates treatment differences at the $(1 - \alpha)100\%$ level of significance if it exceeds $\chi^2_{\alpha,4}$ (App. A.3).

Orthogonal contrasts among the treatments may be estimated unbiasedly from interblock information alone by computing

$$q_k = \sum_{i=1}^{t} c_{ik} \left(\sum_{j(i)}^{r} y_{\cdot j} \right) / (r - \lambda) \tag{6.32}$$

where $\sum_{i=1}^{t} c_{ik} = 0$. The sample variance of q_k is

$$\hat{V}[q_k] = k \sum_{i=1}^{t} c_{ik}^2 (\hat{\sigma}^2 + k\hat{\sigma}_D^2)/(r - \lambda)$$

so the kth contrast is significant if

$$t = q_k / \sqrt{\hat{V}[q_k]} \tag{6.33}$$

is outside the limits $\pm t_{\alpha/2,b-t}$. A combined test using both intrablock and interblock information may be made if one combines the probability of Type I error for this test with that of the corresponding t test described for contrasts in intrablock analysis (previous section) according to (6.31).

6.2.1.4. *Example of Balanced Design*. Consider a study in which a tasting panel is to be convened to score the quality of steaks produced by progeny of $t = 10$ different bulls (treatments). Since a judge's ability to discriminate tastes rapidly diminishes as the number of items is increased, only $k = 5$ samples are to be presented to each judge (block) at one time. Complete balance for this design requires $r = 9$ replications (ratings for each bull), because $\lambda = r(k - 1)/(t - 1) = 4r/9$ and λ must be an integer. Thus there will be $n = tr = 90$ observations to record and $b = n/k = 90/5 = 18$ blocks of 5 steak samples. If $r = 9$, then $\lambda = 4$ comparisons of each pair of bulls in the same block (judge). The minimum efficiency of the design relative to a complete block design is

$$E = (t\lambda/kr) = [(10)(4)/(5)(9)] = 0.89$$

If the judges differ considerably in their notions of average rating or score but are more consistent in rating few samples than many, the actual efficiency quite likely will exceed 100%. Possibly, but not likely, efficiency will be improved if interblock information is recovered (effects of judges considered random). The sample plan, which shows assignment of treatments to blocks (Cochran and Cox 1957) is given in Table 6.2.

Table 6.2. Balanced Design for 10 Treatments in Blocks of 5

Block	Bulls					Block	Bulls					Block	Bulls				
(1)	1	2	3	4	5	(7)	1	4	5	6	10	(13)	2	5	6	8	10
(2)	1	2	3	6	7	(8)	1	4	8	9	10	(14)	2	6	7	9	10
(3)	1	2	4	6	9	(9)	1	5	7	9	10	(15)	3	4	6	7	10
(4)	1	2	5	7	8	(10)	2	3	4	8	10	(16)	3	4	5	7	9
(5)	1	3	6	8	9	(11)	2	3	5	9	10	(17)	3	5	6	8	9
(6)	1	3	7	8	10	(12)	2	4	7	8	9	(18)	4	5	6	7	8

The bulls were assigned treatment numbers from 1 to 10 at random. The judges, assigned block numbers at random, were instructed to score for flavor and tenderness on a scale from 0 to 8, with 8 being best possible. The scores shown in Table 6.3 correspond to the block assignments given for treatments in Table 6.2.

The unadjusted sum of squares for treatments [(6.24)] is

$$ss_T = 9(5.22^2 + 7.11^2 + \ldots + 6.89^2) - 90(5.89^2) = 104.1273$$

Table 6.3. Taste Scores for Balanced Incomplete Block Experiment

Block	Scores					Mean	Block	Scores					Mean	Block	Scores					Mean
(1)	7	8	8	6	7	7.2	(7)	5	5	6	6	7	5.8	(13)	6	4	4	2	5	4.2
(2)	5	7	6	6	7	6.2	(8)	6	5	4	8	8	6.2	(14)	6	5	7	6	6	6.0
(3)	6	8	5	6	7	6.4	(9)	4	4	6	6	6	5.2	(15)	8	6	7	8	8	7.4
(4)	6	7	5	7	4	5.8	(10)	7	7	5	4	8	6.2	(16)	6	4	5	7	6	5.6
(5)	4	5	4	2	4	3.8	(11)	8	8	7	8	8	7.8	(17)	7	5	5	4	6	5.4
(6)	4	6	6	3	6	5.0	(12)	7	4	7	4	6	5.6	(18)	5	6	7	8	5	6.2

Treatment Means: (1) 5.22, (2) 7.11, (3) 6.78, (4) 5.00, (5) 5.44,
(6) 5.56, (7) 7.00, (8) 3.56, (9) 6.33, (10) 6.89; $\bar{y}$=5.89

the unadjusted sum of squares for blocks [(6.14)] is

$$ss_D = 5(7.2^2 + 6.2^2 + \dots + 6.2^2) - 90(5.89^2) = 84.9110$$

and the total sum of squares [(6.13)] is

$$ss_y = (7^2 + 8^2 + \dots + 5^2) - 90(5.89^2) = 187.7110$$

Note that in this case the unadjusted sums of squares for treatments and blocks together exceed the total sum of squares. Adjustments to sums of squares and estimates of treatment effects require computation for each treatment of the total of block means for blocks in which a treatment occurs ($\sum_{j(i)}^{r} \bar{y}_{.j}$). Computations are shown in Table 6.4. As a check on computations,

Table 6.4. Adjustment of Treatments for Block Effects in Balanced Incomplete Block Design

	[A]	[B]	[C]	Adj. Trt. Mean
Treatment	$\bar{y}_{i.}$	$\sum_{j(i)}^{r} \bar{y}_{.j}$	$r[A]-[B]$	$\hat{\mu}+\hat{\tau}_i=\bar{y}+(k/t\lambda)[C]$
1	5.22	51.6	-4.62	5.31
2	7.11	55.4	+8.59	6.96
3	6.78	54.6*	+6.42	6.69
4	5.00	56.6	-11.60	4.44
5	5.44	53.2	-4.24	5.36
6	5.56	51.4	-1.36	5.72
7	7.00	53.0	+10.00	7.14
8	3.56	48.4	-16.36	3.85
9	6.33	52.3	+4.67	6.47
10	6.89	53.8	+8.21	6.92

*Column [B] example: $\sum_{j(3)} \bar{y}_{.j} = 7.2+6.2+3.8+5.0+6.2$ $+7.8+7.4+5.6+5.4 = 54.6$, the total of block means for blocks in which treatment 3 occurs (Tables 6.2, 6.3).

the sum of items in column [B] should equal the overall total ($y_{..}$). The adjusted sum of squares for treatments [(6.15)] is

$$ss_{T(\text{adj})} = [5/(10)(4)](-4.62^2 + 8.59^2 + \dots + 8.21^2) = 93.4498$$

the adjusted sum of squares for blocks [(6.23)] is

$$ss_{D(\text{adj})} = 84.9110 + 93.4498 - 104.1273 = 74.2335$$

the sum of squares for intrablock error [(6.16)] is

$$ss_E = 187.7110 - 84.9110 - 93.4498 = 9.3502$$

and mean square error is 9.3502/63 = 0.1484.

Compare the adjusted treatment means with the unadjusted means (last and first columns of Table 6.4), and note that adjustments not only change the rankings of bulls but comprise as much as 10% of the mean values (see treatment 4). The standard error of each adjusted treatment mean [(6.9)] is

$$s_{\hat{\mu}+\hat{\tau}_i} = \sqrt{(0.1484/90)[1 + (5)(9)(9)/(10)(4)]} = 0.135$$

and the standard error of the difference between two means [(6.11)] is 0.193. The intrablock test of equal treatment effects [(6.15), (6.16)] is $f = (93.4498/9)/(9.3502/63) = 69.96$, which far exceeds $f_{0.001,9,63} = 3.69$, leaving little doubt that treatments (bulls) differ. Tukey's HSD test (Sec. 2.2.4) may be applied to differentiate among adjusted means. Any difference that exceeds

$$s_{\hat{\mu}+\hat{\tau}_i}(q_{0.05,10,63}) = 0.135(4.64) = 0.626 \quad \text{(App. A.8)}$$

is significant at the 95% level. In increasing magnitude, the adjusted means rank the bulls

8 4 1 5 6 9 3 10 2 7

where any two bulls underscored by the same line are not significantly different. It is clear that bulls 4 and 8 are inferior and bulls 2, 3, 7, 9, and 10 are superior. Bull 7 had the highest adjusted mean and is significantly better than all other bulls except 2, 3, and 10.

Since the block (judge) effects are random in this example, one may compute estimates of treatment effects from combined intra- and interblock information [(6.22)] using estimates of the components of variance for blocks and error (see Table 6.1). The analysis is shown in Table 6.5.

The estimated error component of variance is $\hat{\sigma}^2 = 0.1484$, and the estimated component of variance for blocks is

$$\hat{\sigma}_D^2 = (4.3667 - 0.1484)/4.7059 = 0.8964$$

To obtain estimates of treatment means from combined intra- and interblock information [(6.22)], one must first compute

$$\hat{\sigma}_D^2/[(t - 1)\hat{\sigma}^2 + t(k - 1)\hat{\sigma}_D^2] = 0.8964/[9(0.1484) + 40(0.8964)] = 0.0241$$

The remaining computations are displayed in Table 6.6. Note that [B] of Table 6.6 is a function of [B] of Table 6.4. In this example the combined estimates

Table 6.5. Analysis of Variance for Random Effects in Incomplete Block Experiment

Source of Variation	df	*ss*	*ms*	$E[MS]$
Treatments	9	104.1273	...	
Treatments (adj.)	...	[93.4498]	...	
Blocks	...	[84.9110]	...	
Blocks (adj.)	17	74.2335	4.3667	$(\sigma^2+4.7059\sigma_D^2)$*
Intrablock error	63	9.3502	0.1484	σ^2
Total	89	187.7110		

*$[(n-t)/(b-1)] = 80/17 = 4.7059$.

Table 6.6. Combined Intra- and Interblock Estimates of Treatment Means

	$[A]$*	$[B]$†	$[C]$‡	$[D]$§
1	-0.67	-7.0	-0.58	5.31
2	+1.22	+12.0	+1.08	6.97
3	+0.89	+8.0	+0.80	6.69
4	-0.89	+18.0	-1.43	4.46
5	-0.45	+1.0	-0.53	5.36
6	-0.33	-8.0	-0.18	5.71
7	+1.11	0.0	+1.24	7.13
8	-2.33	-23.0	-2.06	3.83
9	+0.44	-3.5	+0.58	6.47
10	+1.00	+4.0	+1.02	6.91

*$[A] = \bar{y}_{i.} - \bar{y}.$

†$[B] = [k(t-1)/r](\sum_{j(i)}^{r} \bar{y}_{.j} - r\bar{y}).$

‡$[C] = [A]+0.0241\{(t-k)[A]-[B]\}.$

§$[D] = \hat{\mu}+\hat{\tau}_c = \bar{y}+[C].$

differ from the corresponding intrablock estimates (Table 6.4) by no more than 0.02.

Although the intrablock test shows little doubt that treatments differ, consider the combined test using both intra- and interblock information. The interblock treatment sum of squares [(6.27)] is

$$ss'_T = 5(-1.4^2 + 2.4^2 + \ldots + 0.8^2)/5 = 48.1300$$

from items in column [B] of Table 6.6, each divided by $k(t - 1)/r$. The remainder of unadjusted sum of squares for blocks is

$$ss'_D = 84.9110 - 48.1300 = 36.7810$$

The interblock test statistic [(6.29)] is $f' = (48.1300/9)(36.7810/8) = 1.16$, which is only a little larger than $f_{0.50,9,8} = 1.01$. In cases such as this, where the number of blocks is only slightly larger than the number of treatments ($b - t = 8$) and block differences are relatively large, the test is likely to have low sensitivity and a combination of probabilities for Type I errors [(6.31)] is not warranted. The intrablock result, based on much more information, should be sustained. The intrablock variance of treatment differences [(6.10)] is

$$(2\hat{\sigma}^2/rE) = 2(0.1484)/[9(0.89)] = 0.03705$$

and the interblock variance [(6.19)] is

$$2(\hat{\sigma}^2 + k\hat{\sigma}_D^2)/[(r - \lambda)/k] = 2[0.1484 + 5(0.8964)]/(5/5) = 9.2608$$

The variance of the combined estimate [see (6.21)] involves $\hat{\phi} = 9.2608/(0.03705 + 9.2608) = 0.996$, and is

$$\hat{\sigma}_C^2 = (0.996^2)(0.03705) + (1 - 0.996^2)(9.2608) = 0.03690$$

Therefore, the estimated efficiency of the combined estimate to the intrablock estimate is (0.03705/0.03690)100% = 100.4%.

Yates (1940) has pointed out that little is gained from recovery of interblock information when block effects are relatively large. Obviously, the gain in efficiency from recovery of interblock information in this example is trivial. Interblock analysis often is like eating an artichoke--you have to wade through so much to get so little!

6.2.2. One-Dimensional Partially Balanced Designs

Partially balanced designs are useful when the number of replications required for complete balance is too large or when block size is restricted (by practical requirements) to some number that does not fit a completely balanced design. The precision of some treatment comparisons may be improved at the expense of others by using partial balance.

6.2.2.1. *General Structure of Partially Balanced Designs.* In partially balanced designs, each treatment occurs more often in the same block with some treatments than with others. Thus, two or more parameters $\{\lambda_g\}$ are required to specify the number of times treatments are paired in the same block, whereas only one λ is required for completely balanced designs. An *associate class* of a particular treatment, say treatment i, is defined as a group of treatments, each of which occurs with treatment i in the same block a constant number of times, i.e., λ_1 times for first associates, λ_2 times for second associates, etc. ($r \geq \lambda_1 > \lambda_2 > \ldots \geq 0$). Because of the inequalities in the numbers of intrablock comparisons, the means of two treatments that are first associates are compared with greater precision than the means of two second associates, which are more precisely compared than third associates, etc. There are $t - 1$ treatments with which treatment i may be compared. If t_g of them are paired with treatment i in λ_g blocks (gth associates of treatment i), $\sum_{g(i)} t_g = t - 1$. Also, there are $k - 1$ pairs of treatments involving treatment i in each of the r blocks in which treatment i appears. Therefore, $\sum_{g(i)} t_g \lambda_g = r(k - 1)$.

For a pair of treatments, say 1 and 2, that are gth associates, let $p^g_{hh'}$ be the number of treatments *common* to the associate classes h of treatment 1 and h' of treatment 2; m associate classes have m^3 constants like $p^g_{hh'}$. Kempthorne (1952) shows that the following conditions are necessary to ensure that the least squares estimation problem is tractable:

1. $p^g_{hh'} = p^g_{h'h}$

2. $t_g p^g_{hh'} = t_h p^h_{gh'}$ if $g < h$

3. $\sum_{h'=1}^{m} p^g_{hh'} = t_h - 1$ if $g = h$

$= t_h$ if $g \neq h$

Given the necessary conditions, only $m(m^2 - 1)/6$ of the m^3 constants are independent. Commonly used designs have only $m = 2$ associate classes with $2^3 = 8$ constants,

$$\{p^1_{11}, p^1_{12}, p^1_{21}, p^1_{22}, p^2_{11}, p^2_{12}, p^2_{21}, p^2_{22}\}$$

If the value for one constant is fixed, values for the other 7 depend on the first value because of the necessary conditions:

1. $p^1_{12} = p^1_{21}$ and $p^2_{12} = p^2_{21}$

2. $t_1 p^1_{21} = t_2 p^2_{11}$ and $t_1 p^1_{22} = t_2 p^2_{12}$

3. $p^1_{11} + p^1_{12} = t_1 - 1 \quad p^2_{21} + p^2_{22} = t_2 - 1 \quad p^2_{11} + p^2_{12} = t_1 \quad p^1_{21} + p^1_{22} = t_2$

Many designs with 2 associate classes have $\lambda_1 = 1$ and $\lambda_2 = 0$, i.e., a particular treatment is compared in the same block once with some treatments and not at all with others. Such designs minimize the number of replications required--the chief advantage of partially balanced over completely balanced designs.

Consider the construction of a design for $t = 9$ treatments in a study of the effects of different rations on nitrogen balance in ruminants. The objectives require expensive digestion stalls and equipment. Suppose only 3 such stalls are available at one time, so blocks based on time are limited in size to $k \leq 3$. For $k = 3$, a completely balanced design requires 4 replicates. For $k = 2$, complete balance requires $r = \lambda(t - 1)/(k - 1) = 8\lambda$, or a minimum of 8 replications of each treatment. For $r = 4$, $n = tr = 36$ and the number of blocks (in time) required is $b = n/k = 36/3 = 12$--too many for most such practical studies. Therefore, consider a partially balanced design for $t = 9$ and $k = 3$, with less replication. Kempthorne has shown that such a design may be constructed with $r = 2$ replicates (6 blocks of 3 animals each) or with $r = 3$ (9 blocks of 3 animals each). Also, Bose et al. (1954) have developed designs with more replicates ($r = 5, 6, \ldots, 10$) for $t = 9$, $k = 3$. If a particular treatment is to be compared one time ($\lambda_1 = 1$) with t_1 treatments and not at all ($\lambda_2 = 0$) with t_2 treatments, then $t_1 + t_2 = 8$ or $t_2 = 8 - t_1$.

For illustration, examine the case with $t_1 = 6$, $t_2 = 2$, or conditions

1. $p^1_{12} = p^1_{21}$ and $p^2_{12} = p^2_{21}$

2. $6p^1_{21} = 2p^2_{11}$ and $6p^1_{22} = 2p^2_{12}$

3. $p^1_{11} + p^1_{12} = 5 \quad p^2_{12} + p^2_{22} = 1 \quad p^2_{11} + p^2_{12} = 6 \quad p^1_{21} + p^1_{22} = 2$

From the third set of conditions it is obvious that one must use either $p^2_{21} = 0$ or $p^2_{22} = 0$. In the latter case $p^2_{21} = 1 = p^2_{12}$ (first conditions) and $p^2_{11} + p^2_{12} = 6$ implies that $p^2_{11} = 5$. However, the second set of conditions requires $6p^1_{21} = 2p^2_{11}$, so $p^2_{11} = 5$ implies $p^1_{21} = 5/3$, which is impossible. Therefore, one must take $p^2_{21} = 0$ and $p^2_{22} = 1$. Now the third set of conditions requires $p^2_{11} + p^2_{12} = 6$ or $p^2_{11} = 6$ because $p^2_{12} = p^2_{21} = 0$. Substitution of $p^2_{11} = 6$ and $p^2_{12} = 0$ into the second set of conditions implies $p^1_{21} = 2 = p^1_{12}$ and $p^1_{22} = 0$. Finally, with $p^1_{12} = 2$, the condition $p^1_{11} + p^1_{12} = 5$ implies $p^1_{11} = 3$. The numbers of first and second associates shared in combinations by two treatments that are first or second associates are summarized by matrices,

$$\mathrm{p}^1_{hh'} = \begin{bmatrix} 3 & 2 \\ 2 & 0 \end{bmatrix} \qquad \mathrm{p}^2_{hh'} = \begin{bmatrix} 6 & 0 \\ 0 & 1 \end{bmatrix}$$

respectively, where h = 1 or 2 for rows and h' = 1 or 2 for columns. That is, two treatments appearing together in the same block will have 3 first and no second associates in common, and two treatments will be first associates of one but second associates of the other. Also, two treatments not appearing together in the same block will have 6 first and 1 second associates in common, and no treatments will be first associates of one but second associates of the other.

Suppose one takes treatments {2, 3, ..., 7} as first associates of treatment 1 and {8, 9} as second associates. Because $p^1_{22} = 0$, any one of the treatments {2, 3, ..., 7} must have {8, 9} as first associates. For treatment 2, choose arbitrarily two treatments from the set {3, 4, ..., 7}, say 6 and 7, as second associates. Because $p^1_{22} = 0$ and treatment 3 is a first associate of treatments 1 and 2, treatments 4 and 5 must be the 2 second associates of treatment 3. Similarly, second associates of all other treatments are fixed. First and second associates of each treatment are listed in Table 6.7.

Table 6.7. Association Scheme for Partially Balanced Design

Treatment	First Associates	Second Associates
1	2 3 4 5 6 7	8 9
2	1 3 4 5 8 9	6 7
3	1 2 6 7 8 9	4 5
4	1 2 6 7 8 9	3 5
5	1 2 6 7 8 9	3 4
6	1 3 4 5 8 9	2 7
7	1 3 4 5 8 9	2 6
8	2 3 4 5 6 7	1 9
9	2 3 4 5 6 7	1 8

Treatments may be properly assigned to blocks by observing the requirements of the association scheme and noting that any 2 first associates must be paired once ($\lambda_1 = 1$) in the same block, and second associates may not appear in the same block ($\lambda_2 = 0$). Beginning with treatment 1, note that first associates are {2, 3, ..., 7}. Let the k = 3 treatments in block 1 be {1, 2, 3}. One cannot use treatments 4 and 5 with treatment 1 for the next block because 5 is a second associate of 4 and $\lambda_2 = 0$. Therefore, let block 2 be {1, 4, 6} and block 3 be {1, 5, 7}, completing the r = 3 replicates of treatment 1. Next examine the remaining first associates of treatment 2, which are {4, 5, 8, 9}. Since 4 and 5 may not appear together nor may 8 and 9, let block 4 be {2, 4, 8} and block 5 be {2, 5, 9}. The remaining first associates of treatment 3 are {6, 7, 8, 9}. Treatments 6 and 7 may not be paired nor may 8 and 9. Therefore, let block 6 be {3, 6, 8} and block 7 be {3, 7, 9}. The remaining first associates of treatment 4 are 7 and 9, but those treatments already have been placed in the same block and may not be paired again because $\lambda_1 = 1$.

Similarly, 6 and 8 may not be placed with treatment 5 because they have been paired in block 6. A solution to this impasse is to change blocks 4 and 5 to {2, 4, 9} and {2, 5, 8} and let blocks 8 and 9 be {4, 7, 8} and {5, 6, 9}. The final block assignments are shown in Table 6.8.

Table 6.8. Partially Balanced Incomplete Block Design

Block:	1	2	3	4	5	6	7	8	9
Treatment	1	1	1	2	2	3	3	4	5
	2	4	5	4	5	6	7	7	6
	3	6	7	9	8	8	9	8	9

For this example in which blocks are based on time, it may be wise to use the blocks in random order, assigning 3 animals as alike as possible to each block. Within each block the treatments should be randomly assigned to the 3 animals.

The somewhat tedious procedure of constructing designs may be avoided in most practical cases by referring to extensive tables of plans provided by Bose et al. (1954) and Clatworthy et al. (1973).

6.2.2.2. *Analysis of Data from Designs with Two Associate Classes*. Intrablock estimators of treatment effects may be obtained by minimizing

$$\sum_{i=1}^{t} \sum_{j=1}^{r} (y_{ij} - \mu - \tau_i - \delta_j)^2$$

The tedious algebra (Kempthorne 1952) is not given here. However, recall from the completely balanced case [(6.8)] that the quantity

$$r\bar{y}_{i'.} - \sum_{j(i')}^{r} \bar{y}_{.j}$$

where $\bar{y}_{i'.}$ is a specific sample treatment mean and the $\bar{y}_{.j}$ are the sample means of the blocks that contain treatment i', was required to obtain $\hat{\tau}_{i'}$. The same quantity and others like it appear in the estimate of a treatment effect in the partially balanced case:

$$\hat{\tau}_{i'} = [(k - c_2)(r\bar{y}_{i'.} - \sum_{j(i')}^{r} \bar{y}_{.j}) + (c_1 - c_2) \sum_{i \neq i'}^{t_1} (r\bar{y}_{i.} - \sum_{j(i)}^{r} \bar{y}_{.j})]/[r(k - 1)] \tag{6.34}$$

where the second term is understood to involve only the first associates of treatment i'. The constants c_1 and c_2 are functions only of the design parameters, where

$$c_1 = k\lambda_1[r(k-1)+\lambda_2] + k(\lambda_1-\lambda_2)(p^1_{12}\lambda_2 - p^2_{12}\lambda_1)/$$
$$\{[r(k-1)+\lambda_1][r(k-1)+\lambda_2] + (\lambda_1-\lambda_2)[r(k-1)(p^1_{12}-p^2_{12}) + (p^1_{12}\lambda_2 - p^2_{12}\lambda_1)]\} \tag{6.35}$$

and c_2 is the same as c_1 except that λ_1 and λ_2 are interchanged in the first term of the numerator. Note that $\lambda_1 > \lambda_2$ implies $c_1 > c_2$. In the common case where $\lambda_1 = 1$ and $\lambda_2 = 0$,

$$c_1 = k\{1 - p^2_{12}/[r(k-1)]\}/\{r(k-1) + 1 + (p^1_{12} - p^2_{12}) - p^2_{12}/[r(k-1)]\} \tag{6.36}$$

and c_2 is the same as c_1 except that unity is deleted from the first term of the numerator.

For designs with $\lambda_1 = 1$, $\lambda_2 = 0$, and $p^2_{12} = 0$, as in the example of the previous section,

$$c_1 = k/[r(k-1) + 1 + p^1_{12}] \tag{6.37}$$

and $c_2 = 0$. (For the example, $c_1 = 1/3$.)

It can be shown that the variance of an estimated treatment mean is

$$V[\hat{\mu} + \hat{\tau}_i] = (\sigma^2/r)\{(1/t) + [1/rk(k-1)][r(k-c_2)^2$$
$$+ (t_1 - 1)(c_1 - c_2)^2\{r + [\lambda_1(k-1)/k][p^1_{11} + 2(k-c_2)(c_1-c_2)]\}]\} \tag{6.38}$$

where $\hat{\mu} = \bar{y}$ as usual. If one substitutes ms_E for σ^2, the square root of (6.38) becomes the standard error of an estimated treatment mean.

The variance of the difference between two estimated treatment means is

$$V[\hat{\tau}_1 - \hat{\tau}_2] = (2\sigma^2/r)[(k - c_a)/(k-1)] \tag{6.39}$$

where $c_a = c_1$ if the two treatments are first associates or $c_a = c_2$ if the two treatments are second associates. When $\lambda_1 > \lambda_2$, then $c_1 > c_2$; first associates are compared with better precision (smaller variance) than are second associates because first associates are compared under more homogeneous conditions (in the same block for λ_1 of the r units involved in each mean).

The minimum efficiency of a partially balanced design relative to a randomized complete block design (RCBD) with the same amount of replication is

$$E_a = (k-1)/(k-c_a) \tag{6.40}$$

where $c_a = c_1$ for first associates or $c_a = c_2$ for second associates. In the example given previously, $c_1 = 1/3$, $c_2 = 0$, and $k = 3$, so $E_1 = 0.75$ and $E_2 = 0.67$. If the smaller blocks of the partially balanced design permit greater homogeneity of animals in the same block, the error variance may be reduced sufficiently that E_1, and possibly E_2 as well, may exceed unity for the results of a particular experiment. The error variance may be estimated by

$$ms_E = (ss_y - ss_D - ss_{T(\text{adj})})/(n - t - b + 1) \tag{6.41}$$

where ss_y and ss_D are the usual sums of squares for total data and blocks [see (6.13), (6.14)] and $ss_{T(\text{adj})}$ is the sum of squares for treatments adjusted for block differences,

$$ss_{T(\text{adj})} = \sum_{i=1}^{t} \hat{\tau}_i (r\bar{y}_{i.} - \sum_{j(i)}^{r} \bar{y}_{.j}) \tag{6.42}$$

Recovery of interblock information may be worthwhile for cases involving blocks with random effects, but the computations are rather tedious if not fully adapted to a high-speed computer (Kempthorne 1952; Federer 1955; Cochran and Cox 1957; Kendall and Stuart 1966). Experience has shown that interblock information is not likely to add significantly to efficiency if blocks number no more than 12; for $t \leq 12$ treatments that implies $r > k$, since $tr = bk$, i.e., the replication per treatment should exceed block size. For many experiments this requirement is too stringent because the main reason for choosing partial over complete balance is the desire to use low replication.

6.2.2.3. *Example of Partially Balanced Design*. Consider the example of Sec. 6.2.2.1 ($t = 9$, $r = 3$, $k = 3$) with block assignments as in Table 6.8. The corresponding data (coded) with treatment numbers in parentheses are shown in Table 6.9. Treatment effects may be estimated with (6.34). Computations required to obtain the estimates and adjusted treatment means are summarized in Table 6.10. As a computational check, note that the total of column [B] equals r times the total of [A], or $y_{..}$; and columns [C], [D], and [$\hat{\tau}_i$] sum to zero. Also note that each entry in [D] is the sum of $t_1 = 6$ entries in [C], where each treatment has 6 first associates.

The sum of squares of treatments adjusted for blocks [(6.42)] may be computed by multiplying [C] and [$\hat{\tau}_i$] from Table 6.10. A summary of the analysis of variance is shown in Table 6.11. Although the f test of overall treatment differences is not highly significant, well-designed orthogonal contrasts or other specific comparisons might bring out differences between certain treatments with good confidence.

As shown in Table 6.10, the standard error of an adjusted treatment mean is ±2.36 [see (6.38)]. Equation (6.39) may be used to compute standard errors for the adjusted mean differences between first and second associate treatments by substituting ms_E for σ^2 and taking the square root of the result. Note that $(k - c_a)/(k - 1) = 1/E_a$, where $E_1 = 0.75$ and $E_2 = 0.67$ minimum efficiencies for the design in this example. Therefore, the standard error of the adjusted mean difference between 2 first associates is

Table 6.9. Responses and Treatment Numbers for 9 Incomplete Blocks in Partially Balanced Design*

Block:	1	2	3	4	5	6	7	8	9
	33.72(1)	38.58(1)	34.55(1)	42.95(2)	45.12(2)	43.04(3)	40.64(3)	38.53(4)	36.40(5)
	37.80(2)	45.39(4)	34.82(5)	45.35(4)	36.36(5)	45.22(6)	30.49(7)	34.58(7)	33.28(6)
	42.25(3)	47.75(6)	38.29(7)	48.84(9)	40.58(8)	37.26(8)	36.34(9)	40.81(8)	38.46(9)
Means ($\bar{y}_{.j}$):									
	37.92	43.91	35.89	45.71	40.69	41.84	35.82	37.97	36.05

*t=9, k=3, r=3, λ_1=1, λ_2=0, c_1=1/3, c_2=0, p^1_{11}=3.

Table 6.10. Adjustment of Treatments for Blocks in Partially Balanced Design

Treatment	[A] $\bar{y}_{i.}$	[B] $\sum_{j(i)}^{3} \bar{y}_{.j}$	[C] $r[A]-[B]$	First Associates	[D] First Assoc. Only*	Trt. Effects $[\hat{\tau}_i]$†	Adj. Trt. Means $(\hat{\mu}+\hat{\tau}_i)$‡
1	35.62	117.72	-10.86	{234567}	+6.66	-5.06	34.47
2	41.96	124.32	+1.56	{134589}	+0.33	+0.80	40.33
3	41.98	115.58	+10.36	{126789}	-6.99	+4.79	44.32
4	43.09	127.59	+1.68	{126789}	-6.99	+0.45	39.98
5	35.86	112.63	-5.05	{126789}	-6.99	-2.91	36.62
6	42.08	121.80	+4.44	{134589}	+0.33	+2.24	41.77
7	34.45	109.68	-6.33	{134589}	+0.33	-3.15	36.38
8	39.55	120.50	-1.85	{234567}	+6.66	-0.56	38.97
9	41.21	117.58	+6.05	{234567}	+6.66	+3.40	42.93
Total	355.80	1067.40	0	...	0	0	355.77

* $\sum_{i}^{6} (r\bar{y}_{i.} - \sum_{j(i)}^{3} \bar{y}_{.j})$.

† $\{(k-c_2)[C]+(c_1-c_2)[D]\}/[r(k-1)]$.

‡ Standard error of an adjusted treatment mean is ±2.36 [see (6.38)].

Table 6.11. Analysis of Variance for Experiment in Partially Balanced Design

Source of Variation	df	*ss*	*ms*	*f* Ratio
Treatments (adj.)	8	172.767	21.596	2.326*
Blocks	8	324.625	40.578	
Error	10	92.856	9.286	
Total	26	590.248		

*P[Type I error] ≃ 0.10.

$$\sqrt{2ms_E/(rE_1)} = \sqrt{2(9.286)/[3(0.75)]} = 2.87$$

and the corresponding result for second associates is

$$\sqrt{2ms_E/(rE_2)} = \sqrt{2(9.286)/[3(0.67)]} = 3.04$$

Suppose each treatment is to be compared with the control (treatment 1), and one-sided comparisons are assumed to be valid (treatments expected to be at least as good as control). One may use Dunnett's procedure (Sec. 2.2.5), with critical value $t_{D,0.05,8,10} = 2.76$ (App. A.9.2). Any first associate of the control (treatments 2, 3, ..., 7) is significantly better if the adjusted mean difference exceeds (2.87)(2.76) = 7.92, and any second associate of the control (treatments 8, 9) is significantly better if the difference exceeds (3.04)(2.76) = 8.39. The adjusted treatment means in Table 6.10 show that only treatments 3 and 9 are better than the control at a 95% significance level.

In this experiment incomplete blocks were necessary because of physical limitations. It may be of interest to estimate the efficiency of the design used relative to a complete block design. One may adjust block means for treatments as follows:

$$\hat{\mu} + \hat{\beta}_j = \bar{y}_{\cdot j} - \left(\sum_{i(j)}^{3} \hat{\tau}_i/3\right)$$

where the last term is the average of the estimated effects of the 3 treatments included in a given block (Table 6.10). The resulting adjusted block means are

{37.74, 44.70, 39.60, 44.16, 41.58, 39.68, 34.14, 39.06, 35.14}

Assuming that complete blocks could have been formed from relatively homogeneous incomplete blocks, one may average the 3 highest, 3 middle, and 3 lowest block means to obtain 43.48, 39.45, and 35.67, respectively. Using these results, the sum of squares for complete blocks is 274.576, which is 50.049 less than the sum of squares for incomplete blocks (Table 6.11). Therefore, the estimated sum of squares for error in the RCBD would be increased by 50.049 and the degrees of freedom for error would increase by 6, since complete blocks would use only 2 df (instead of 8). The estimated error variance in the complete block design is

$$\hat{\sigma}^2_{RCBD} = (92.856 + 50.049)/16 = 8.932$$

which is slightly smaller than $\hat{\sigma}^2 = ms_E = 9.286$ for the incomplete blocks. Thus in experiments with few degrees of freedom, smaller blocks do not necessarily guarantee decreased experimental error variance. The estimated efficiency of the partially balanced incomplete block design (PBIB) relative to a complete block design is

$$E_{PBIB:RCBD} = (E_a \hat{\sigma}^2_{RCBD}/\hat{\sigma}^2)100\%$$

which is 72% for first associates and 64% for second associates. The failure to achieve 100% efficiency is the price paid (in sensitivity) for the convenience of using limited facilities in small blocks. Despite the relative inefficiency of incomplete blocks in this example, it is rather obvious that blocking was effective relative to a completely randomized design because the block mean square was more than 4 times as large as the error mean square (Table 6.11).

6.2.2.4. *Chain Block Designs*. Chain blocks (*Youden linked blocks*) are a special type of partially balanced design. Many treatments can be accommodated with small block size (k) and very few replications ($r < k$). For example, 10 treatments in blocks of 4 or 15 treatments in blocks of 5 require only 2 replicates. Three replicates are required for 35 treatments in blocks of 7. A design with 10 treatments is shown in Table 6.12. Note that the linking be-

Table 6.12. Chain Block Design for 10 Treatments in Blocks of 4

Block:	1	2	3	4	5
Treatment	1	3	5	7	9
	2	4	6	8	10
	3	5	7	9	1
	4	6	8	10	2

tween blocks is cyclic with each block containing 2 of the 4 treatments found in the previous block. For details see Youden and Connor (1953) or Cochran and Cox (1957).

6.2.3. Multidimensional Designs

Multidimensional incomplete block designs permit the removal of a sum of squares for replications from the total sum of squares because the incomplete blocks can be grouped into complete replicates (or groups of replicates).

Most of the designs are lattices, so called because the plans may be represented as a lattice of lines with treatment numbers at the intersections. However, some nonlattice plans exist that must be constructed by the methods described in Sec. 6.2.1.1. Because of greater efficiency, multidimensional designs generally are preferred over one-dimensional designs when both options are available.

6.2.3.1. *Nonlattice Designs*. Relatively few nonlattice incomplete block plans that can be grouped into complete replicates are available. A few more in which groups of replicates may be formed have been found (Cochran and Cox 1957). A list of both types of completely balanced plans of practical interest is given in Table 6.13. Note that t/k must be an integer. Reasonably good designs are available for most treatment numbers in the range $6 \leq t \leq 16$.

Table 6.13. Some Configurations for Nonlattice, Multidimensional, Balanced, Incomplete Block Designs for t Treatments in Blocks of k, Replicated r Times

Designs in Replicates				Designs in Groups of Replicates			
t	k	r	n	t	k	r	n
6	2	5	30	6	4	10	60
6	3	10	60	7	2	6	42
8	2	7	56	9	2	8	72
8	4	7	56	9	4	8	72
10	2	9	90	9	5	10	90
15	3	7	105	9	6	8	72
21	3	10	210	10	3	9	90
28	4	9	252	11	2	10	110
				13	3	6	78
				16	6	6	96
				19	3	9	171
				25	4	8	200
				41	5	10	410

Note that most of the designs require fairly large numbers of experimental units. Partially balanced designs should be used if lower replication is necessary (Cochran and Cox).

The designs in complete replicates can be analyzed as if they were complete block designs without resulting in biased estimates of treatment means (Yates 1939), but the intrablock adjustments should provide more precise estimates in most cases. Neither type of design should be appreciably less efficient than corresponding complete block designs for experiments of the same size (Yates 1940) and therefore should be preferred over one-dimensional incomplete block designs when the option is available. They are analyzed in the manner of balanced one-dimensional designs except for the removal of a sum of squares for replicates from the error sum of squares.

6.2.3.2. *General Structure of Two-Dimensional Lattices.* Two-dimensional lattices are special cases of balanced incomplete block designs in which $t = k^2$; the number of treatments must be the square of the block size k, which is restricted to numbers that are primes or powers of primes (2, 3, 4, 5, 7, 8, 9, ...). They share with other multidimensional designs the property that complete replicates may be formed and accounted for in the analysis, usually resulting in better efficiency than one-dimensional designs of similar size. Unfortunately, useful square two-dimensional lattices exist only for t = 9, 16, 25, 49, 64, or 81 treatments, unless $t \geq 120$. (See Sec. 6.2.3.4 for rectangular lattices that permit other numbers of treatments.) For this reason they are used rather rarely in most research disciplines, although agronomic variety trials and industrial chemistry are outstanding exceptions.

Two-dimensional lattice plans may be constructed with pseudofactors and methods of partial confounding used in the construction of factorial incomplete block plans (Sec. 6.5.1.2), and the lattice designs have been referred to as quasifactorial incomplete block designs (Kempthorne 1952). The basic steps of construction are:

1. Select arbitrary correspondence between the treatments to be used and combinations of pseudofactors A and B, each at k levels, as if the experiment were to be a k^2 factorial.
2. Represent the $k + 1$ orthogonal factorial components of $k - 1$ df each by the symbols A, B, (AB), (AB^2), ..., (AB^{k-1}).
3. Arbitrarily select a different pseudofactor component to confound with the k blocks of each replicate (Secs. 6.3, 6.4, 6.5).

As in one-dimensional designs, the minimum number of replications required for complete balance is $r = \lambda(t - 1)/(k - 1)$, where λ is the number of times a treatment occurs in the same block with each of the other treatments. For lattices in which $t = k^2$, minimum replication is $r = \lambda(k + 1)$ or $r = k + 1$ because the minimum value of λ is unity. Since the minimum number of replications exactly equals the number of orthogonal factorial components, each component of the pseudofactorial may be confounded with the $k - 1$ df for blocks in one replicate and thus achieve balanced partial confounding of all treatments. If one uses minimum r, the number of experimental units or subjects required is $n = rt = (k + 1)k^2$. Also, since $bk = rt$, the number of blocks required is $b = (k + 1)k$ or $t + k$.

Consider an experiment to be performed with t = 9 treatments in blocks of k = 3. The minimum amount of replication for complete balance is r = 4 and b = 12 blocks are required. One may arbitrarily assign the set of treatment numbers {1, 2, ..., 9} to a corresponding set of pseudofactor symbols $\{A_0B_0, A_0B_1, A_0B_2, A_1B_0, A_1B_1, A_1B_2, A_2B_0, A_2B_1, A_2B_2\}$ or in shorthand notation {00, 01, 02, ..., 22}. Using the methods of Sec. 6.4.1, one may confound orthogonal components A, B, (AB), and (AB^2), each having 2 df, with the blocks of replicates 1, 2, 3, and 4, respectively. Then one obtains blocks for the first replicate: {00, 01, 02}, {10, 11, 12}, and {20, 21, 22} that correspond to {1, 2, 3}, {4, 5, 6}, and {7, 8, 9}. The entire design is shown in Table 6.14. Tables of plans for balanced lattices have been given by Cochran and Cox (1957).

If the number of replicates required for complete balance is too large

Table 6.14. Balanced Lattice Design for 9 Treatments in Blocks of 3

Replicate 1		2		3		4	
Block	Treatments	Block	Treatments	Block	Treatments	Block	Treatments
(1)	1 2 3	(4)	1 4 7	(7)	1 6 8	(10)	1 5 9
(2)	4 5 6	(5)	2 5 8	(8)	2 4 9	(11)	3 4 8
(3)	7 8 9	(6)	3 6 9	(9)	3 5 7	(12)	2 6 7

for the resources available in a given experimental situation, one may turn to partially balanced lattices that usually involve 4 or fewer replicates. The designs are termed *double* (or *simple*) lattice if $r = 2$, *triple* lattice if $r = 3$, and *quadruple* lattice if $r = 4$. Construction may be accomplished as for completely balanced lattices except that only 2, 3, or 4 of the $k + 1$ orthogonal pseudofactor components are confounded with block effects. The partially balanced design in Table 6.8 is actually a lattice design because the blocks may be grouped into complete replicates. (That fact was ignored in the discussion of Sec. 6.2.2, which relates to the general problem of partially balanced incomplete block designs.)

6.2.3.3. *Analysis of Lattices*. Since lattices are special cases of balanced and partially balanced incomplete block designs, the analyses of Secs. 6.2.1 and 6.2.2 are pertinent. See Cochran and Cox (1957) for certain simplifications. The only complication is an additional computation of a sum of squares of replicates, with the result that the quantity computed does not become part of the intrablock experimental error (residual sum of squares). See Federer (1955) and Winer (1962) for numerical examples. Kempthorne (1952) has shown the general correspondence of treatment effects to pseudofactorial effects.

In Sec. 6.2.1.2, the minimum efficiency of intrablock information in a balanced design relative to a complete block design with the same replication is shown to be $E = (t\lambda/kr)$. For completely balanced two-dimensional lattices having the minimum number of replicates ($r = k + 1$ with $\lambda = 1$), the minimum efficiency becomes

$$E = [k^2(1)/k(k + 1)] = k/(k + 1)$$

The resulting efficiencies, for experiments with t = 9, 16, 25, 49, 64, or 81 treatments, are E = 0.75, 0.80, 0.83, 0.88, 0.89, or 0.90, respectively. Actual efficiencies may exceed 100% if small blocks permit more homogeneity of subjects within each block with correspondingly reduced σ^2. Also, recovery of interblock information may improve efficiency in some cases with random effects of blocks.

6.2.3.4. *Miscellaneous Lattices*. *Rectangular* lattices are two-dimensional, partially balanced designs developed by Harshbarger (1946, 1949) to accommodate numbers of treatments intermediate to those required for other lattices. In rectangular designs, $t = k(k + 1)$ $(k \geq 2)$, so the number of treatments must be 12, 20, 30, 42, 56, 72, 90, etc. Excepting t = 30 or 90, all the useful designs ($t \leq 90$) may be constructed from balanced lattices with $t = (k + 1)^2$. For $t = 30$ or 90, Latin square designs of size 6×6 or 10×10 can be used as a basis for construction (Kempthorne 1952). See Cochran and Cox (1957) for specific plans and for the method of analysis, essentially the same as for other lattices.

Cubic lattices are three dimensional, i.e., the number of treatments is the cube of the block size $t = k^3$. The number of replicates typically is $r = 3$ (or some multiple of 3) with k^2 blocks per replicate. The designs permit a drastic reduction in block size when the number of treatments is large, although only $t = 8, 27, 64, 125$, etc., are permitted. Construction of the designs requires 3 pseudofactors as discussed in Sec. 6.2.3.2. Since $\lambda = r(k - 1)/(t - 1) = r/(k^2 + k + 1)$, the minimum number of replicates for complete balance is $k^2 + k + 1$ or 7, 13, 21, etc. Usually, few replicates are desired, so most useful cubic lattices are only partially balanced.

See Federer (1955, p. 315) for a classification of rare and unusual varieties of lattices.

6.3. SYMMETRICAL TWO-LEVEL FACTORIAL EXPERIMENTS

Symmetrical designs are designs in which each factor is applied at the same number of levels. When several factors are to be studied simultaneously in the early stages of examining a particular response variable, the number of levels of each factor is commonly limited to 2 (often presence and absence). Also, more factors are likely to be studied in the initial experiment than in later studies. When the number of treatment combinations is large or experimental materials or subjects are quite variable, the investigator should consider using incomplete block designs so that the experimental units within each block will be relatively homogeneous and experimental error will be reduced. In other cases block size is restricted by practical considerations--it may be impossible to apply all combinations and to measure responses under homogeneous conditions relative to time, space, etc. Occasionally, even one full replicate of the desired treatment combinations is too much for resources available, so one must turn to fractionally replicated designs (usually a single incomplete block).

6.3.1. Construction of Incomplete Blocks

As in nonfactorial experiments in incomplete blocks (Sec. 6.2), the factorial design problem involves developing a plan for the balanced assignment of treatments (treatment combinations in the factorial case) to blocks. Deliberate *confounding* is a device to allocate combinations to incomplete blocks within a given replicate in such a manner that contrasts or comparisons for factorial effects of little or no importance are identical with (and inseparable from) differences between blocks. Complete confounding of one or more factorial effects occurs when the same effects are confounded with block differences in every replicate or the entire design consists of a single or fractional replicate. Partial confounding occurs when different factorial effects are selected for confounding in different replicates; then all factorial effects are estimable, but those involved in partial confounding are estimated less precisely because information about them is available only in replicates where they are not confounded with blocks.

6.3.1.1. *Defining Contrasts*.

The essential notation for symmetrical 2-level experiments with n factors (2^n treatment combinations) is described in Sec.

2.4.6. Yates's method of analysis is shown to be a convenient elaboration of the orthogonal contrast principle for factorial effects (Sec. 2.4.3.1). Summarizing briefly, note that the main effects and interaction of 2 factors A and B may be derived from the following orthogonal contrasts among treatment combination totals:

$$q_A = y_{11.} + y_{10.} - y_{01.} - y_{00.}$$

$$q_B = y_{11.} - y_{10.} + y_{01.} - y_{00.}$$

$$q_{AB} = y_{11.} - y_{10.} - y_{01.} + y_{00.}$$

where 0 and 1 designate low and high levels of each factor. Equivalent notation for symmetrical 2-level experiments uses letters for high levels only and (1) for control:

$$q_A = ab + a - b - (1) \qquad q_B = ab - a + b - (1) \qquad q_{AB} = ab - a - b + (1)$$

Alternatively, using vectors of ordered levels for symbols,

$$q_A = \{11\} + \{10\} - \{01\} - \{00\}$$

$$q_B = \{11\} - \{10\} + \{01\} - \{00\}$$

$$q_{AB} = \{11\} - \{10\} - \{01\} + \{00\} \tag{6.43}$$

Deliberate confounding dictates the selection of a factorial effect (preferably of trivial interest and effect) to be sacrificed for the purpose of forming incomplete blocks; i.e., some factorial contrast, termed the *defining contrast* (DC), must be used to define the different blocks.

A method of allocating combinations to blocks that works well with all symmetrical factorials in which the number of levels of each factor is a prime (p = 2, 3, 5, 7, etc.) involves addition performed to the number base p. For example, in base 2, 0 + 0 = 0, 0 + 1 = 1, 1 + 0 = 1, and 1 + 1 = 10. For 1 + 1 = 10 it is understood that "10" means $(1 \times 2^1) + (0 \times 2^0)$, just as "43" means $(4 \times 10^1) + (3 \times 10^0)$ in base 10 arithmetic. Alternatively, one may perform addition to the usual base 10 and reduce the result modulo p (divide the result by p and save the remainder or residue). The base 2 sum (1 + 1), computed in base 10, modulo 2, is (1 + 1)/2 = 2/2 = 1, the quotient, with 0 remainder (i.e., 10 in base 2). The additions are performed on certain elements of the vectors of ordered levels of each treatment combination. Let x_i be the value of the ith element in the ordered vector of levels denoting a particular treatment combination, i.e., each x_i may take on values 0, 1, 2, ..., p - 1 in a p-level factorial design. The selected elements to be summed in each vector are designated by the defining contrast.

It is not practical to form incomplete blocks for the 2^2 factorial but the method may be illustrated most simply for that case. For all 2-level factorials (p = 2), the x_i (i = 1, 2, ..., f factors) take on values 0 or 1 only, denoting low level and high level of a factor in a particular treatment combination. For only 2 factors, the 4 vectors of x_i are those shown in each equation of (6.43). For any treatment combination vector, the x_i to be summed are designated by the defining contrast. If the main effect of factor *A* is selected as a defining contrast, only x_1 is involved; if the main effect of *B* is selected, only x_2 is involved; or if *AB* interaction is selected, both x_1 and x_2 are involved. In general, for a defining contrast with *f* factors involved, the computation $\sum_{i=1}^{f} x_i/2$ will yield an integer quotient *Q* plus a remainder or residue ($R = 0, 1$), which is the same as the least significant digit of a binary sum. The results for some treatment combinations will yield $R = 0$; others will yield $R = 1$. On this basis the entire set of treatment combinations is divided into 2 incomplete blocks. The same residue classes also define the 2 sides (+, -) of the orthogonal contrast for the effect chosen as defining contrast, thus confounding that effect with the difference between 2 incomplete blocks. For example, suppose the defining contrast selected is the interaction *AB*. Then $(x_1 + x_2)/2$ produces $R = 0$ for {00} and {11} and $R = 1$ for {10} and {01}. Thus treatment combinations (1) and *ab* should be used in one block and *a* and *b* in the other. Alternatively, in binary addition, {11}, for example, produces the binary sum 10, where the first digit corresponds to *Q* and the second to *R* from (1 + 1)/2 in base 10 arithmetic. The proper allocations of treatment combinations to blocks for each possible defining contrast in a 2^2 experiment (*A*, *B*, *AB*) are shown in Table 6.15. Note that the allocations to blocks are exactly the same as the corresponding allocations to (-, +) or (+, -) in the contrasts for effects of *A*, *B*, and *AB* [(6.43)].

Table 6.15. Allocations of Treatment Combinations to Incomplete Blocks, with Effects of *A*, *B*, or *AB* Confounded with Blocks

DC	Vector Sums	Block 0 (*R*=0)	Block 1 (*R*=1)
A	$x_1/2$	[{00}, {01}] or [(1), *b*]	[{10}, {11}] or [*a*, *ab*]
B	$x_2/2$	[{00}, {10}] or [(1), *a*]	[{01}, {11}] or [*b*, *ab*]
AB	$(x_1+x_2)/2$	[{00}, {11}] or [(1), *ab*]	[{10}, {01}] or [*a*, *b*]

6.3.1.2. *Two Incomplete Blocks per Replicate*. Although any factorial effect may be selected as the defining contrast to establish 2 incomplete blocks, the highest ordered interaction usually is chosen because it is least likely to have effects of significant magnitude and usually is the effect of least interest (see split plots in Chap. 8 for exceptions). However, when more than one replicate is to be used, unless one interaction is known to have trivial effect the investigator may choose to confound a different interaction in each replicate so that no factorial effect is totally confounded with blocks throughout the experiment.

With rare exceptions, the smallest practical factorial for incomplete block designs is a 2^3 experiment (a study with 8 treatment combinations).

Suppose it is desired to have $r = 4$ replicates so that $4(2^3) = 32$ subjects are involved, 4 in each of 8 blocks. Let the defining contrasts for the 4 replicates be *AB*, *AC*, *BC*, and *ABC*, respectively. The vector symbols for treatment combinations are {000}, {100}, etc. For the first replicate, with DC = *AB*, the pertinent vector sums are $(x_1 + x_2)/2$ and the residues are $R = 0$ and $R = 1$. Considering each combination in turn, one finds that {000}, {001}, {110}, and {111}, or (1), *c*, *ab*, and *abc*, each produces $R = 0$; and collectively these form the *intrablock subgroup* or *principal block*. The other 4 combinations (*a*, *b*, *ac*, *bc*) produce $R = 1$ and form the *coset* or *alternate block*. In the other 3 replicates the vector sums for *AC*, *BC*, and *ABC* are $(x_1 + x_3)/2$, $(x_2 + x_3)/2$, and $(x_1 + x_2 + x_3)/2$, respectively. The entire design is shown in Table 6.16.

Table 6.16. A 2^3 Factorial in 2 Incomplete Blocks per Replicate with Partial Confounding

Replicate:	1	2	3	4
DC:	*AB*	*AC*	*BC*	*ABC*
Block 0	(1),*c*,*ab*,*abc*	(1),*b*,*ac*,*abc*	(1),*a*,*bc*,*abc*	(1),*ab*,*ac*,*bc*
Block 1	*a*,*b*,*ac*,*bc*	*a*,*c*,*ab*,*bc*	*b*,*c*,*ab*,*ac*	*a*,*b*,*c*,*abc*

Four relatively homogeneous subjects should be assigned to the same group; each group of 4 should be randomly assigned to one of the 8 blocks; and within each block the 4 subjects should be randomly assigned, one to each of the 4 treatment combinations.

Some mathematical tricks may be used to simplify the process of identifying all the treatment combinations that belong in a certain block. The methods are especially useful in designs for larger factorial experiments with many combinations in each block. By enforcing an artificial rule of commutative multiplication, such that

$$(1) = a^2 = b^2 = c^2 = \ldots = 1$$

the symbols in the principal block form an Abelian group with elements of period $p = 2$. Then the product of any two symbols in the principal block is another symbol in the same block; for example, in the principal block of replicate 1 in Table 6.16, $(c)(ab) = abc$, or $(ab)(abc) = a^2b^2c = c$. Elements in the alternate block can be found by multiplying each element in the principal block by any element in the alternate block. For example, if an experimenter had already determined the elements of the principal block in the first replicate of Table 6.16 and had determined only the element *a* for the alternate block, then $(a)(c) = ac$, $(a)(ab) = a^2b = b$, and $(a)(abc) = a^2bc = bc$ provide the remaining elements in the alternate block.

If the plan for an incomplete block experiment is reported in the literature or otherwise specified without naming the defining contrast, the DC can be found by trial and error. Select a logical possibility as a trial DC and find the residue of the vector sum for each treatment combination of the principal block [the block containing the treatment combination (1), or 000...].

If all the residues are zero, the trial DC is the correct one. If not, select a different trial DC and check residues again.

An outline of the analysis of variance for the 2^3 design in 4 replicates is shown in Table 6.17. The design provides full information on the main effects of A, B, and C but only three-fourths as much information about interactions, each of which is confounded with blocks in one of the 4 replicates.

Table 6.17. Sources of Variation for 2^3 Factorial Experiment in Incomplete Block Design with 4 Replicates

Source of Variation	Degrees of Freedom
Replicates	3
Blocks/replicates	4
A	1
B	1
C	1
AB	1
AC	1
BC	1
ABC	1
Error	17
Total	31

Therefore, the sum of squares for blocks within replicates (see Sec. 2.4.2 for formulation) is identical with

$$(ss_{AB})_1 + (ss_{AC})_2 + (ss_{BC})_3 + (ss_{ABC})_4$$

where the numerical subscripts indicate the replicate to which an interaction computation is restricted. Because of the partial confounding, computation of the unconfounded portions of sums of squares for interactions cannot be accomplished in the usual manner as for factorials in completely randomized or complete block designs [(2.167)-(2.170)]. Suppose an interaction used as a defining contrast, say $\mathrm{DC}_{i'}$, is confounded in replicate i', i.e., the difference between the two blocks in that replicate also measures the interaction. Let $y_{(i')j.}$ be the total of $2^{n-1}(r - 1)$ responses, in the $r - 1$ replicates where $i \neq i'$, of the 2^{n-1} treatment combinations listed in block j of replicate i'. Then the unconfounded portion of the sum of squares for that interaction is

$$ss_{\mathrm{DC}_{i'}} = \sum_{j=0}^{1} y^2_{(i')j.}/[2^{n-1}(r - 1)] - (\sum_{j=0}^{1} y_{(i')j.})^2/[2^n(r - 1)] \tag{6.44}$$

For example, consider the data in Table 6.18, corresponding to the design of Table 6.16. In the first replicate, *AB* is the defining contrast DC_1. Treatment combinations (1), *c*, *ab*, and *abc* appear in the principal block of that replicate. Summing responses to those combinations in the other 3 replicates, one obtains

Table 6.18. Data for 2^3 Factorial in Incomplete Block Design with Partial Confounding

Replicate:		1	2	3	4
DC:		*AB*	*AC*	*BC*	*ABC*
Block 0:	treatment	(1),*c*,*ab*,*abc*	(1),*b*,*ac*,*abc*	(1),*a*,*bc*,*abc*	(1),*ab*,*ac*,*bc*
	response	32,31,41,40	25,26,30,32	26,29,24,33	27,35,33,28
Block 1:	treatment	*a*,*b*,*ac*,*bc*	*a*,*c*,*ab*,*bc*	*b*,*c*,*ab*,*ac*	*a*,*b*,*c*,*abc*
	response	32,29,32,28	32,29,33,26	29,28,35,30	29,24,26,33

$$y_{(1)0.} = (25 + 29 + 33 + 32) + (26 + 28 + 35 + 33) + (27 + 26 + 35 + 33) = 362$$

Similarly, for combinations *a*, *b*, *ac*, and *bc* that appear in the alternate block of the first replicate, one obtains from the remaining replicates

$$y_{(1)1.} = (32 + 26 + 30 + 26) + (29 + 29 + 30 + 24) + (29 + 24 + 33 + 28) = 340$$

Then, in accordance with (6.44),

$$ss_{DC_1} = ss_{AB} = (362^2 + 340^2)/12 - [(362 + 340)^2/24] = 20.17$$

Sums of squares for the other interactions may be computed similarly. Sums of squares for replicates, blocks within replicates, and the main effects of *A*, *B*, and *C* all follow standard computations. The error (residual) sum of squares may be obtained by subtracting all other sums of squares from the total sum of squares. The error sum of squares could be decomposed into interactions of replicates with the main effects *A*, *B*, and *C* (unconfounded in any replicate and therefore providing 3 df each) plus the interactions of replicates with *AB*, *AC*, *BC*, and *ABC* (each confounded in one replicate and providing only 2 df each).

If interactions are significant, the experimenter likely will be interested in treatment combination means and will want to make specific comparisons among those as suggested in Sec. 2.2. Unfortunately, in experiments with partial confounding the treatment combination means require adjustment so that they are based only on the particular replications of effects that are not confounded with blocks. Designate a particular adjusted combination mean as

$\bar{y}'_{.k'l'm'}$, where k', l', and m' are either 0 or 1 depending on the levels of factors A, B, and C, respectively. Also, designate $y_{.k'..}$, $y_{..l'.}$, and $y_{...m'}$ as the totals of $2^{n-1}r$ observations (over r replicates) for the particular level of each factor involved. The adjusted combination mean may be computed from

$$\bar{y}'_{.k'l'm'} = (y_{.k'..} + y_{..l'.} + y_{...m'})/[2^{n-1}r] + \sum_{i=1}^{r} y_{(i)j.}/[2^{n-1}(r - 1)] - (n + r - 1)y_{..}/[2^n r] \qquad (6.45)$$

where $y_{..}$ is the total of all $2^n r$ observations and the $y_{(i)j.}$ are totals of treatment combinations listed in block j of replicate i [as in (6.44) for the computation of sums of squares for interactions]. The block number (j = 0 or 1) to look at for the list of treatment combinations is in each case determined by the confounding pattern. Consider the design of Table 6.18, with AB, AC, BC, and ABC confounded in replicates 1, 2, 3, and 4, respectively:

for $i = 1$, $j = k' + l' \pmod 2$ for $i = 2$, $j = k' + m' \pmod 2$

for $i = 3$, $j = l' + m' \pmod 2$ for $i = 4$, $j = k' + l' + m' \pmod 2$

For example, the *unadjusted* mean $\bar{y}_{.111}$ for treatment combination (abc) of Table 6.18 is (40 + 32 + 33 + 33)/4 = 34.5. The total $y_{.1..}$ includes 16 observations for combinations including the high level of factor A, i.e., all symbols with (a) present. The result is

$$(41 + 40 + 32 + 32) + (30 + 32 + 32 + 33) + (29 + 33 + 35 + 30) + (35 + 33 + 29 + 33) = 529$$

Similarly, $y_{..1.} = 496$ for (b), $y_{...1} = 483$ for (c), and $y_{..} = 967$ for all 32 observations. The totals $y_{(i)j.}$ will involve combinations listed in block j = 0 of replicates 1, 2, and 3 because $k' + l' = k' + m' = l' + m' = 0 \pmod 2$ for the combination $k'l'm' = 111$. The total $y_{(4)j.}$ will involve combinations listed in block $j = 1$ of replicate 4 because $k' + l' + m' = 1 \pmod 2$. The total $y_{(1)0.} = 362$ was determined earlier in the computation of ss_{AB}. Similarly, $y_{(2)0.} = 368$, $y_{(3)0.} = 364$, and $y_{(4)1.} = 370$ are quantities one would normally compute in the course of finding ss_{AC}, ss_{BC}, and ss_{ABC}, respectively. Now, one may compute the *adjusted* mean for combination (abc):

$$\bar{y}'_{.111} = [(529 + 496 + 483)/16] + [(362 + 368 + 364 + 370)/12] - [6(967)/32]$$

$$= 94.25 + 122.00 - 181.32 = 34.93$$

It differs from the unadjusted mean by 0.43 (a little more than 1% of the mean in this case). Of course, these adjustments and the complex computations for sums of squares of interactions are not required for experiments with complete confounding, i.e., when the same interaction (such as *ABC*) is confounded with blocks in all replicates. In that case one merely ignores the completely confounded interaction in the analysis, and no adjustments to combination means are required if other interactions are significant.

Assigning more than one subject per combination to the same block is rarely done because it counteracts the purpose of forming incomplete blocks (to permit greater homogeneity of subjects within each block). If it is done and the effects of blocks are random, intra- and interblock estimates of interactions may be combined (Winer 1962, p. 397).

For experiments with a rather large number of treatment combinations, say $2^5 = 32$ or more, it may be feasible to utilize only a single replication if one is willing to assume that the higher ordered interactions (say those involving 3 or more factors) are negligible in effect relative to the magnitude of main effects and first order interactions. For example, a 2^5 factorial requires 5 df for main effects and 10 df for first order interactions. If the 5-factor interaction *ABCDE* is used as a defining contrast and thus is confounded with blocks (1 df), the remaining 15 df (10 df for 3-factor interactions plus 5 df for 4-factor interactions) may be pooled for use as experimental error. See Holms and Berettoni (1969) for a pooling strategy and the average probability of Type I error involved. For experiments with 6 or more factors (64 or more treatment combinations), the block size may be too large (32 or more subjects) to permit relative uniformity of subjects within the same block. In such cases one may consider the possibility of providing a design with 4 or more blocks per replicate.

6.3.1.3. *More than Two Blocks per Replicate*. Although a 2^3 factorial in 4 incomplete blocks per replicate is rarely practical, it will serve as a simple illustration of difficulties encountered in obtaining smaller blocks than would be used in a design with only 2 blocks per replicate. If one selects 2 of the 4 interactions *AB*, *AC*, *BC*, and *ABC* as defining contrasts, combinations of residue groups (R = 0, 1) for each defining contrast produce 4 groupings (blocks) of treatment combinations instead of 2 (see Table 6.16). Unfortunately, a third effect is automatically confounded with block differences in the process. Consider that 4 blocks have 3 df but the 2 effects selected as defining contrasts have but 1 df each, leaving another degree of freedom confounded as well. The effect automatically confounded is known as the *generalized interaction* (GI) of the 2 DC.

For example, if one selects $DC_1 = AB$ and $DC_2 = BC$ from a 2^3 factorial, $(x_1 + x_2)/2$ produces a residue of $R_1 = 0$ or 1 and $(x_2 + x_3)/2$ produces a residue of $R_2 = 0$ or 1 for each treatment combination. Combinations that produce $R_1 = 0$ and $R_2 = 0$ [i.e., (1) and *abc*] form the principal block (say block 1). Blocks 2, 3, and 4 are defined by $R_1 = 0$, $R_2 = 1$ (*c*, *ab*); $R_1 = 1$, $R_2 = 0$ (*a*, *bc*); and $R_1 = 1$, $R_2 = 1$ (*b*, *ac*), respectively. Consider a full set of 3 orthogonal contrasts among the 4 block effects. If the effects of the respective blocks are denoted $\{\delta_1, \delta_2, \delta_3$, and $\delta_4\}$, one can see that $(\delta_1 + \delta_2) - (\delta_3 + \delta_4)$ equals the effect of interaction *AB* (both *a* and *b* or neither versus one of them), $(\delta_1 + \delta_3) - (\delta_2 + \delta_4)$ equals the effect of *BC* (both *b* and *c* or neither versus one of them), and $(\delta_1 + \delta_4) - (\delta_2 + \delta_3)$ equals the effect of AC (both a and c or neither versus one of them). Obviously, AC is the third effect automatically confounded; it is the generalized interaction of defining contrasts AB and BC. Note that the coefficients of the block effects (δs) in the linear function confounded with AC are products of the coefficients of corresponding δs in the linear functions confounded with AB and BC. This result is parallel to the situation in Sec. 2.4 where the coefficients of treatment totals in the linear function describing an interaction contrast are products of the coefficients in orthogonal contrasts describing the main effects of the factors involved in the interaction.

A convenient method for quickly determining the generalized interaction of 2 DC is to reduce the product of the factorial symbols for the defining contrasts according to the rule $A^2 = B^2 = C^2 = \ldots = 1$. For example, $(AB)(BC) = AB^2C = AC$.

Designs with 4 blocks per replicate are not practical for most 2^n experiments with $n < 5$ factors unless partial confounding is used (different defining contrasts in different replicates), because one cannot avoid confounding at least one 2-factor interaction in any given replicate. For experiments with 5 or more factors, careful choice of defining contrasts will avoid confounding of 2-factor interactions. In general one should choose 2 interactions having as few factors in common as possible as defining contrasts.

For example, for a 2^6 experiment with 4 blocks in a single replicate, one might choose $DC_1 = ABCD$ and $DC_2 = CDEF$. Then,

$$GI = (ABCD)(CDEF) = ABC^2D^2EF = ABEF$$

i.e., the 4-factor interaction $ABEF$ is automatically confounded with blocks. Sixteen treatment combinations having zero residues for both $\sum_{i=1}^{4} x_i/2$ and $\sum_{i=3}^{6} x_i/2$ would be allocated to the principal block. For example, combination cd, or {001100}, belongs to the principal block, but combination cde, or {001110}, belongs in the block designated by $R_1 = 0$, $R_2 = 1$. The 16 combinations in each block should be randomly assigned to the 16 relatively homogeneous subjects allotted to that block. Experimental error normally would be composed of the 6-factor interaction plus six 5-factor interactions plus 12 unconfounded 4-factor interactions (19 df in all), on the assumption that the true effects of such interactions are trivial relative to the magnitude of the lower ordered factorial effects and random error.

For experiments with a large number of treatment combinations or very heterogeneous subjects, designs that have more than 4 incomplete blocks per replicate may be constructed. In general, for a 2^n experiment one may construct $b = 2^s$ incomplete blocks per replicate ($s < n$). Such designs have $k = 2^{n-s}$ subjects per block. Any s factorial effects may be selected as defining contrasts provided that none selected is a generalized interaction of any subset of the others. Since 2^s blocks will have $2^s - 1$ df, then $2^s - 1$ factorial effects will be confounded with blocks. Effects confounded are the s independent defining contrasts plus $\sum_{i=2}^{s} C(s, i)$ generalized interactions among 2 or more of the defining contrasts, where $C(s, i)$ represents the number of combinations of s things taken i at a time. For example,

$$C(3, 2) = [3!/(2!)(1!)] = (3)(2)(1)/[(2)(1)(1)] = 3$$

and $C(3, 3) = 1$, for a total of 4 generalized interactions when $s = 3$ DC are used to form $2^3 = 8$ blocks per replicate. For $s \geq 3$ DC, one or more of the effects automatically confounded cannot be found by taking direct products of just 2 of the symbols for the defining contrasts but must be found by forming generalized interactions among 3 or more symbols.

Fisher (1942) showed that if the number of factors is $n = 2^m - 1$ $(m \geq 2)$, a design can be formed in a single replicate for which no main effects or 2-factor interactions will be confounded if the block size $k = 2^{n-s} > n$. For example, if $n = 7$ factors, $k = 2^{7-4} = 8$ subjects in each of $2^4 = 16$ blocks will permit estimation of all main effects and 2-factor interactions.

Consider a 2^7 experiment on rats ($n = 7$ factors and 128 treatment combinations). Suppose it is desirable to block the 128 rats that will be used into groups of 16. Therefore $k = 2^{7-s} = 16$, and $s = 3$ DC are required to form $2^3 = 8$ incomplete blocks. Let the 3 DC be 3-way interactions ABC, CDE, and EFG, no two of which have more than one factor in common. Since there will be 8 blocks with 7 df, then 4 additional effects will be automatically confounded. Generalized interactions among pairs of defining contrast symbols produce

$$(ABC)(CDE) = ABC^2DE = ABDE \quad (ABC)(EFG) = ABCEFG \quad (CDE)(EFG) = CDE^2FG = CDFG$$

Note that the product of any two of these results produces the third. The remaining confounded effect can be obtained by taking the generalized interaction of one of the three results with the defining contrast not involved in that result, which is the same as the triple product of the defining contrasts:

$$[(ABC)(CDE)](EFG) = (ABDE)(EFG) = ABDFG$$

Note that if either of the remaining GI ($ABCEFG$ or $CDFG$) is multiplied by the third DC (EFG), the result reproduces one of the other DC. Two residue groups ($R = 0, 1$) of 64 treatment combinations can be formed with respect to each of the 3 DC so that the 8 combinations of R_1, R_2, and R_3 values represent the 8 blocks of 16 treatment combinations each. The 8 groups of 16 homogeneous rats should be randomly assigned block numbers, and the rats should be randomly assigned, one to each treatment within each block. There will be 7 main effects and $C(7, 2) = 21$ two-factor interactions. All the $C(7, 3) = 35$ three-factor interactions (except the 3 used as defining contrasts) may be examined if that seems desirable. The $C(7, 4) - 2 = 33$ unconfounded 4-way interactions, $C(7, 5) - 1 = 20$ unconfounded 5-way interactions, $C(7, 6) - 1 = 6$ unconfounded 6-way interactions, and the one 7-way interaction may be used for error (60 df in all).

6.3.2. Analysis by Yates's Method

A simplified method of computing estimates of mean factorial effects via orthogonal contrasts is displayed in Sec. 2.4.6 for completely randomized symmetrical experiments. That procedure may be summarized briefly: (1) construct a table with treatment combination totals in a definite order {(1), *a*, *b*, *ab*, *c*, *ac*, *bc*, *abc*, ...} on the left margin, (2) allocate n columns for computing (when there are n factors) plus a final column to record estimates of factorial effects, (3) fill the top half of each column with sums of pairs of totals from the entire previous column, (4) fill the bottom half of each column with differences of the same pairs (second total minus first in each pair), and (5) compute factorial effects (Δs) in column $n + 1$ by dividing each result except the first in column n (qs) by $r2^{n-1}$. The same procedure may be used for factorials in incomplete blocks with complete confounding except then the estimates for certain factorial effects are known to be confounded with block effects. The sum of squares for any factorial effect can be calculated in the manner used for orthogonal contrasts:

$$ss_{q_k} = [q_k^2/r\Sigma c_{ik}^2] = [q_k^2/2^n r] \tag{6.46}$$

where q_k is the kth factorial contrast total found in the nth column of computations (by Yates's method) for n factors in a design with r replicates. The sum of squares for error is

$$ss_E = ss_y - ss_D - \Sigma_k(ss_{q_k}) \tag{6.47}$$

where ss_y is the total sum of squares, ss_D is the sum of squares for br blocks (ignoring replicates if any), and $\Sigma_k(ss_{q_k})$ is the total of the contrast sums of squares for unconfounded factorial effects. The degrees of freedom for error are $(2^n - b)(r - 1)$ for replicated experiments (the product of degrees of freedom for unconfounded factorial effects and degrees of freedom for replicates). For unreplicated experiments, one begins Yates's analysis with individual observations--the observations themselves are the treatment combination "totals." Then the sum of squares for error is composed of the sums of squares for as many unconfounded higher ordered interactions as are not of interest and thought to be negligible.

The standard error of any estimated factorial effect, $\hat{\Delta}_k = q_k/(2^{n-1}r)$, is

$$s_{\hat{\Delta}_k} = \sqrt{2ms_E/(2^{n-1}r)} = \sqrt{ms_E/(2^{n-2}r)} \tag{6.48}$$

A particular effect is significant if the confidence interval (CI) does not include zero (ν_E = df for error):

$$(1 - \alpha)100\%(\text{CI})(\Delta_k) = \hat{\Delta}_k \pm t_{\alpha/2,\nu_E} s_{\hat{\Delta}_k} \tag{6.49}$$

If interactions are significant, interpretation may be aided by separating various effects. For example, consider a 2^3 experiment with DC = ABC for 2 blocks in a single replicate. If the AB interaction is significant one may wish to examine the average effect of A separately for high and low levels of B. The factorial contrast means $(\hat{\Delta}_k)$ can be combined conveniently to do that. If one represents the single observation for each treatment combination by the corresponding symbol, then

$$\hat{\Delta}_A = \{[a + ab + ac + abc] - [(1) + b + c + bc]\}/4$$

$$\hat{\Delta}_{AB} = \{[(1) + c + ab + abc] - [a + b + ac + bc]\}/4$$

Note that

$$\hat{\Delta}_A + \hat{\Delta}_{AB} = [(2ab + 2abc) - (2b + 2bc)]/4 = [(ab + abc)/2] - [(b + bc)/2]$$

which is the average effect of A at the high level of B. Similarly,

$$\hat{\Delta}_A - \hat{\Delta}_{AB} = [(a + ac)/2] - [(1) + c]/2$$

which is the average effect of A at the low level of B.

For experiments with partial confounding see the analysis in Sec. 6.3.1.2.

In factorial experiments in incomplete blocks one may encounter results in which one or more cells are missing. (If 2 or more are missing one may employ a covariance technique or turn to a generalized least squares computer program.) If only one cell is missing one may easily estimate the missing value from formulas derived by minimizing mean square error. In complete confounding, the missing value may be computed from

$$\hat{y}_{i'j'm'} = [ry'_{i'j'.} + ky'_{.jm'} - \sum_{i}^{r}(\sum_{m(jm')} y_{ijm})]/[(r - 1)(k - 1)] \quad (6.50)$$

where $y'_{i'j'.}$ is the total of the $k - 1$ observations in block j' of replicate i' (the one with a value missing), $y'_{.jm'}$ is the total of the $r - 1$ observations resulting from application of treatment combination m' (the one with a value missing), and $\sum_{i=1}^{r}(\sum_{m(jm')} y_{ijm})$ is the total of $rk - 1$ observations in all the blocks (j) in which treatment combination m' was applied. After the estimated value is placed with the actual observations, analysis proceeds as usual except that degrees of freedom for error are reduced by one and the approximate standard error of any factorial effect becomes

$$S_{\hat{\Delta}} \simeq \sqrt{ms_E\{[1/(2^{n-1}r - 1.5)] + (1/2^{n-1}r)\}} \quad (6.51)$$

In the case of partial confounding, the missing value may be computed from

$$\hat{y}_{i'j'm'} = [tr(r - 1)y'_{i'j'.} + tk(r - 1)y'_{.jm'} - kry'_{i'j.} + ky'_{.j.}$$

$$- tr \sum_{i \neq i'}^{r-1} \left(\sum_{m(jm')}^{k} y_{ijm} \right) - t(r - 1) \sum_{i=1}^{r} \left(\sum_{m \neq m'(i'j'm')}^{k-1} y_{ijm} \right) \tag{6.52}$$

$$+ t \sum_{h \neq i'}^{r-1} \left(\sum_{i=1}^{r} \sum_{m \neq m'(hjm')}^{k-1} y_{ijm} \right)] / \{ (r - 1)[t(r - 1)(k - 1) - (t - k)] \}$$

where $y'_{i'j'.}$ and $y'_{.jm}$ are as defined for (6.50), $y'_{i'j.}$ is the total of $t - 1$ observations for replicate i' (the one with a value missing), $y'_{.j.}$ is the total of all $tr - 1$ observations in the experiment, $\sum_{i \neq i'}^{r-1} (\sum_{m(jm')}^{k} y_{ijm})$ is the total of $k(r - 1)$ observations in the $r - 1$ complete replicates in the blocks j having treatment combination m', $\sum_{i=1}^{r} (\sum_{m \neq m'(i'j'm')}^{k-1} y_{ijm})$ is the total of $r(k - 1)$ observations over all replicates for the $k - 1$ treatment combinations applied with combination m' in block j' of replicate i', and $(\sum_{i=1}^{r} \sum_{m \neq m'(hjm')}^{k-1} y_{ijm})$ is the total of $r(k - 1)$ observations over all replicates for the $k - 1$ treatment combinations that appear with combination m' in the same block j of one of the complete replicates $(h \neq i')$.

An exact appraisal of the efficiency of a randomized incomplete block design (RIBD) relative to a complete block design (RCBD) is difficult. However, if one is willing to assume that the replicates of the RIBD are equivalent to the blocks of the RCBD, a simple comparison is possible. From the results of an experiment performed in a RIBD with b blocks of k subjects in each of r replicates, one obtains $\hat{\sigma}^2_{\text{RIBD}} = ms_E$ and

$$\hat{\sigma}^2_{\text{RCBD}} = [ss_{D/R} + rb(k - 1)ms_E]/[r(bk - 1)] \tag{6.53}$$

where $ss_{D/R}$ is the sum of squares for blocks within replicates, which may be computed by subtracting the sum of squares for replicates from the sum of squares for blocks ignoring replicates (ss_D) as indicated for (6.47). Ignoring the trivial correction for difference in degrees of freedom, one obtains the relative efficiency,

$$E_{\text{RIBD:RCBD}} = (\hat{\sigma}^2_{\text{RCBD}}/\hat{\sigma}^2_{\text{RIBD}})100\% \tag{6.54}$$

The efficiency of incomplete blocks may be overrated if (1) the differences between complete blocks in the RCBD could be made larger than the differences between replicates in the RIBD or (2) the interactions confounded with blocks in the RIBD are not negligible.

6.3.3. Example (four blocks per replicate)

An experiment was designed to investigate the possible effects of 5 factors, each at 2 levels (absence or presence), on serum nonurea nitrogen. Because of heterogeneity of animals available, the investigator decided to reduce block size from 32 animals (complete block) to 8 animals in each of 4 blocks. Two 3-factor interactions of little interest with only one factor in common were selected as defining contrasts: *ADE* and *BCD*. The generalized interaction

$$(ADE)(BCD) = (ABCD^2E) = (ABCE)$$

was automatically confounded with blocks. Residues R_1 from $(x_1 + x_4 + x_5)/2$ and R_2 from $(x_2 + x_3 + x_4)/2$ were computed for each treatment combination, with $\{R_1, R_2\} = \{0, 0\}$, $\{0, 1\}$, $\{1, 0\}$, and $\{1, 1\}$ being designated blocks 1, 2, 3, and 4, respectively. For example, the combination with all factors present, *abcde* (or 11111), produces $(1 + 1 + 1)/2$ with $R_1 = 1$ and $(1 + 1 + 1)/2$ with $R_2 = 1$ and belongs in block 4. Combination *abc* (or 11100) produces $(1 + 0 + 0)/2$ with $R_1 = 1$ and $(1 + 1 + 0)/2$ with $R_2 = 0$ and belongs in block 3. Two replicates of 32 animals each were used, with the same effects confounded in both replicates (i.e., complete confounding). The complete design and corresponding responses (mg/100 mL) are shown in Table 6.19.

The total sum of squares for the experiment is

$$ss_y = (7.40^2 + 11.84^2 + \ldots + 18.05^2) - (773.74^2/64) = 428.60$$

and the sum of squares among the 8 blocks (ignoring replicates) is

$$ss_D = (84.12^2 + 92.12^2 + \ldots + 104.69^2)/8 - (773.74^2/64) = 87.16$$

Yates's method may be used to compute orthogonal contrasts for the factorial effects. Computations are shown in Table 6.20. Remember that results in the top half of columns labeled 1-5 are sums of pairs of results in the entire previous column, and results in the bottom half are differences of the same pairs.

The sum of squares for each of the 28 unconfounded factorial effects [(6.46)] is $q_k^2/64$, and the total of those results is

$$\sum_k^{28} (ss_{q_k}) = (42.86^2 + 61.70^2 + \ldots + 6.88^2)/64 = 331.84$$

The residual or error sum of squares [(6.47)] is

$$ss_E = 428.60 - 87.16 - 331.84 = 9.60$$

and has $28(r - 1) = 28$ df, so $ms_E = 9.60/28 = 0.343$.

The standard error of each estimated factorial effect is $\sqrt{0.343/16} = 0.146$, and the t values for 95% and 99% confidence are 2.048 and 2.763, respectively (App. A.4, with 28 df). Therefore, any effect that has absolute magnitude of $(0.146)(2.048) = 0.30$ or more is significant at the 95% level, and any effect that exceeds $(0.146)(2.763) = 0.40$ in magnitude is significant

Table 6.19. Serum Nonurea Nitrogen (mg/100 mL) for 2^5 Experiment

	Replicate I							
	Block 1		2		3		4	Total
(1)	7.40	*b*	12.48	*a*	7.80	*d*	8.88	
ae	11.84	*c*	7.84	*e*	11.08	*ab*	11.04	
bc	8.96	*ad*	9.64	*bd*	10.08	*ac*	10.00	
abd	9.96	*de*	10.08	*cd*	6.92	*be*	11.76	
acd	8.60	*abe*	13.76	*abc*	10.08	*ce*	11.56	
bde	12.84	*ace*	13.76	*bce*	13.28	*ade*	13.12	
cde	9.84	*bcde*	13.56	*abde*	13.24	*bcd*	8.44	
abce	14.68	*abcd*	11.00	*acde*	12.92	*abcde*	15.08	
Total	84.12		92.12		85.40		89.88	351.52
	Replicate II							
	Block 1		2		3		4	
(1)	10.01	*b*	14.36	*a*	10.96	*d*	9.33	
ae	14.67	*c*	9.56	*e*	14.56	*ab*	12.86	
bc	11.66	*ad*	11.88	*bd*	11.26	*ac*	10.60	
abd	12.75	*de*	11.63	*cd*	8.99	*be*	13.42	
acd	11.18	*abe*	16.98	*abc*	11.09	*ce*	13.70	
bde	15.88	*ace*	15.09	*bce*	14.70	*ade*	15.27	
cde	12.25	*bcde*	17.67	*abde*	15.86	*bcd*	11.46	
abce	16.65	*abcd*	12.83	*acde*	15.06	*abcde*	18.05	422.22
Total	105.05		110.00		102.48		104.69	Σy=773.74

at the 99% level. Highly significant main effects are *A*, *B*, and *E*, with factor *E* having the largest average effect. Significant 2-factor interactions are *AB*, *AC*, *AD*, *AE*, and *CE*, i.e., factor *A* interacts with each of the other 4 factors and *C* interacts with *E*. Note that each of the interactions involving *A* is no more than approximately 35% of the magnitude of the main effect of *A*, and the *CE* interaction is only about 20% as large as the main effect of *E*. Three higher order interactions involving *E* (*BCE*, *BDE*, and *ABDE*) also are significant but may be difficult to interpret in a practical manner. Interactions *AB*, *ABD*, and *ABDE* all have negative effects, suggesting that factors *A* and *B* are antagonistic to each other; interactions involving *C* as well as *A* and *B* are positive but not significant. The effect of *A* in the absence of *B* may be estimated by

$$\hat{\Delta}_A - \hat{\Delta}_{AB} = 1.34 - (-0.46) = 1.80$$

and the effect of *A* in the presence of *B* by

Table 6.20. Estimates of Factorial Effects Determined by Yates's Method (2^5 experiment in incomplete blocks)

Trt.	Total	1	2	3	4	n=5 (q_k)	Fact.	Mean Effect ($\hat{\Delta}=q_k/32$)
(1)	17.41	36.17	86.91	166.70	329.90	(773.74)		($\bar{y}$=773.74/64=12.09)
a	18.76	50.74	79.79	163.20	443.84	42.86	*A*:	1.34
b	26.84	38.00	83.78	221.49	14.64	61.70	*B*:	1.93
ab	23.90	41.79	79.42	222.35	28.22	-14.66	*AB*:	-0.46
c	17.40	39.73	108.07	2.16	30.72	0.38	*C*:	0.01
ac	20.60	44.05	113.42	12.48	30.98	9.70	*AC*:	0.30
bc	20.62	35.69	107.92	13.37	-8.82	0.94	*BC*:	0.03
abc	21.17	43.73	114.33	14.85	-5.84	1.02	*ABC*:	0.03
d	18.21	52.15	-1.59	18.36	-11.48	-2.64	*D*:	-0.08
ad	21.52	55.92	3.75	12.36	11.86	11.80	*AD*:	0.37
bd	21.34	54.11	4.68	8.97	8.46	7.04	*BD*:	0.22
abd	22.71	59.31	7.80	22.01	1.24	-9.68	*ABD*:	-0.30
cd	15.91	50.10	6.43	-6.94	-7.06	3.92	*CD*:	0.12
acd	19.78	57.82	6.94	-1.88	8.00	-2.00	*ACD*:	-0.06
bcd	19.90	50.07	7.06	4.45	3.64	[19.64]	*BCD*:	[0.61]*
abcd	23.83	64.36	7.79	-10.62	-2.62	7.60	*ABCD*:	0.24
e	25.65	1.35	14.57	-7.12	-3.50	113.94	*E*:	3.56
ae	26.51	-2.94	3.79	-4.36	0.86	13.58	*AE*:	0.42
be	25.18	3.20	4.32	5.35	10.32	0.26	*BE*:	0.01
abe	30.74	0.55	8.04	6.51	1.48	2.98	*ABE*:	0.09
ce	25.26	3.31	3.77	5.34	-6.00	23.34	*CE*:	0.73
ace	28.85	1.37	5.20	3.12	13.04	-7.22	*ACE*:	-0.23
bce	27.98	3.87	7.72	0.51	5.06	15.06	*BCE*:	0.47
abce	31.33	3.93	14.29	0.73	-14.74	[-6.26]	*ABCE*:	[-0.20]*
de	21.71	0.87	-4.29	-10.78	2.76	4.36	*DE*:	0.14
ade	28.39	5.56	-2.65	3.72	1.16	[-8.84]	*ADE*:	[-0.28]*
bde	28.72	3.59	-1.94	1.43	-2.22	19.04	*BDE*:	0.60
abde	29.10	3.35	0.06	6.57	0.22	-19.80	*ABDE*:	-0.62
cde	22.09	6.68	4.69	1.64	14.50	-1.60	*CDE*:	-0.05
acde	27.98	0.38	-0.24	2.00	5.14	2.44	*ACDE*:	0.08
bcde	31.23	5.89	-6.30	-4.93	0.36	-9.36	*BCDE*:	-0.29
abcde	33.13	1.90	-3.99	2.31	7.24	6.88	*ABCDE*:	0.22

*These effects are completely confounded with blocks.

$$\hat{\Delta}_A + \hat{\Delta}_{AB} = 1.34 + (-0.46) = 0.88$$

The standard error of each of these estimates is equal to $\sqrt{2}$ times the standard error of any $\hat{\Delta}$, or (1.414)(0.146) = 0.206. Factor E has the largest main effect (+3.56), enhanced considerably by the presence of B and C together:

$$\hat{\Delta}_E + \hat{\Delta}_{BE} + \hat{\Delta}_{CE} + \hat{\Delta}_{BCE} = 3.56 + 0.01 + 0.73 + 0.47 = 4.77$$

However, the composite of effects represented by treatment combination means should be considered in view of the number of significant interactions. The 10 highest means all involve factor E plus at least 2 other factors, one of which is A or B. The standard error of a treatment combination mean is

$$\sqrt{ms_E/r} = \sqrt{0.343/2} = 0.414$$

Then, the half-width of Tukey's 95% interval (Sec. 2.2) is

$$(0.414)q_{0.05,32,28} = (0.414)(5.93) = 2.46$$

and any two treatment combinations that differ by more than that are significantly different. The 10 highest combinations all are significantly different from the 15 lowest combinations. Further research, using at least 3 factors, perhaps at 3 levels instead of 2, may be necessary to clarify these preliminary results.

The estimated efficiency of the incomplete block design (RIBD) relative to a complete block design (RCBD) requires computation of the sum of squares for replicates,

$$ss_R = (351.2^2 + 422.2^2)/32 - (773.74^2/64) = 78.10$$

Then, the sum of squares for blocks within replicates and the estimated error for complete blocks [(6.53)] are

$$ss_{D/R} = ss_D - ss_R = 87.16 - 78.10 = 9.06 \qquad \hat{\sigma}^2_{\text{RCBD}} = [9.06 + 56(0.343)]/62 = 0.456$$

The estimated efficiency is

$$E_{\text{RIBD:RCBD}} = (0.456/0.343)100\% = 133\%$$

i.e., forming blocks of 8 animals appears to be approximately 33% more efficient than forming blocks of 32 animals for this 2^5 experiment.

6.3.4. Fractional Replication

Fractional factorial experiments were formally introduced by Finney (1945). In exploratory research one may wish to examine a large number of factors simultaneously to look for interactions. For some experiments a complete replication is too costly or too large for facilities available. Experience in many disciplines suggests that the effects of high order interactions are small relative to main effects and first order interactions and to the magnitude of experimental error commonly encountered in replicated experiments. Fractionally replicated designs, which reduce the size of the experiment but permit estimation of main effects and first order interactions, may be very useful in such circumstances. In a few cases cost is so great or facilities are so restricted that fractional designs seem desirable for the estimation of main effects alone in the first experiment. However, one should not use fractional designs unless the risk of confusion from confounded effects is likely to be small.

The degree of resolution (ϕ) of a fractional design is defined (Webb 1965) for designs that permit estimation of factorial effects up to order $m < \phi/2$ when all interactions of order $\phi - m$ and higher are negligible ($m = 1$ for main effects, $m = 2$ for 2-factor interactions, etc.). Designs with resolution $\phi <$ III are of no use, and most of those with $\phi >$ VI are either impossible to construct or not economical for practical experiments. Characteristics of designs of resolution III, IV, V, and VI are described in Table 6.21.

Table 6.21. Estimability of Effects for Commonly Used Classes of Fractional Designs

ϕ, Resolution	m, Highest Order of Estimable Effects	$\geq(\phi - m)$, Order of Effects Assumed Negligible
III	1, main effects only	≥2, all interactions
IV	1, main effects only	≥3, interactions among 3 or more factors
V	2, 2-factor interactions	≥3, interactions among 3 or more factors
VI	2, 2-factor interactions	≥4, interactions among 4 or more factors

Fractional designs may be constructed by using the principles involved in the formation of incomplete blocks. For example, a half-replicate design for a 2^n experiment uses only one of the 2 blocks formed by choosing 1 DC and a quarter-replicate uses only one of 4 blocks formed when 2 DC are selected.

6.3.4.1. *Half-Replicate Designs*. Because only one of the 2 blocks formed by a defining contrast is actually used in a half-replicate experiment, the overall mean for the experiment is equivalent to the mean for one side of the contrast for the effect selected as defining contrast. This equivalence is known as the *fundamental identity I*. To illustrate the problems encountered, consider a 2^3 experiment (simple but impractical) for which one wishes to restrict the number of subjects or experimental units to 4 instead of the 8 required for a full replicate. Suppose ABC is selected as the defining contrast to form 2 blocks, {(1), ab, ac, bc} and {a, b, c, abc}, as in replicate 4 of Table 6.16. Selecting at random one of the 2 blocks for a half-replicate experiment implies that the fundamental identity $I = |\Delta_{ABC}|/2$ (or $I = ABC$ to simplify notation). Suppose the principal block {(1), ab, ac, bc} is selected; the only available data will be $\{y_{000}, y_{110}, y_{101}, y_{011}\}$. Examine the contrast totals q_k for a 2^3 experiment (e.g., see Chap. 2), and note the signs of the combinations (1), *ab*, *ac*, and *bc*. Each factorial effect is estimated by $\hat{\Delta}_k = q_k/(r2^{n-1})$, where $r = 1/2$ and $2^{n-1} = 4$. Therefore, $\hat{\Delta}_k = q_k/2$ and the overall mean is estimated by $\hat{\mu} = y_{...}/4$ for a half-replicate of a 2^3 experiment. Equations for estimating effects, using only the information available in the half-replicate, are:

$$4\hat{\mu} = y_{...} = +y_{011} + y_{101} + y_{110} + y_{000}$$

$$2\hat{\Delta}_A = q_A = -y_{011} + y_{101} + y_{110} - y_{000}$$

$$2\hat{\Delta}_B = q_B = +y_{011} - y_{101} + y_{110} - y_{000}$$

$$2\hat{\Delta}_{AB} = q_{AB} = -y_{011} - y_{101} + y_{110} + y_{000}$$

$$2\hat{\Delta}_C = q_C = +y_{011} + y_{101} - y_{110} - y_{000}$$

$$2\hat{\Delta}_{AC} = q_{AC} = -y_{011} + y_{101} - y_{110} + y_{000}$$

$$2\hat{\Delta}_{BC} = q_{BC} = +y_{011} - y_{101} - y_{110} + y_{000}$$

$$2\hat{\Delta}_{ABC} = q_{ABC} = -y_{011} - y_{101} - y_{110} - y_{000}$$

From these equations, the following identities are obvious:

$$2\hat{\mu} = -\hat{\Delta}_{ABC} \quad \hat{\Delta}_A = -\hat{\Delta}_{BC} \quad \hat{\Delta}_B = -\hat{\Delta}_{AC} \quad \hat{\Delta}_C = -\hat{\Delta}_{AB}$$

That is, each main effect is completely confounded with a 2-factor interaction. Therefore a half-replicate design for a 2^3 experiment is of resolution III, i.e., only main effects are estimable and then only if all interactions are negligible.

Two or more factorial effects that are completely confounded are referred to as *aliases*. Using concise notation, one may say (for the 2^3 experiment in a half-replicate) that the fundamental identity is $I = ABC$ and the alias structure is $A = BC$, $B = AC$, and $C = AB$. The same alias structure would occur if one selected the alternate block as the half-replicate except that the aliased effects would be related positively (e.g., $\hat{\Delta}_A = +\hat{\Delta}_{BC}$) instead of negatively. The alias of any effect can be determined easily by finding the generalized interaction of that effect with the fundamental identity, using the rule $A^2 = B^2 = C^2 = ... = 1$. For example, if $I = ABC$, the alias of A is $(A)\cdot(ABC) = A^2BC = BC$.

A 2^3 experiment in a half-replicate design usually is not practical even for studying main effects in the absence of all interactions because there are no degrees of freedom for experimental error. In some cases a prior estimate of error may be available. Total degrees of freedom are $(1/2)2^3 - 1 = 3$, all of which are required to estimate the 3 main effects.

A 2^4 experiment, with $I = ABCD$, has alias structure

$$A = BCD \quad B = ACD \quad C = ABD \quad D = ABC \quad AB = CD \quad AC = BD \quad AD = BC$$

That is, the design is of resolution IV, permitting estimation of main effects if interactions among 3 or more factors are negligible. There are only $(1/2)2^4 - 1 = 7$ df in all, with 4 df required for main effects and 3 df left for experimental error. These 3 df represent the 3 alias pairs of 2-factor interactions. Obviously, the experiment would have very low sensitivity even if the interactions were negligible.

A 2^5 experiment, with $I = ABCDE$, is of resolution V because the aliases of main effects are 4-factor interactions, and the aliases of 2-factor interactions are 3-factor interactions (e.g., $A = BCDE$ and $AB = CDE$). Five of the $(1/2)2^5 - 1 = 15$ df represent main effects, and the remaining 10 df represent the $C(5, 2) = 10$ two-factor interactions. Therefore, although the design permits estimation of the effects of 2-factor interactions, there are no degrees of freedom for error unless one pools the sums of squares for some of those interactions for use as experimental error.

Now it is apparent that half-replicates of 2^n experiments with $n = 5$ or fewer factors are not very useful unless the 2-factor interactions are likely to be trivial. However, a 2^6 experiment with $I = ABCDEF$ is of resolution VI, permitting estimation of main effects and 2-factor interactions if their respective alias interactions of 5 and 4 factors are negligible. The $(1/2)2^6 - 1 = 31$ df are divided into 6 df for main effects, $C(6, 2) = 15$ df for 2-factor interactions, and $31 - 6 - 15 = 10$ df for error, which is composed of 10 alias pairs of 3-factor interactions ($ABC = DEF$, $ABD = CEF$, etc.).

Half-replicates with 7 or more factors all have resolution $\phi > \text{VI}$ but may be too large [$(1/2)2^n \geq 64$ subjects for $n \geq 7$] or the required number of subjects may be rather heterogeneous. In the former case one may turn to smaller fractions of a replicate (Secs. 6.3.4.2, 6.3.4.3), and in the latter case it may be possible to form 2 or more blocks within the half-replicate to permit greater homogeneity of subjects in the same block.

Consider $(1/2)2^6$ in 2 blocks of 16 subjects each, with $I = ABCDEF$. Any effect selected as a defining contrast to form the 2 blocks will automatically confound its alias as well. Therefore, suppose the defining contrast is the alias pair $ABC = DEF$. If the half-replicate selected consists of the 32 treatment combinations for which $\sum_{i=1}^{6} x_i/2$ produces a zero residue, block 1 will contain 16 of the combinations for which $\sum_{i=1}^{3} x_i/2$ produces a zero residue and block 2 will contain the remaining 16 combinations:

Block 1: (1), *ab*, *ac*, *bc*, *de*, *df*, *ef*, *abde*, *abdf*, *abef*, *acde*, *acdf*, *acef*, *bcde*, *bcdf*, *bcef*
Block 2: *ad*, *ae*, *af*, *bd*, *be*, *bf*, *cd*, *ce*, *cf*, *abcd*, *abce*, *abcf*, *adef*, *bdef*, *cdef*, *abcdef*

Error consists of the remaining 9 alias pairs of 3-factor interactions (i.e., all except $ABC = DEF$). It is not possible to form 4 blocks of 8 subjects each within the half-replicate without confounding one of the 2-factor interactions, because the generalized interaction of any 2 of the 3-factor alias pairs selected as defining contrasts automatically confounds a 2-factor effect with blocks. For example, if $DC_1 = ABC = DEF$ and $DC_2 = ABD = CEF$, then

$$(ABC)(ABD) = A^2B^2CD = CD$$

which has alias *ABEF*.

Consider $(1/2)2^7$ in 4 blocks of 16 units each, with $I = ABCDEFG$, $DC_1 = ABCD = EFG$, and $DC_2 = ABEF = CDG$. The generalized interaction,

$$(ABCD)(ABEF) = A^2B^2CDEF = CDEF$$

which has alias *ABG*, is automatically confounded with blocks. The major partitions of variation are shown in Table 6.22. It is possible to construct $(1/2)2^7$ in 8 blocks without confounding any of the 2-factor interactions with

Table 6.22. Partition of Variation in Half-Replicate Design (2^7 experiment in 4 blocks of 16 units each)

Source of Variation	df
Blocks (*ABCD*=*EFG*, *ABEF*=*CDG*,*CDEF*=*ABG*)	3
Main effects	7
Two-factor interactions	*C*(7,2)=21
Error (unconfounded alias pairs of 3- and 4-factor interactions)	*C*(7,3)-3=32
Total	$(1/2)2^7$-1=63

blocks by using $I = ABCDEFG$ and 3 DC; for example, $ABCD = EFG$, $CDEF = ABG$, and $ACEG = BDF$. Since there are 7 df among 8 blocks, an additional 4 alias pairs are automatically confounded with blocks:

$(ABCD)(CDEF) = ABEF$ with alias *CDG*

$(ABCD)(ACEG) = BDEG$ with alias *ACF*

$(CDEF)(ACEG) = ADFG$ with alias *BCE*

$(ABCD)(CDEF)(ACEG) = BCFG$ with alias *ADE*

The analysis of data from half-replicate designs is surprisingly easy.

f any one of the factors in $(1/2)2^n$ is ignored, the remaining combinations orm a complete 2^{n-1} factorial experiment; computations may then be made as or a 2^{n-1} factorial in a completely randomized design (Chap. 2) or a block esign (Sec. 6.3.3). For example, if one ignores f in treatment combinations ound in the 2 blocks of the $(1/2)2^6$ experiment mentioned, the 32 treatment ombinations are labeled as they would be in a full replicate of a 2^5 experi-ent in 2 incomplete blocks. Using Yates's method (as in the example of Sec. .3.3), one would obtain mean estimates $\hat{\Delta}_k$ for the 31 factorial effects. The ffect labeled $\hat{\Delta}_{ABCDE}$ must be interpreted, via the alias structure imposed by $= ABCDEF$, to be the same as $\hat{\Delta}_F$. Similarly, the 5 effects involving 4 of the actors A, B, C, D, and E must be interpreted to be 2-factor interactions in-olving factor F. For example, $\hat{\Delta}_{ABCD} = \hat{\Delta}_{EF}$, $\hat{\Delta}_{ABCE} = \hat{\Delta}_{DF}$, etc. Of course, $_{ABC}$ would be confounded with blocks and the remaining nine 3-factor effects ould be pooled for experimental error (see example of Sec. 6.3.4.4).

.3.4.2. *Quarter-Replicate Designs*. If the number of subjects required for a alf-replicate design is too large for the resources available, one may use a maller fraction. Consider a 2^7 experiment that requires 64 subjects for a alf-replicate design. Suppose the experimenter wishes to restrict the size f the investigation to $(1/4)2^7 = 32$ subjects. To obtain a quarter-replicate ne will have to form 4 incomplete blocks and select one. Two defining con-rasts must be selected to form the 4 blocks; and the generalized interaction f the defining contrasts also will be lost as it too becomes part of the fun-amental identity I. Let $DC_1 = ABCDE$ and $DC_2 = CDEFG$. The generalized inter-ction is $(ABCDE)(CDEFG) = ABFG$. Thus the fundamental identity for this quar-er-replicate design is

$$I = ABCDE = CDEFG = ABFG$$

very estimable factorial effect will have 3 aliases instead of 1. The alias-s may be determined by obtaining the generalized interactions of each effect ith the 3 effects in the fundamental identity. For example, the main effect f factor A will have aliases $A(ABCDE) = BCDE$, $A(CDEFG) = ACDEFG$, and $A(ABFG)$ $= BFG$. Also, AB has aliases CDE, $ABCDEFG$, and FG, i.e., some of the 2-factor nteractions are confounded among themselves ($AB = FG$, $AF = BG$, $AG = BF$). All ther 2-factor interactions have aliases that contain at least 3 factors. The esign therefore has only resolution IV, although 15 of the 21 two-factor in-eractions are estimable if interactions involving 3 or more factors are neg-igible. In some experiments it may be possible judiciously to assign the ymbols A, B, F, and G to factors least likely to interact among themselves. Also, see Sec. 6.3.4.5 for an irregular fraction of resolution V.) Treatment ombinations having zero residues for both $\sum_{i=1}^{5} x_i/2$ and $\sum_{i=3}^{7} x_i/2$ and the major artitions of variation are shown in Table 6.23.

Experiments with 8 or more factors, of resolution V or higher in quarter-eplicate designs, still require many subjects [$(1/4)2^8 = 64$ or more] that may

Table 6.23. Design and Analysis for Quarter-Replicate of 2^7 Experiment

I=ABCDE=CDEFG=ABFG

Treatment combinations:

(1),*ab*,*cd*,*ce*,*de*,*fg*,*acf*,*acg*,*adf*,*adg*,
aef,*aeg*,*bcf*,*bcg*,*bdf*,*bdg*,*bef*,*beg*,*abcd*,
abce,*abde*,*abfg*,*cdfg*,*cefg*,*defg*,*acdef*,
acdeg,*bcdef*,*bcdeg*,*abcdfg*,*abcefg*,*abdefg*

Source of Variation	df
Main effects	7
Unconfounded 2-factor interactions	15
AB=FG,*AF=BG*,*AG=BF*	3*
Error (higher order alias quartets)	6
Total	31

*These effects may be tested and pooled with error if clearly nonsignificant.

be rather heterogeneous. Experimental error may be reduced by blocking within the quarter-replicate. Consider $(1/4)2^7$ in 2 blocks of 16 units each, with *I* = *ABCDE* = *CDEFG* = *ABFG* as before. The defining contrast required for blocking consists of an alias quartet, e.g., *ADG* = *BCEG* = *ACEF* = *BDF*. Sixteen treatment combinations from Table 6.23 that have zero residue for $(x_1 + x_4 + x_7)/2$ will appear in one block, and the remaining combinations will appear in the other block. The alias structure is unchanged by blocking, but the degrees of freedom for error are reduced by one.

Consider $(1/4)2^8$ in 4 blocks of 16 units each, with *I* = *ABCDE* = *DEFGH* = *ABCFGH*. Formation of 4 blocks requires the confounding of 2 DC and their generalized interaction with the 3 df for blocks. For example, if

$$\text{DC}_1 = ADH = BCEH = AEFG = BCDFG \qquad \text{DC}_2 = BEG = ACDG = BDFH = ACEFH$$

then (*ADH*)(*BEG*) = *ABDEGH* = *CGH* = *ABF* = *CDEF*, which is also confounded with blocks. All 8 main effects and *C*(8, 2) = 28 two-factor interactions are estimable, leaving 3 df for blocks and 24 df for error.

The analysis of data from quarter-replicate designs is similar to that for half-replicates (Sec. 6.3.4.1). If 2 factors in $(1/4)2^n$ are ignored, the remaining combinations form a complete 2^{n-2} factorial. Main effects and 2-factor interactions involving the ignored factors may be obtained by renaming the appropriate aliases. For example, if one ignores *a* and *g* in the treatment combinations of Table 6.23, the remaining symbols describe a complete replicate of a 2^5 experiment. After obtaining estimates of all effects, note that $\hat{\Delta}_A = \hat{\Delta}_{BCDE}$, $\hat{\Delta}_G = \hat{\Delta}_{CDEF}$, $\hat{\Delta}_{AC} = \hat{\Delta}_{BDE}$, etc., by taking generalized interactions of desired effects involving *A* and *G* with the fundamental identity *I*. Draper and Stoneman (1964) have described a method of estimating 1, 2, or 3 missing values to simplify the analysis when data are incomplete because of mortality or other reasons.

6.3.4.3. *Construction of Any Regular Fraction.* A 2^{-s} fraction (s = 1, 2, 3, ..., but $s < n$) is one of 2^s incomplete blocks formed by selecting s independent interactions as defining contrasts. These defining contrasts and their generalized interactions make a total of $2^s - 1$ factorial effects that are part of the fundamental identity *I*. Therefore, each estimable factorial effect will have $2^s - 1$ aliases, the generalized interactions of the estimable effect with *I*.

For example, $(1/8)2^8 = 32$ subjects and treatment combinations. Three DC and their 4 GI make the fundamental identity, and every estimable effect has 7 aliases. The design has resolution IV, but a few of the 2-factor interactions are estimable (Cochran and Cox 1957, p. 286).

A 2^{-s} fraction may be divided into 2^u blocks $[(s + u) \leq n]$ of 2^{n-s-u} units each by selecting as defining contrasts u alias sets of 2^s effects each. For $u \geq 2$, $2^u - u - 1$ additional alias sets will be automatically confounded with blocks. The additional sets are the generalized interactions among the defining contrasts.

For example, 1/16 of 2^8 is 16 units, i.e., $s = 4$ because $2^{-4} = 1/16$. If one wishes to divide the 16 units into 4 blocks of 4 each, then $u = 2$ ($2^2 = 4$ blocks of $2^{8-4-2} = 4$ units each). If one takes the fundamental identity and defining contrasts,

$$I = ABCD = ABEF = ABGH = ACEH \quad (+\ 11\ \text{GI})$$

$$\text{DC}_1 = AB = CD = EF = GH = BCEH = \ldots$$

$$\text{DC}_2 = AC = BD = BCEF = BCGH = EH = \ldots$$

then the generalized interaction,

$$\text{GI} = (AB)(AC) = BC = AD = ACEF = ACGH = \ldots$$

also is confounded with blocks. The design has resolution IV, but only 4 df are available for error (including all 16 two-factor interactions not confounded with blocks) because the 8 main effects and 4 blocks use 11 of the 15 total df.

Cochran and Cox (1957, pp. 276-89) and the U.S. National Bureau of Standards (1957) have provided extensive lists of plans for regular fractions of factorial experiments.

6.3.4.4. *Example with Fractional Replication*. Suppose an experimenter wishes to study the mechanisms of utilization of 7 amino acids for the growth of pigs; the main effects are known to be nontrivial, so the first order interactions are of chief interest. A full replicate of $2^7 = 128$ combinations is too extensive for the initial study. Therefore a half-replicate design is proposed, to be constructed with 8 blocks (litters) of 8 pigs each. The fundamental identity selected is $I = ABCDEFG$, with treatment combinations to be used including the "negative" control (1) and all other combinations containing an even number of letters (zero residue from $\sum_{i=1}^{7} x_i/2$).

The alias structure indicates 6-factor aliases of main effects, 5-factor aliases of first order interactions, and 4-factor aliases of 3-factor interactions, although the latter must be used for experimental error because the 7 main effects $C(7, 2) = 21$ first order interactions and 8 blocks use 35 of the 63 df.

The 3 DC required for blocking are alias pairs of 3- and 4-factor interactions, selected so that their generalized interactions are not confounded with any main effects or first order interactions. Selected are $DC_1 = ABC = DEFG$, $DC_2 = ADE = BCFG$, and $DC_3 = BDF = ACEG$. Effects automatically confounded with blocks are

$$(ABC)(ADE) = BCDE = AFG \qquad (ABC)(BDF) = ACDF = BEG$$

$$(ADE)(BDF) = ABEF = CDG \qquad (ABC)(ADE)(BDF) = (AFG)(BDF) = ABDG = CEF$$

Each treatment combination in the half-replicate produces residues R_1, R_2, and $R_3 = 0$ or 1 from $(x_1 + x_2 + x_3)/2$, $(x_1 + x_4 + x_5)/2$, and $(x_2 + x_4 + x_6)/2$. For example, *abcd* (or 1111000) produces $(1 + 1 + 1)/2$, $(1 + 1 + 0)/2$, and $(1 + 1 + 0)/2$ so that $R_1 = 1$, $R_2 = 0$, and $R_3 = 0$. The 8 blocks {1, 2, ..., 8} are formed from residue combinations {000, 001, 010, 011, 100, 101, 110, 111}. For example, *abcd* has residue combination 100 and is assigned to block 5. The complete design and hypothetical observed responses (coded) are shown in Table 6.24.

Ignoring *g* on treatment combinations utilized, one may compute mean estimates of the 63 resulting factorial effects ($\hat{\Delta}_k$) by Yates's method. Effects symbolized by 4 or more of the letters *A*-*F* must be renamed by finding the generalized interaction of the nominal symbol with the fundamental identity $I = ABCDEFG$. For example, *ABCD* becomes *EFG*, *ABCDE* becomes *FG*, *ABCDEF* becomes *G*, etc. After converting symbols, the sums of squares for all 3-factor effects except the 7 effects confounded with blocks may be pooled for experimental error [$C(7, 3) - 7 = 28$ df] as follows [see (6.46)]:

$$ss_E = (1/2)2^{n-2} \sum_{k}^{28} \hat{\Delta}_k^2 = 16 \sum_{k}^{28} \hat{\Delta}_k^2$$

In similar fashion, sums of squares for the 7 effects confounded with blocks may be pooled for computing the block sum of squares. Since the standard error of any factorial effect [(6.48)] is $\sqrt{ms_E/[(1/2)2^{n-2}]}$, the simplest way to

Table 6.24. Amino Acid Combinations in Half-Replicate of 2^7 Factorial Experiment (growth response of pigs)

Block:	1		2		3		4		5		6		7		8
(1)	32	*bc*	41	*ac*	48	*ab*	47	*ae*	48	*ad*	63	*ag*	35	*af*	65
abdg	52	*de*	43	*df*	52	*dg*	57	*bf*	58	*bg*	57	*bd*	36	*be*	53
abef	56	*fg*	37	*eg*	44	*ef*	54	*cg*	55	*cf*	56	*ce*	31	*cd*	59
acdf	60	*abdf*	71	*abde*	66	*acde*	58	*abcd*	63	*abce*	64	*abcf*	48	*abcg*	48
aceg	38	*abeg*	43	*abfg*	58	*acfg*	68	*adfg*	60	*aefg*	72	*adef*	62	*adeg*	71
bcde	49	*acdg*	48	*bcdg*	65	*bcdf*	69	*bdeg*	58	*bdef*	67	*befg*	41	*bdfg*	66
bcfg	39	*acef*	46	*bcef*	45	*bceg*	55	*cdef*	51	*cdeg*	63	*cdfg*	42	*cefg*	58
defg	46	*bcdefg*	50	*acdefg*	68	*abdefg*	83	*abcefg*	71	*abcdfg*	85	*abcdeg*	49	*abcdef*	89

I=ABCDEFG. Confounded with blocks *ABC*,*ADE*,*BDF*,*AFG*,*BEG*,*CDG*,*CEF*.

compute the standard error is

$$s_{\hat{\Delta}_k} = \sqrt{\sum_{k}^{28} \hat{\Delta}_k^2 / 28}$$

Any unconfounded effect that exceeds $t_{\alpha/2,28} s_{\hat{\Delta}_k}$ is significant at the $(1 - \alpha) \cdot$ 100% level. Estimates of mean effects ($\hat{\Delta}_k$) are shown in Table 6.25.

Four of the previously established main effects are confirmed with high significance (*A*, *B*, *D*, *F*), but estimates of the other 3 (*C*, *E*, *G*) are not large enough to show significance. Interactions of *A* with *D*, *E*, and *F* and of *B* with *D* are significant. The *CE* interaction is very near the 95% significance level. Supposedly the amino acid labeled *A* (arginine) has important influence on growth directly and also in synergy with at least 3 other amino acids. Note that 5 of the 7 block effects were "significant," indicating that blocking is important in reducing experimental error.

6.3.4.5. *Irregular Fractions*. An irregular fractional replicate cannot be constructed by the simple methods of defining contrasts used for regular fractions 2^{-s}, such as 1/2 or 1/4. Irregular fractions often require fewer experimental units than do regular fractions to estimate a specified set of effects, but most irregular plans do not permit completely orthogonal estimation of the desired effects. Therefore the use of such plans should be restricted to experiments in which the experimental units are very expensive or nearly unique and limited in number, studies that require extensive valuable resources of other kinds, or investigations that may endanger lives of human volunteers.

Most irregular plans presently known are of resolution IV or V (see Table 6.21). Webb (1968) and Margolin (1969a) proved that the smallest fractional plan of resolution IV (permitting estimation of main effects only, but not assuming first order interactions negligible) for a 2^n experiment contains $2n$ treatment combinations. For example, an experiment with 5 factors must contain 10 of the 32 treatment combinations to have resolution IV. Section 6.3.4.1 shows that a half-replicate of 2^5 (16 treatment combinations) is of resolution V but provides no degrees of freedom for error. Also, $(1/4)2^5 = 8$ combinations provide resolution III. Box and Wilson (1951) showed that a plan of resolution IV can be constructed by using treatment combinations of a plan of resolution III and their mirror images--the *foldover* technique. Margolin (1969c) later proved that the smallest plan of resolution IV must be a foldover plan. For example, the 2^5 in 10 combinations may be obtained from a 2^4 in 5 combinations (the smallest 2^4 plan of resolution III) by adding a fifth factor to either the 2^4 plan or its mirror:

2^4 in 5 = {(1), *abc*, *abd*, *acd*, *bcd*}

mirror = {*abcd*, *d*, *c*, *b*, *a*}

2^5 in 10 = {(1), *abc*, *abd*, *acd*, *bcd*, *abcde*, *de*, *ce*, *be*, *ae*}

Table 6.25. Estimates of Factorial Effects in Half-Replicate of 2^7 Experiment (utilization of amino acids by pigs)

Main Effects		Two-Factor Interactions						Blocks		Error							
A	8.56**	*AB*	1.06	*BD*	2.44*	*CG*	-0.56	(*ABC*)	4.88	*ABD*	0.88	*ACG*	0.38	*BCF*	-0.62	*CDF*	0.94
B	4.75**	*AC*	-0.88	*BE*	0.62	*DE*	-0.06	(*ADE*)	1.50	*ABE*	2.06	*ADF*	0.06	*BCG*	1.19	*CEG*	-0.06
C	0.81	*AD*	2.38*	*BF*	1.44	*DF*	-0.38	(*AFG*)	-10.12	*ABF*	1.50	*ADG*	-0.19	*BDE*	-1.69	*CFG*	-1.88
D	9.69**	*AE*	2.44*	*BG*	0.19	*DG*	0.69	(*BDF*)	8.75	*ABG*	0.44	*AEF*	0	*BDG*	-0.56	*DEF*	-0.12
E	1.62	*AF*	5.88**	*CD*	0.12	*EF*	-0.06	(*BEG*)	5.50	*ACD*	-1.06	*AEG*	-0.31	*BEF*	-1.19	*DEG*	0.69
F	7.94**	*AG*	1.31	*CE*	-2.19	*EG*	-0.75	(*CDG*)	5.00	*ACE*	0	*BCD*	1.25	*BFG*	-0.19	*DFG*	1.62
G	-0.44	*BC*	0.31	*CF*	-1.00	*FG*	0.75	(*CEF*)	2.00	*ACF*	2.06	*BCE*	0.50	*CDE*	-1.12	*EFG*	-1.94

Standard error of effect = $\sqrt{[0.88^2+2.06^2+\ldots+(-1.94)^2]/28}=1.10$, $t_{0.025,28}=2.048$, $t_{0.005,28}=2.763$.

*P(Type I error) < 0.05, **P(Type I error) < 0.01.

Plans of resolution IV in $2n$ combinations with n = 4, 8, or 16 factors permit orthogonal estimation of the main effects. Estimation is nonorthogonal for other values of n. Margolin (1969a) has listed nonorthogonal plans of resolution III that may be used to construct plans of resolution IV. One should remember that the plan with the smallest number of treatment combinations possible is not always the most desirable.

All orthogonal plans of resolution V (permitting estimation of first order interactions) require far more subjects than the number of main effects and first order interactions to be estimated [excepting the $(1/2)2^5$ design that provides no degrees of freedom for error]. For example, a 2^7 experiment has 7 main effects and $C(7, 2) = 21$ first order interactions, but the smallest orthogonal design of resolution V is the half-replicate that includes 64 subjects. Webb (1965) developed *saturated* plans of resolution V (plans having no degrees of freedom for error) that are nonorthogonal for $n > 5$ factors. Such plans minimize cost but are not desirable statistically unless one has a prior estimate of experimental error. Addelman (1969) produced nonorthogonal plans of resolution V for as many as 17 factors. The most practical of those plans are $(5/16)2^7$ and $(3/16)2^8$, because the regular $(1/4)2^7$ requires confounding of 6 of the 21 first order interactions and the regular $(1/4)2^8$ requires 64 subjects, while the 3/16 fraction requires only 48 subjects for resolution V. John (1961) showed how to obtain resolution V for 2^4 and 2^5 experiments by using 3/4 replicates. In 1962 he developed the general case of 3/4 replicates for 2^n experiments.

Webb (1971) published a short catalogue of small 2^n designs ($3 \leq n \leq 9$) that do not require more than 20 subjects. Most of the designs are of resolution III or IV, although a design of resolution V for a 2^4 experiment requires only 11 subjects (one fewer than a 3/4 replicate).

6.3.4.6. *Sequences of Fractions*. Ordinary factorial experiments must be completed before any estimates of effects can be made. When subjects can or must be treated in sequence, as in some clinical trials, some conclusions can be made earlier. Although such conclusions may be of limited validity and precision, they may be worthwhile for one or more reasons:

1. To determine large effects quickly and terminate the experiment as soon as conclusive results are obtained (saving resources)
2. To recognize quickly that the range of doses applied should be narrowed, expanded, or shifted
3. To decide whether more subjects should be added to improve precision of estimated effects or to permit estimation of additional parameters

Davies and Hay (1950) were early advocates of evaluating each fraction of a sequence of fractional replicates as soon as it is completed, and Daniel (1956) discussed the general problem of designing second stage experiments to clarify the interpretation of results from fractional factorial designs.

The various sequential strategies may be classified into (1) sequences of incomplete blocks (fractions) of treatment combinations, (2) sequences that introduce one factor at a time, and (3) sequences robust against time trends. Daniel (1962) discussed some sequences of fractions, and Addelman (1969) presented a catalogue of sequences involving as many as 11 factors. For example, a sequence of four 1/64 replicates of a 2^{10} experiment (16 subjects per fraction) was devised by using the defining contrasts, {*ABCDEF*, *ABCGHJ*, *ABDGK*, *AEG*, *ABH*, *ACD*}. (The letter *I* is not used to designate a factor because of possible confusion with the fundamental identity.) Treatment combinations specified by the following groups of residue combinations $\{R_1R_2...R_6\}$ for the 6 DC are suggested as the 4 sequential fractions: {001100}, {001101}, {001110}, and {001011}. The design permits estimation of all main effects and 5 first order interactions after the first stage. Additional interactions (15 more and 29 more) can be estimated after the second and third stages, and all 2-factor interactions can be estimated after the fourth stage (resolution V). The average variance of the estimates is only half as much after the second stage as after the first, but only small reductions in variance (especially for interactions) are achieved by adding the third and fourth stages.

Daniel (1958) discussed the advantages of introducing one factor at a time (assuming low level to represent absence of a factor), and Webb (1965) devised such designs with the minimal number of subjects that permit estimation of main effects and first order interactions. The minimal number for a 2^n experiment is $(n^2 + n + 2)/2$ (e.g., 4 factors require 11 subjects). Suppose the treatment combinations of a 2^3 experiment were run in the order {(1), *a*, *b*, *ab*, *c*, *ac*, *bc*, *abc*}. After responses were obtained from the first 2 subjects one could estimate Δ_A; after 3 subjects one could estimate Δ_B; after 4, Δ_{AB}; after 5, Δ_C; after 6, Δ_{AC}; after 7, Δ_{BC}; and after 8, Δ_{ABC}. Obviously, only the first 7 subjects are required for the estimation of main effects and first order interactions. However, one must have a prior estimate of experimental error if any probability statement is to be made. The effect of a fourth factor Δ_D and all 2-factor interactions involving *D* could be estimated by adding only 4 treatment combinations to the first 7, namely {*d*, *ad*, *bd*, *cd*}. The factors should be introduced in descending order of importance (if indeed prior conviction suggests they differ in importance). Unfortunately, the predetermined ordering of the treatment combinations precludes the possibility of randomizing combinations with respect to time (although in some cases the subjects can be randomly ordered). Therefore, such designs are not robust against time trends.

Daniel and Wilcoxon (1966) proposed designs providing estimates that are orthogonal or nearly orthogonal to linear and quadratic trends in time. For example, the proper sequence of treatment combinations of 2 replicates of a 2^2 experiment is {(1), *ab*, *b*, *a*, (1), *ab*, *b*, *a*}. Draper and Stoneman (1968b) produced designs, limited to 8 subjects, that are robust against linear time trends and also tend to minimize the number of level changes from one subject to the next. The latter characteristic is desirable when the administration of the treatments is complex--for example, changing of several levels at one time may waste much time in resetting equipment involved or may lead to confusion and errors. Consider a 2^3 experiment. The sequence {(1), *ab*, *abc*, *c*, *ac*, *bc*, *b*, *a*} is free of linear time trend for all 3 main effects but involves 11 changes of level in administration of the treatments to 8 subjects. The interchanging of *c* and *ac* in the sequence would reduce the number of changes in level to 9 but would partially confound a linear time trend with the main effect of factor *A*.

A special case of sequential design is the repetition of the same fraction. Such designs are not saving of resources with respect to minimal estimation of effects but they possess an advantage missing from other fractional plans--provision of a "pure" estimate of experimental error (an estimate not dependent on triviality of higher ordered interactions for its validity). The "pure" error may be confounded with time effects, however. One may test the significance of the set of interactions commonly pooled for error and pool with the "pure" error only those that are clearly trivial.

Be cautious in interpreting tests of significance for combined stages of experimentation, especially when decisions to perform later stages are based on results observed in earlier stages. The amount of bias introduced in the significance level for the combined test is unknown. However, many experimenters regard the risk of bias as a reasonable price to pay for the advantages and economies gained.

6.3.5. Sequential Paired Trials for All-or-None Responses

In clinical trials a disease may be sufficiently rare that a study must be conducted over a lengthy period to obtain enough patients to make reasonably precise comparison of the curative powers of 2 treatments. The treatments may be drugs, diets, operative techniques, etc. One treatment will be a *control* (placebo or absence of treatment or standard form of treatment) labeled "old," and the other treatment, experimental in nature, labeled "new." As patients come into the experiment they are "paired," i.e., each set of 2 patients is placed on the old or new treatment randomly.

Such experiments are not factorial and are fractional only in the sense that a complete block (pair) is split into halves by the individual patients arriving at different times. Despite this anomaly of classification, it is reasonable to discuss the subject in this chapter with sequential fractions.

Sequential analysis was developed by Wald (1947) for military purposes, but its use spread to industrial applications (Jackson 1960) and medical trials (Armitage 1960, 1971) and has developed as a branch of statistics in its own right (Wetherill 1966; Ghosh 1970). Sequential experimentation offers three advantages over the usual fixed-sample procedures: (1) perhaps most important, it may permit significant reduction in the amount of data required to reach statistically valid conclusions (Chow et al. 1971); (2) it permits analysis of data as they arrive instead of waiting until all results are available; and (3) in some cases all computation is eliminated.

Bross (1952) developed a simple sequential technique for evaluating *all-or-none* responses in paired trials. For example, "lived-died" and "cured-not cured" are all-or-none classifications. Two consecutive patients constitute a pair or patients may be paired by age, sex, severity of disease, etc. Obviously, the same procedure may be applied to other experimental units, such as animals, in vitro cultures, etc.

The old and new treatments should be randomly assigned within each pair. For a given pair, if both treatments lead to the same conclusion (both cured or both not cured), that pair provides no information as to which treatment is superior. However, such ties should be recorded if finally one wishes to make a proper estimate of the proportion of patients cured by either treatment. Obviously, a decision can be made at any stage in the sequence of pairs. The experimenter can stop the experiment and make a recommendation ("use the new," "use the old," "use either new or old") or postpone the recommendations and continue the experiment, adding other pairs. Of course, the appropriate decision will depend on what has happened up to that point in the sequence of pairs. Has one of the treatments "won" in every pair, most of the pairs, or a majority of the pairs? Or has each treatment "won" about an equal number of times? The length of the series required for good confidence will depend on the data obtained, but on the average a sequential plan will require smaller total sample size than the usual fixed-sample experiment (Chap. 1).

Let ω_1 be the probability of cure by the old treatment and let ω_2 be the probability of cure by the new treatment. Then, considering pairs of patients, one obtains $\omega_1(1 - \omega_2)$ as the probability that the patient given the old treatment is cured and the patient given the new treatment is not. Similarly, one obtains $(1 - \omega_1)\omega_2$ as the probability that the patient given the old treatment is not cured and the patient given the new treatment is cured. Limiting the discussion to pairs of patients for which a treatment wins, one obtains the probability that the new treatment will win:

$$\omega^* = (1 - \omega_1)\omega_2/(\omega_1 + \omega_2 - 2\omega_1\omega_2) \tag{6.55}$$

When the treatments are truly equal in curative ability ($\omega_1 = \omega_2$), then $\omega^* = 0.5$. However, if the new treatment is better than the old, $\omega^* > 0.5$. Suppose one defines $\omega^* = 0.6$ as a "small" advantage for the new treatment, and $\omega^* = 0.7$ as an "important" advantage. The proportion of patients cured by the new treatment (ω_2) necessary to obtain each of these defined advantages depends on the probability of cure by the old treatment (ω_1). Some of the relationships are tabulated in Table 6.26.

Note that an old treatment that is only "moderately" effective (say $0.3 \leq \omega_1 \leq 0.5$) requires the definitions of "small" and "important" advantages of the new treatment to represent approximately 10% and 20% more of the total patients that will be cured. For an old treatment that has "low" (say $\omega_1 < 0.3$) or "high" efficacy, the same definitions imply smaller percentages of the total patients. However, for low ω_1 the percent improvement is the largest, as one would think it should be. For example, at $\omega_1 = 0.05$, an important advantage implies $\omega_2 = 0.11$ (only 6% more of the total patients cured); but the change in rate of cure is more than twofold, since the improvement (0.06) represents 120% of the old rate of cure. Conversely, if the old treatment has a high rate of cure, say $\omega_1 = 0.90$, an important advantage of the new treatment implies $\omega_2 = 0.95$, which is 5% more of the total patients but an improvement of only 5.5% over the old rate of cure.

Bross devised two charts (Fig. 6.1) that enable the experimenter to make reasonable decisions about the treatments without making any computations. Figure 6.1.1 was constructed so that the probability of Type I error is approximately 0.10 for one-sided alternatives ($\bar{H}$:new > old) or 0.20 for two-sided alternatives ($\bar{H}$:new ≠ old). The power of the procedure is approximately 0.9, 0.75, or 0.5 when $\omega^* = 0.7, 0.65$, or 0.6, respectively. That is, if the new treatment actually has an important advantage ($\omega^* = 0.7$), the probability of detecting the advantage is 0.9. Figure 6.1.2 was constructed with 0.05 probability of Type I error (or 0.10 for two-sided alternatives) and approximately 0.85, 0.65, or 0.40 power to detect $\omega^* = 0.7, 0.65$, or 0.6, respectively.

Ordinary graph paper may be used to draw in the barriers according to

Table 6.26. Improvement in Proportion Cured Represented by Important and Small Advantages of New Treatment

Probability of Cure by Old Treatment (ω_1)	Proportion Cured by New Treatment (ω_2) for Important Advantage (ω^*=0.7)	Small Advantage (ω^*=0.6)
0.05	0.11	0.07
.10	.21	.14
.15	.29	.21
.20	.37	.27
.25	.44	.33
.30	.50	.39
.35	.56	.45
.40	.61	.50
.45	.66	.55
.50	.70	.60
.55	.74	.65
.60	.78	.69
.65	.81	.74
.70	.84	.78
.75	.88	.82
.80	.90	.86
.85	.93	.89
.90	.95	.93
0.95	0.98	0.97

Fig. 6.1.1 or 6.1.2, depending on the experimenter's view of the relative risks of errors of the first and second kind. For example, if the old treatment is thought to cure about 70% of the patients, $\omega^* = 0.7$ indicates the new treatment would cure about 84% of the patients (Table 6.26). If one wishes to have 90% chance to detect the 14% advantage and is willing to accept a 10% risk that a conclusion stating "new is better than old" may be erroneous, the barriers of Fig. 6.1.1 should be used. However, if one desires risk of Type I error smaller than 10%, the barriers of Fig. 6.1.2 may be used; but the power of detecting the 14% improvement, if it exists, is reduced to 85%. Having drawn the appropriate barriers, one should observe the following plotting rules:

1. The outcome of each pair of patients is plotted only if one treatment wins (the new treatment cures and old does not or vice versa). Ties are recorded but not plotted.

2. If the new treatment wins, mark an x in the square *above* the blackened square in the lower left corner or *above* the previous x if other wins have been recorded.

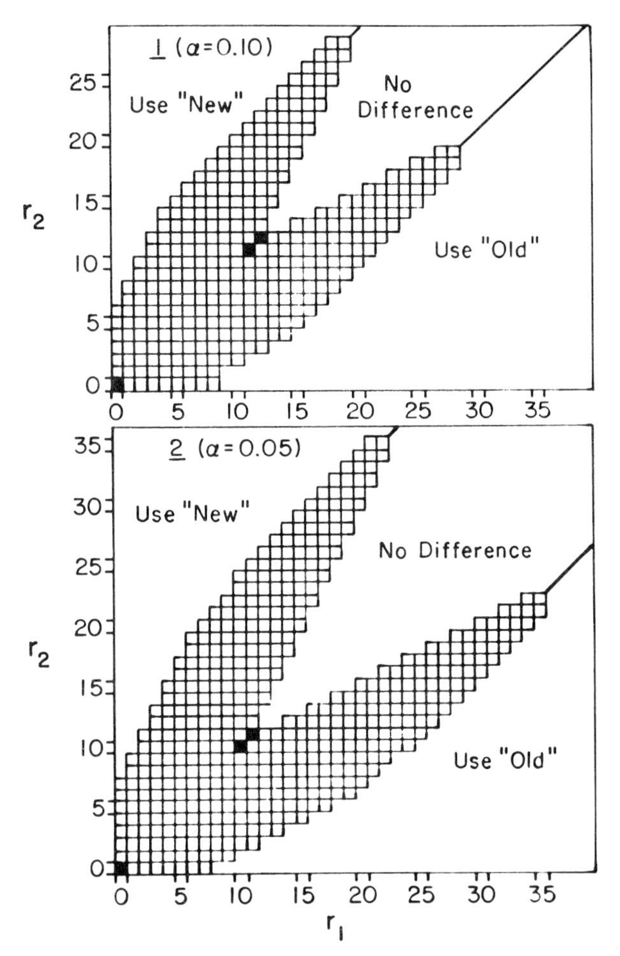

Fig. 6.1. Decision charts for sequential paired trials; number of "wins" by "old" treatment (r_1); number of "wins" by "new" treatment (r_2).

Reproduced by permission from I. D. J. Bross, *Biometrics* 8(1952): 197-98.

3. If the old treatment wins, mark an x in the square to the *right* of the blackened square in the lower left corner or to the *right* of the previous x if other wins have been recorded.

Note three barriers in each figure: top, middle, and bottom. Normally, the experiment is continued (additional pairs observed) until one barrier is crossed. Decision rules are outlined in Table 6.27.

If one wishes to estimate the proportion of patients cured by either treatment, the results of pairs that were tied also must be considered. Nearly unbiased estimates for old and new treatments are

$$\hat{\omega}_1 = (r_1 + r_3) / \sum_{i=1}^{4} r_i \tag{6.56}$$

Table 6.27. Decision Rules for Sequential Paired Trials (see Fig. 6.1.1)

Barrier Crossed	Appropriate Conclusion	Action
Top	new treatment superior	use new treatment
Middle	no major difference	use preferred treatment*
Bottom	old treatment superior	use old treatment

*Experiments in which one treatment is more convenient or economical (preferred) may be terminated without crossing a barrier as soon as the path of *x*s enters the region between the middle and the top barrier (if new is preferred) or the bottom barrier (if old is preferred).

$$\hat{\omega}_2 = (r_2 + r_3) / \sum_{i=1}^{4} r_i \tag{6.57}$$

where r_1, r_2, r_3, and r_4 represent the numbers of pairs in which old won, new won, both cured, and both failed to cure, respectively. Note that $\sum_{i=1}^{4} r_i$ is the total number of pairs observed (or half of n, the total number of patients observed).

For Fig. 6.1.1 (10% Type I error), the longest possible path to a decision is 48 squares or 96 patients if no ties occur. The experiment may require more patients to conclude the longest path if ties occur or fewer patients if a barrier is crossed by a shorter path. Bross has shown that the expected maximum number of patients is twice the longest possible path times the ratio $\omega^* / [(1 - \omega_1)\omega_2]$ or $67.2/[(1 - \omega_1)\omega_2]$ for the 10% chart with $\omega^* = 0.7$ (probability that the new treatment will win). For example, the expected maximum number of patients is 192 for $\{\omega_1 = 0.50, \omega_2 = 0.70\}$ (see Table 6.26). If both treatments are rarely successful or nearly always successful, a very large experiment may be required to reach a decision. For example, the expected maximum number of patients is 644 for $\{\omega_1 = 0.05, \omega_2 = 0.11\}$ or 1372 for $\{\omega_1 = 0.95, \omega_2 = 0.98\}$. Figure 6.1.2 (5%) has a maximum path of 58 squares (pairs not tied) or 116 patients. The expected maximum number of patients becomes $69.6/[(1 - \omega_1)\omega_2]$ with $\omega^* = 0.6$. The expected maximum is much greater than for the 10% chart when the proportion cured by the old treatment is low, 1046 patients for $\{\omega_1 = 0.05, \omega_2 = 0.07\}$, but only slightly greater when ω_1 is larger, 232 patients for $\{\omega_1 = 0.5, \omega_2 = 0.6\}$.

It may be more practical to consider the median number of patients required. Bross has shown that the median length of path is 29 for 10% and 28 for 5% error when $\omega_1 = \omega_2$ ($\omega^* = 0.5$) but is only 23 for 10% and 28 for 5% if

$\omega^* = 0.7$. That is, the experiment is expected to be shorter for the 10% test when the new treatment has an important advantage than when the treatments do not differ. Such an improvement is not obtained for the 5% test. Bross's comparison of fixed sample size and average sequential sample size, for $\omega^* = 0.5$, is shown in Table 6.28. Generally the sequential plans require only about two-thirds as many patients (or fewer if the cure rate is high) as fixed-sample plans with the same specifications, although it is possible that in a particular experiment the sequential scheme may require more observations than the fixed-sample plan.

Table 6.28. Expected Number of Patients Required in Fixed and Sequential Samples

	10% Error (Fig. 6.1.1)		5% Error (Fig. 6.1.2)	
ω_1	Fixed	Average Sequential	Fixed	Average Sequential
0.10	280	176	322	180
.20	176	114	202	116
.50	150	96	172	96
0.90	728	338	842	340

Armitage (1960), Spicer (1962), Alling (1966), Choi (1968), and Breslow (1970) also have discussed sequential procedures for a binomial sequence. As with Bross's method, the designs are "closed" only in terms of the number of pairs not tied, as tied pairs are discarded. That represents uncertainty about maximum length of the trial, especially if not much is known about the effectiveness of the 2 treatments. Aroian and Öksoy (1970) considered sequential procedures that provide for tied pairs, thereby permitting plans with a definite maximum sample size. Elfring and Schultz (1973) have extended that idea for situations in which it is convenient to work with more than one pair at a time, as when patients are confined to a clinic or metabolic ward.

6.4. SYMMETRICAL THREE-LEVEL FACTORIAL EXPERIMENTS

Factors found to be significant either as main effects or in interactions in exploratory 2-level experiments may be utilized in further stages of experimentation. If 3 levels of each factor are used, one may examine simple nonlinear response to dose and begin to look for optimum combinations of doses of the various significant factors. Ultimately one may turn to response surface designs (Chap. 9), but in the early stages one may have many treatment combinations to examine and the available subjects may be sufficiently heterogeneous that incomplete block designs are warranted to reduce experimental error. Practical considerations also may restrict block size or prevent using more than a fraction of the desired treatment combinations.

Special notation for symmetrical 3-level factorials is explained in Chap. 2. Briefly, a 3^n experiment involves n factors each at 3 levels designated 0,

1, 2. For example, a 3^2 experiment involves factors A and B at 3 levels each (usually equally spaced doses) with 9 treatment combinations designated {00, 01, 02, 10, 11, 12, 20, 21, 22}. Because each main effect has 2 df and interactions involving f factors have 2^f df, simple contrasts are not sufficient to describe each factorial effect as is possible for 2-level factorials.

Consider the total response to increases in level of factor A,

$$(y_{2..} - y_{1..}) + (y_{1..} - y_{0..}) = y_{2..} - y_{0..} \tag{6.58}$$

Equation (6.58) is a measure of the total *linear* effect of A. Next consider the difference in response to increases in A in different parts of the range of A:

$$(y_{2..} - y_{1..}) - (y_{1..} - y_{0..}) = y_{2..} - 2y_{1..} + y_{0..} \tag{6.59}$$

Equation (6.59) is a measure of the *nonlinear* (quadratic) effect of A. Note that (6.58) and (6.59) are orthogonal contrasts. The coefficients of 0, 1, and 2 levels, {-1, 0, +1} and {+1, -2, +1}, derived here heuristically, are the same coefficients obtained by the theory of orthogonal polynomials (Chap. 2 and App. A.7). Similar contrasts may be made among totals for different levels of factor B.

The 4 df for AB interaction are orthogonal to the main effects. Those, too, may be partitioned orthogonally into contrasts with 1 df each by taking products of the coefficients ξ'_{ij} for the linear or quadratic effects of the 2 factors. The entire array of contrasts is shown in Table 6.29. As usual, one may obtain the sum of squares for any contrast from

Table 6.29. Orthogonal Polynomial Coefficients ξ'_{ij} for Linear (L) and Quadratic (Q) Trends (3^2 experiment with equally spaced levels)

Contrast Totals	Treatment Combination Totals								
	$y_{00.}$	$y_{01.}$	$y_{02.}$	$y_{10.}$	$y_{11.}$	$y_{12.}$	$y_{20.}$	$y_{21.}$	$y_{22.}$
$q_1(A_L)$	-1	-1	-1	0	0	0	+1	+1	+1
$q_2(A_Q)$	+1	+1	+1	-2	-2	-2	+1	+1	+1
$q_3(B_L)$	-1	0	+1	-1	0	+1	-1	0	+1
$q_4(B_Q)$	+1	-2	+1	+1	-2	+1	+1	-2	+1
$q_5(A_LB_L)$	+1	0	-1	0	0	0	-1	0	+1
$q_6(A_LB_Q)$	-1	+2	-1	0	0	0	+1	-2	+1
$q_7(A_QB_L)$	-1	0	+1	+2	0	-2	-1	0	+1
$q_8(A_QB_Q)$	+1	-2	+1	-2	+4	-2	+1	-2	+1

$$ss_{q_k} = q_k^2/[r\Sigma(\xi'_{ij})^2] \tag{6.60}$$

where there are r replicates of each treatment combination.

6.4.1. Construction of Incomplete Blocks

As for symmetrical 2-level factorial experiments, the design problem involves proper allocation of a subset of treatment combinations to each block. The use of deliberate confounding of certain treatment effects with block effects to accomplish that goal is common to 2- and 3-level experiments, but the techniques of developing a proper design are necessarily different.

6.4.1.1. *Orthogonal Components with Two Degrees of Freedom.* In a symmetrical 2-level experiment every factorial effect has only 1 df, so the choice of effect to confound with the difference between 2 blocks is arbitrary. However, in 3-level experiments each main effect has 2 df, and an interaction involving f factors has 2^f df. It is necessary to deliberately confound 2 df for treatment effects with 2 df among 3 incomplete blocks. The chief difficulty is to find a way to partition interactions into orthogonal components that have 2 df each. For example, AB has 2 such components, ABC has 3, etc. Obviously, the orthogonal polynomial contrasts are of no help, as they have only 1 df each.

Consider a 3^2 factorial, unreplicated. The 9 treatment combinations have 8 df that may be partitioned into 4 orthogonal components of 2 df each: 2 main effects, A and B, and 2 components of the interaction, symbolized (AB) and (AB^2), that have no practical interpretation in the analysis of data. A convenient system for obtaining the orthogonal partition is related to the Latin square design. For t treatments or treatment combinations, a Latin square of t rows and t columns is designed to ensure that each treatment occurs exactly once in each row and each column. For example, suppose the 3 levels of factor A, {0, 1, 2}, are allocated to a 3 × 3 square:

A square		Column		
	Row	0	1	2
	0:	0	1	2
	1:	1	2	0
	2:	2	0	1

Two Latin squares are orthogonal if, when superimposed, each treatment of the first square appears exactly once with each treatment of the second square somewhere in the same cell. For example, consider the necessary arrangement of levels of factor B, {0, 1, 2}, in a second square orthogonal to the one above:

B square		Column		
	Row	0	1	2
	0:	0	1	2
	1:	2	0	1
	2:	1	2	0

When the A and B squares are superimposed, one obtains symbols for the 9 treatment combinations in the 9 cells:

AB square	Row	Column 0	Column 1	Column 2
	0:	00	11	22
	1:	12	20	01
	2:	21	02	10

Furthermore, both A and B are orthogonal to rows and columns. The structure is equivalent to that of a Graeco-Latin square design (Chap. 7), i.e., each level of A occurs exactly once in each row and column, each level of B occurs exactly once in each row and column, and each combination of A and B occurs exactly once in the entire square. Now there are 4 arrangements (A, B, rows, columns) that are orthogonal and have 2 df each.

The modulo-p arithmetic for defining relations in p-level factorials (p = prime number) is introduced in Sec. 6.3.1.1. A defining relation (DR) serves the same purpose as a defining contrast (DC), but for $p > 2$ it is not a contrast because it has 2 or more df, whereas all contrasts have only 1 df. For $p = 3$ levels and $n = 2$ factors, one may obtain orthogonal partitions of treatment combinations identical to the AB square above by computing for each treatment combination the vector sum,

$$\Sigma a_i x_i / 3 = Q + R \tag{6.61}$$

where x_i is the level of the ith factor in the treatment combination and a_i = 0 for any factor not involved in the 2 df selected as a defining relation but equals the superscript (exponent), 1 or 2, on a factor that is involved (Table 6.30). The symbols Q and R represent the quotient and remainder (residue) after division by 3. Only the residues, which may take on values 0, 1, or 2, are of interest because they designate 3 groups of treatment combinations that make up 3 incomplete blocks. The partition of treatment combinations corresponding to each choice of defining relation is shown in Table 6.30. Note that rows 0, 1, 2 of the AB square correspond exactly to residue groups R = 0, 2, 1 when the (AB^2) component is used as a defining relation and columns 0, 1,

Table 6.30. Possible Allocations of Treatment Combinations to Blocks (3^2 experiment)

DR	Vector Sum	Treatment Combinations in Blocks R=0	R=1	R=2
A	$x_1/3$	{00,01,02}	{10,11,12}	{20,21,22}
B	$x_2/3$	{00,10,20}	{01,11,21}	{02,12,22}
(AB)	$(x_1+x_2)/3$	{00,12,21}	{22,01,10}	{11,20,02}
(AB^2)	$(x_1+2x_2)/3$	{00,11,22}	{21,02,10}	{12,20,01}

2 correspond to R = 0, 2, 1 when the (AB) component is used.

The vector sum $(2x_1 + 2x_2)/3$ produces exactly the same allocation of treatment combinations as does $(x_1 + x_2)/3$. Likewise, $(2x_1 + x_2)/3$ and $(x_1 + 2x_2)/3$ produce the same blocks. That is, component symbol (A^2B^2) is identical with (AB) and symbol (A^2B) is identical with (AB^2). Clearly, some rule is required to determine values of the superscripts (exponents) a_i such that a unique set of symbols defines a complete orthogonal partition of any interaction into components with 2 df each. The convention adopted requires that the *first* factor listed in any component must *not* have exponent $a_i = 2$. In general, a 3^n factorial has $(3^n - 1)/2$ components of treatment effects with 2 df each, and n of these are main effects. An interaction involving f factors has 2^f df, so there are 2^{f-1} orthogonal components with 2 df each. For such an interaction there are 2^f symbols $(A^{a_1}B^{a_2}\ldots)$, a_i = 1 or 2, but only 2^{f-1} symbols are required to identify the components uniquely. For example, in a 3^3 experiment the 3-factor interaction has 8 df or 4 components of 2 df each. Possible symbols are

$$(ABC),\ (ABC^2),\ (AB^2C),\ (AB^2C^2),\ (A^2BC),\ (A^2BC^2),\ (A^2B^2C),\ (A^2B^2C^2)$$

By observing the convention that A may not have exponent 2, one eliminates the last 4 symbols; the first 4 symbols uniquely identify the 4 orthogonal components. In the formation of more than 3 blocks per replicate, 2 or more DR are required and 1 or more GI will be automatically confounded with blocks. In finding the generalized interaction one may encounter products of symbols with individual exponents > 2 on some factors. Those should be reduced, modulo 3; i.e., divide by 3 and retain only the residue. When the residue is zero, drop the symbol for that factor. If the resulting component still has $a_i = 2$ for the first factor in the component, one can obtain the corresponding symbol from the permissible set by squaring the component and again reducing the exponents, modulo 3. For example,

$$(AB^2C^2)(ABC^2) = (A^2B^3C^4) = (A^2C)$$

but $(A^2C)^2 = (A^4C^2) = (AC^2)$, which is a permissible symbol for the generalized interaction. The uniqueness principle is necessary to ensure that selected defining relations are independent.

6.4.1.2. *Three Incomplete Blocks per Replicate*. The division of 3^n treatment combinations into 3 incomplete blocks requires just 1 DR with 2 df. The 4 possible partitions for a 3^2 experiment are shown in Table 6.30. Four of the 8 df for a single replicate are required for estimation of main effects A and B (2 df each) and blocks have 2 df, leaving only 2 df for error, the component of the AB interaction not used as a defining relation. Therefore, in the few cases when a 3^2 experiment in incomplete blocks is necessary, it is advisable to use at least 2 replicates and partial confounding (confound different effects in different replicates). For example, with 4 replicates one could confound A, B, (AB), and (AB^2) in 1 replicate each so that equal 3/4 information would be available for all effects. Or one could confound (AB) in 2 replicates and (AB^2) in the other 2, retaining full information on the 2 main effects but only 1/2 information on the interaction. Sources of variation and degrees of freedom are shown in Table 6.31. Experimental error consists of interactions of the effects of replicates with treatment effects for repli-

Table 6.31. Partition of Variation for 3^2 Experiment in 4 Replicates of 3 Blocks Each

Source of Variation	df
Replicates	3
Blocks/replicates	8
A	2
B	2
AB	4
Error	16
Total	35

cates in which the treatment effects are not confounded with blocks. For example, if (AB) and (AB^2) are confounded in 2 replicates (R) each, the 16 df for error are composed of 6 df for RA, 6 df for RB, and 4 df for RAB. There are only 4 df for RAB (instead of 12) because AB can be estimated in only 2 of the 4 replicates (1 df for the 2 replicates).

Computation of separate sums of squares for replicates and blocks within replicates is unnecessary because the combined sums of squares equal the sum of squares for the 12 blocks (ignoring replicates). Sums of squares for main effects follow routine computation if the main effects have not been confounded in any replicate; but the sum of squares for interaction must be computed in two parts, $ss_{(AB)}$ and $ss_{(AB^2)}$, within the replicates in which those components are not confounded. Suppose a particular component, say (AB^2), has been used as a defining relation in 2 replicates, i' and i''. Let $y_{(i',i'')j.}$ be the total of $3^{n-1}(r - 2)$ responses in $r - 2$ replicates (where $i \neq i'$, $i \neq i''$) of the 3^{n-1} treatment combinations listed in block j of replicate i' or i''. Then the unconfounded portion of the sum of squares for that interaction component is

$$ss_{(AB^2)} = \sum_{j=0}^{2} y^2_{(i',i'')j.}/[3^{n-1}(r - 2)] - (\sum_{j=0}^{2} y_{(i',i'')j.})^2/[3^n(r - 2)] \quad (6.62)$$

Similar computations apply for obtaining $ss_{(AB)}$. The complete interaction sum of squares is

$$ss_{AB} = ss_{(AB)} + ss_{(AB^2)} \tag{6.63}$$

See (6.44) and the discussion following for the parallel case in 2^n experiments. The error sum of squares may be obtained as usual by subtracting the other sums of squares from the total sum of squares.

If interaction is significant, the experimenter's interest likely will be focused on treatment combination means and specific comparisons among them. In experiments with partial confounding, the combination means require adjustment for block effects. Let $\bar{y}'_{.k'l'}$ be a particular adjusted mean for a 3^2 experiment, where k' and l' are 0, 1, or 2 depending on the levels of factors A and B, respectively. Also, designate $y_{.k'.}$ and $y_{..l'}$ as the totals of $3^{n-1}r$ observations (over all r replicates) for the particular level of each factor involved. The adjusted combination mean may be computed from

$$\bar{y}'_{.k'l'} = (y_{.k'.} + y_{..l'})/[3^{n-1}r] + (y_{(1,2)j.} + y_{(3,4)j.})/[3^{n-1}(r-2)] - y_{..}/[3^{n-1}r] \tag{6.64}$$

where $y_{..}$ is the total of all $3^n r$ observations; $y_{(1,2)j.}$ is the total of $3^n(r - 2)$ observations, in the $r - 2$ replicates other than 1 and 2 in which component (AB^2) is confounded, of the 3^{n-1} treatment combinations listed in block j of replicate 1 or 2 [as in (6.62)]; and $y_{(3,4)j.}$ is a comparable total with respect to component (AB), which is confounded in replicates 3 and 4. The block number (j = 0, 1, or 2) to look at for the list of treatment combinations is in each case determined by the confounding pattern. If (AB^2) is confounded in replicates 1 and 2 and (AB) is confounded in replicates 3 and 4, the treatment combinations are assigned to blocks as in the last two lines of Table 6.30. Then, for (1, 2), $j = k' + 2l'$ (mod 3), and for (3, 4), $j = k' + l'$ (mod 3). For example, consider the treatment combination {21}, the highest level of A and medium level of B. For $y_{(1,2)j.}$, $j = 2 + 2(1) = 1$ (mod 3), and for $y_{(3,4)j.}$, $j = 2 + 1 = 0$ (mod 3). The adjusted mean for the combination is

$$\bar{y}'_{.21} = [(y_{.2.} + y_{..1})/12] + [(y_{(1,2)1.} + y_{(3,4)0.})/6] - (y_{..}/12)$$

where $y_{.2.}$ is the sum of 12 responses for subjects receiving the highest level of A; $y_{..1}$ is the sum of 12 responses for subjects receiving the medium level of B; $y_{(1,2)1.}$ is the total of 6 responses for subjects in replicates 3 and 4 that received the treatment combinations listed in block 1 of replicates 1 and 2, i.e., {21, 02, 10}; $y_{(3,4)0.}$ is the total of 6 responses for subjects in replicates 1 and 2 that received the treatment combinations listed in block 0 of replicates 3 and 4, i.e., {00, 12, 21}; and $y_{..}$ is the total of all 36 responses.

For a 3^3 experiment, one of 4 components of the 3-factor interaction, $\{(ABC), (ABC^2), (AB^2C), (AB^2C^2)\}$, may be selected as a defining relation. If the ABC interaction must be examined in its entirety, a different component could be confounded in each replicate. Then, however, adjustment of treatment combination means for block effects is quite complicated (Federer 1955, pp. 244-53; Kirk 1968, p. 360). Suppose one decides to confound (AB^2C^2) with blocks in every replicate. Then the treatment combinations are to be allocated to 3 blocks according to residues, R = 0, 1, or 2, for the vector sum $(x_1 + 2x_2 + 2x_3)/3$. For example, treatment combination {222}, containing the highest level of each of the 3 factors, produces the sum, [2 + 2(2) + 2(2)]/3 = 10/3, with residue = 1; but {221} produces [2 + 2(2) + 2(1)]/3 = 8/3, with residue = 2. The complete assignment of treatment combinations is shown in Table 6.32. Note that the same pattern of levels for A and B occurs in each

Table 6.32. A 3^3 Factorial Experiment in 3 Incomplete Blocks, (AB^2C^2) Confounded

Block 0	Block 1	Block 2
000	002	001
012	011	010
021	020	022
101	100	102
110	112	111
122	121	120
202	201	200
211	210	212
220	222	221

block. Therefore only the third digit (level of C) must be determined with reference to the particular component confounded.

Allocation of treatment combinations to blocks can be very time consuming in designing experiments with many factors. A convenient rule that simplifies the process after one has identified all the elements of the principal block [the one containing {000...}] and at least one element in each of the other blocks is: Add the "starting" element of any other block to each of the elements of the principal block and reduce each digit of each sum, modulo 3, to obtain the remaining elements. For example, in Table 6.32 suppose one had identified all elements of block 0 but only the element {112} of block 1. Then,

{112} + {000} = {112} {112} + {012} = {124} = {121}

{112} + {021} = {133} = {100} etc.

If an incomplete block plan and results are reported but the defining relation is not named, one can determine which component is confounded by trial and error. For example, look at the principal block of Table 6.32. Suppose (ABC^2) is proposed as a possible defining relation. If correct, every treatment combination in the principal block should produce a residue of zero from $(x_1 + x_2 + 2x_3)/3$. However, immediately one discovers that {012} produces residue = 2, so (ABC^2) is not correct. Eventually, when one tries (AB^2C^2), all combinations produce zero residues.

The 3^3 experiment has only 6 df for error if only 1 replicate is used; therefore, commonly 2 or more replicates are used. Sources of variation and degrees of freedom for an experiment with 2 replicates are shown in Table 6.33. The 3 unconfounded components of *ABC* interaction may be pooled with error or one may compute sums of squares and test them jointly if desired. Computations required are

Table 6.33. Partition of Variation for 3^3 Experiment in 2 Replicates of 3 Blocks Each

Source of Variation	df
Replicates	1
Blocks/replicates (AB^2C^2)	4
Main effects	6
Two-factor interactions	12
$(ABC)+(ABC^2)+(AB^2C)$	6
Error	24
Total	53

$$ss_{(ABC)} = \sum_{j=0}^{2} \Big(\sum_{k+l+m=j}^{9} y_{.klm} \Big)^2 / [3^{n-1}r] - (y_{..}^2/3^n r) \tag{6.65}$$

$$ss_{(ABC^2)} = \sum_{j=0}^{2} \Big(\sum_{k+l+2m=j}^{9} y_{.klm} \Big)^2 / [3^{n-1}r] - (y_{..}^2/3^n r) \tag{6.66}$$

$$ss_{(AB^2C)} = \sum_{j=0}^{2} \Big(\sum_{k+2l+m=j}^{9} y_{.klm} \Big)^2 / [3^{n-1}r] - (y_{..}^2/3^n r) \tag{6.67}$$

where $y_{.klm}$ is the total of r observations (over r replicates) of subjects given treatment combination klm (k, l, m = 0, 1, 2), and the various sums indicate groupings of combination totals that would have formed blocks if a different component (the one for which the sum of squares is being computed) had been used as a defining relation.

Experiments with 4 or more factors often are performed in a single replicate ($3^n \geq 81$ subjects, $n \geq 4$), because all 2- and 3-factor interactions as well as main effects may be estimated and one or more higher order interactions are available for use as experimental error. For example, a 3^4 experiment in 3 blocks, with $(ABCD^2)$ as the defining relation, has 2 df for blocks, 8 df for the 4 main effects, 24 df for the six 2-factor interactions, 32 df for the four 3-factor interactions, and 14 df for error. Error consists of the 7 unconfounded components of $ABCD$ interaction. However, a problem of heterogeneity of subjects often arises when 4 or more factors are used, because 27 or more subjects must be placed in each of the 3 incomplete blocks. In many cases greater homogeneity may be possible if the experiment is designed with more than 3 blocks per replicate.

6.4.1.3. *More than Three Blocks per Replicate*. To obtain smaller block size than is possible with only 3 blocks per replicate, one must use 2 or more DR to obtain 9 or more blocks per replicate. Consider a 3^3 experiment in 9 blocks of 3 subjects each. The 9 blocks will have 8 df, but 2 DR have only 4 df. It is obvious that 2 additional components of 2 df each will be automatically confounded with blocks. The additional components may be determined easily by finding the generalized interactions of the symbol for one defining relation with the first two powers of the symbol for the other, reducing both results if necessary by rules for uniqueness established in Sec. 6.4.1.1. For example, if one selects defining relations (AB^2C^2) and (ABC^2), the generalized interactions are

$$(AB^2C^2)(ABC^2) = (A^2B^3C^4) = (A^2C) = (A^4C^2) = (AC^2)$$

$$(AB^2C^2)(ABC^2)^2 = (A^3B^4C^6) = B$$

Obviously, that is not a good plan because the main effect of B is confounded. One can avoid confounding of main effects by selecting only 1 DR that is a component of ABC interaction and choosing a second component from one of the 2-factor interactions. If one chooses (AB^2C^2) and (BC^2), then

$$(AB^2C^2)(BC^2) = (AB^3C^4) = (AC) \qquad (AB^2C^2)(BC^2)^2 = (AB^4C^6) = (AB)$$

are automatically confounded. Such a plan usually requires replication for two reasons: (1) to use partial confounding so that all components of 2-factor interactions can be estimated and (2) to obtain greater sensitivity. A single replicate provides only 6 df for error [the components (ABC), (ABC^2), and (AB^2C)] unless one also uses (AB^2), (AC^2), and (BC) and attempts to estimate main effects only. The 9 blocks are formed by combinations of residues, R_1 = 0, 1, 2 from $(x_1 + 2x_2 + 2x_3)/3$ and R_2 = 0, 1, 2 from $(x_2 + 2x_3)/3$. For example, treatment combination {021} produces $R_1 = 0$ and $R_2 = 1$ and will be used in the block designated by $\{R_1R_2\} = \{01\}$.

A 3^4 experiment in 9 blocks of 9 subjects each permits estimation of all main effects and 2-factor interactions with reasonable sensitivity from a single replicate if one uses 2 components of 3-factor interactions as the defining relations. For example, (ABC^2) and (BCD^2) have generalized interactions

$$(ABC^2)(BCD^2) = (AB^2C^3D^2) = (AB^2D^2) \qquad (ABC^2)(BCD^2)^2 = (AB^3C^4D^4) = (ACD)$$

Treatment combinations having residue combinations $\{R_1R_2\} = \{00, 01, 02, 10, 11, 12, 20, 21, 22\}$ will be assigned to blocks $\{1, 2, \ldots, 9\}$. For example, treatment combination {1212} has residues $R_1 = 2$ for $(x_1 + x_2 + 2x_3)/3$ and $R_2 = 1$ for $(x_2 + x_3 + 2x_4)/3$ and is assigned to block 8. Sources of variation and degrees of freedom are shown in Table 6.34. The error term consists of the

Table 6.34. Partition of Variation for 3^4 Experiment in 9 Incomplete Blocks

Source of Variation	df
Blocks: $(ABC^2),(BCD^2),(AB^2D^2),(ACD)$	8
Main effects	8
Two-factor interactions	24
Error	40
Total	80

4-factor interaction (16 df) plus 12 components of 3-factor interactions. If one is unsure of the triviality of 3-factor interactions, a partial check could be made by testing the unconfounded components against $ABCD$ used as error.

Experiments with 5 or more factors often are too large and expensive (243 or more experimental units) or do not permit good homogeneity of units in 9 blocks (27 or more units per block). In the former case, one may turn to fractionally replicated designs; in the latter case, more than 9 blocks may be used in a single, full replicate.

In general, one may construct 3^s incomplete blocks in a replicate by selecting s independent defining relations ($s < n$). The number of units per block will be 3^{n-s}; the number of degrees of freedom for blocks will be $3^s - 1$. Therefore, the number of confounded components (2 df each) will be $(3^s - 1)/2$ including the s components selected as defining relations. Automatically confounded with blocks will be $(3^s - 2s - 1)/2$ GI of the defining relations. For example, consider a 3^6 experiment in 27 blocks of 27 mice each. In this case $s = 3$, so 3 components must be selected as defining relations. The number of generalized interactions automatically confounded is $[3^3 - 2(3) - 1]/2$

= 10, so 13 components in all are confounded with the 26 df for blocks. Let defining relations be $(ABCD^2)$, $(BCDE^2)$, and $(CDEF^2)$; generalized interactions are:

$$(ABCD^2)(BCDE^2) = (AB^2C^2E^2) \qquad (ABCD^2)(BCDE^2)^2 = (ADE)$$

$$(ABCD^2)(CDEF^2) = (ABCEF^2) \qquad (ABCD^2)(CDEF^2)^2 = (ABDE^2F)$$

$$(BCDE^2)(CDEF^2) = (BC^2D^2F^2) \qquad (BCDE^2)(CDEF^2)^2 = (BEF)$$

$$(ABCD^2)(BCDE^2)(CDEF^2) = (AB^2DF^2)$$

$$(ABCD^2)(BCDE^2)(CDEF^2)^2 = (AB^2CD^2EF)$$

$$(ABCD^2)(BCDE^2)^2(CDEF^2) = (ACD^2E^2F^2)$$

$$(ABCD^2)(BCDE^2)^2(CDEF^2)^2 = (AC^2F)$$

Three residue groups (R_i = 0, 1, 2) of treatment combinations can be found for each defining relation, according to $(x_1 + x_2 + x_3 + 2x_4)/3$, $(x_2 + x_3 + x_4 + 2x_5)/3$, and $(x_3 + x_4 + x_5 + 2x_6)$, respectively, for $(ABCD^2)$, $(BCDE^2)$, and $(CDEF^2)$. The residue values can be combined in 27 combinations to form the 27 blocks. Within each block the 27 homogeneous mice should be assigned randomly, one to each of the treatment combinations allocated to that block. In general, with s residue groupings for s defining relations the various values of R_1, R_2, ..., R_s can be combined in 3^s ways to form 3^s blocks.

6.4.2. Example (three blocks per replicate)

An experiment was performed to investigate the effects of 3 factors, each at equally spaced levels (coded 0, 1, 2), on the concentration of cholesterol in the urine (μg/100 mL). Incomplete blocks were formed to reduce block size to 9 animals because of heterogeneity of those available. Four replicates, 108 animals in all, were used. The design was developed using the component (ABC^2) as a defining relation in all replicates, i.e., treatment combinations were divided according to residue R = 0, 1, or 2 from the vector sum $(x_1 + x_2 + 2x_3)/3$. Combination {220} produces [2 + 2 + 2(0)]/3 with R = 1, {221} produces [2 + 2 + 2(1)]/3 with R = 0, {222} produces [2 + 2 + 2(2)]/3 with R = 2, etc. Therefore, the 3 treatment combinations that involve the highest levels of both A and B are assigned to different blocks. The complete plan and data are shown in Table 6.35.

It is convenient for computing sums of squares to summarize totals of responses in 2-way arrangements according to each pair of the 3 factors. Such a summary is shown in Table 6.36.

The total sum of squares is ss_y = 62,041 and the sum of squares for replicates is

Table 6.35. A 3^3 Experiment in 3 Blocks per Replicate (cholesterol in urine, μg/100 mL)

Block within Replicate	Treatment Combination	Replicate 1	2	3	4	Total
1	000	60	51	49	90	250
	011	75	68	65	60	268
	022	112	80	49	69	310
	101	78	103	108	122	411
	112	148	95	84	96	423
	120	83	80	77	96	336
	202	69	101	86	123	379
	210	68	102	65	88	323
	221	117	102	102	131	452
Subtotal		810	782	685	875	
2	002	50	55	69	55	229
	010	42	55	59	64	220
	021	64	67	77	59	267
	100	60	99	69	81	309
	111	97	118	75	107	397
	122	92	81	84	91	348
	201	93	84	134	111	422
	212	103	93	105	88	389
	220	104	93	107	99	403
Subtotal		705	745	779	755	
3	001	60	73	32	41	206
	012	50	50	40	51	191
	020	45	55	43	85	228
	102	112	104	87	107	410
	110	86	67	87	91	331
	121	35	81	96	109	321
	200	96	37	63	88	284
	211	110	86	90	96	382
	222	103	84	118	101	406
Subtotal		697	637	656	769	
Total		2212	2164	2120	2399	8895

$$ss_R = [(2212^2 + 2164^2 + 2120^2 + 2399^2)/27] - (8895^2/108) = 1674$$

The sum of squares for blocks within replicates is

$$ss_{D/R} = [(810^2 + 705^2 + \ldots + 769^2)/9] - [(2212^2 + \ldots + 2399^2)/27] = 4021$$

The sums of squares for main effects and 2-factor interactions may be computed from totals in Table 6.36 in the manner used for factorials in completely randomized designs (Chap. 2). The error, or residual sum of squares, may be obtained by subtraction as usual. Note that it includes the unconfounded por-

Table 6.36. Total Responses to Different Factor Levels and Combinations of Levels of 2 Factors (see Table 6.35)

		A 0	A 1	A 2	Total $(B,C,\Sigma y)$
B	0	685	1130	1085	2900
	1	679	1151	1094	2924
	2	805	1005	1261	3071
C	0	698	976	1010	2684
	1	741	1129	1256	3126
	2	730	1181	1174	3085
		B 0	B 1	B 2	
C	0	843	874	967	2684
	1	1039	1047	1040	3126
	2	1018	1003	1064	3085
Total $(A,\Sigma y)$		2169	3286	3440	8895

tions of the *ABC* interaction. Sums of squares for orthogonal polynomial partitions of main effects and interactions may be obtained by applying coefficients as in Table 6.29. Results of the analysis of variance are summarized in Table 6.37.

Both linear and quadratic components of *A* and *C* were significant, indicating declining response for successive increments of dose of treatment. The main effect of *B* was not significant but the *AB* interaction was significant, indicating that the responses to increments of factor *A* are not parallel for each level of factor *B*. The polynomial partitions of *AB* interaction indicate that most, if not all, of the effect can be expressed by interaction of the quadratic for *A* with the linear for *B*, i.e., A_QB_L. See Fig. 6.2. In Fig. 6.2.1 the mean response to increments of *A* is plotted for each level of *B* and for average *B*. The latter response curve is clearly nonlinear, as indicated by the significance of A_Q. The suggestion that the three curves for different levels of *B* are not parallel is supported by significance of the *AB* interaction. Figure 6.2.2 shows that the response to *B* for average *A* is trivial if not zero; hence, the nonsignificance of the main effect of *B*. However, the responses to *B* for different levels of *A* appear to be nonparallel, further supporting *AB* interaction. The three response lines appear to be nearly linear, suggesting A_QB_L in conjunction with Fig. 6.2.1. However, there is a suggestion of nonlinearity, or A_QB_Q, that is supported (but not strongly) by the size of the mean square for A_QB_Q in Table 6.37 (P[Type I error] < 0.10). Finally, Fig. 6.2.1 suggests that the highest level of *A* produces the highest

Table 6.37. Analysis of Variance, Including Linear (L) and Quadratic (Q) Effects (3^3 experiment in incomplete blocks) (see Table 6.35)

Source of Variation	df		Sums of Squares		Mean Squares
Replicates	3		1,674		558
Blocks/replicates	8		4,021		464
A	2		26,730		13,365**
A_L		1		22,437	22,437**
A_Q		1		4,293	4,293**
$B^{\#}$	2		476		238
C	2		3,313		1,656**
C_L		1		2,233	2,233**
C_Q		1		1,080	1,080*
AB	4		3,042		760*
A_LB_L		1		65	65
A_LB_Q		1		5	5
A_QB_L		1		2,071	2,071**
A_QB_Q		1		901	901
$AC^{\#}$	4		1,278		320
$BC^{\#}$	4		390		98
Error	78		21,117		271
Total	107		62,041		

*P(Type I error) < 0.05.

**P(Type I error) < 0.01.

#Polynomial partitions not shown for nonsignificant main effects and interactions.

amount of cholesterol in the urine only in the presence of the highest level of B.

6.4.3. Fractional Replication

As in the case of 2-level factorial experiments, fractional replication may be used for 3-level experiments to reduce resources required without deleting important factors from the study. However, fractional replication generally is less satisfactory for 3^n than for 2^n experiments. Fractions of 3^n are used primarily to estimate main effects of quantitative factors unconfounded with any first order interactions. Such designs have resolution IV (Table 6.21) and permit estimation of second degree polynomial response. In some areas of investigation, 3^n fractions are useful in sequential experimentation in the process of seeking and refining estimates of optimum treatment combinations (Chap. 9).

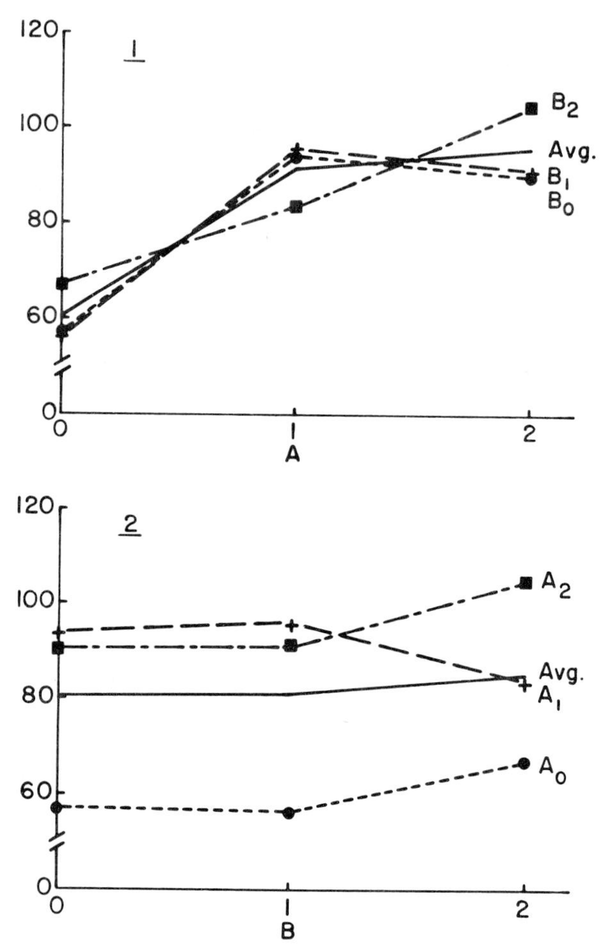

Fig. 6.2. Interaction of 2 factors affecting cholesterol in urine (see Table 6.35).

Fractional designs may be constructed by the mechanics of forming incomplete blocks. Usually, the principal block (containing the control {000...}) of the set forming a complete replicate is used as the fractional plan. Because the mean of the results of the fractional design (*I*) is equivalent to one of the 3 means that would make up the effect of the defining relation used to form 3 incomplete blocks, every estimable effect will have 2 aliases. The aliases of any estimable effect may be determined by finding the generalized interactions of the symbol for the effect with the symbols for the defining relation and its square.

6.4.3.1. *One-Third Replicate Designs*. The smallest 3-level factorial experiment that permits estimation of main effects from 1/3 replicate involves 3 factors or $(1/3)3^3 = 9$ treatment combinations. Suppose one selects component (ABC^2) as the defining relation to create 3 blocks, one of which will serve as the 1/3 replicate. If only the principal block is used, only the 9 treatment combinations having zero residue for $(x_1 + x_2 + 2x_3)/3$ will be used: 000,

11, 022, 101, 112, 120, 202, 210, 221. If the fundamental identity is ABC^2), aliases of A are

$$A(ABC^2) = (A^2BC^2) = (A^2BC^2)^2 = (A^4B^2C^4) = (AB^2C)$$

$$A(ABC^2)^2 = (A^3B^2C^4) = (B^2C)^2 = (B^4C^2) = (BC^2)$$

he aliases of B are

$$B(ABC^2) = (AB^2C^2)$$

$$B(ABC^2)^2 = (A^2B^3C^4) = (A^2C)^2 = (A^4C^2) = (AC^2)$$

nd the aliases of C are

$$C(ABC^2) = (ABC^3) = (AB)$$

$$C(ABC^2)^2 = (A^2B^2C^5) = (A^2B^2C^2)^2 = (A^4B^4C^4) = (ABC)$$

his design has only resolution III (Table 6.21) because each main effect has ne component of a 2-factor interaction as an alias. Also, only 2 df are vailable for error because the remaining 3 components that are not aliases of ain effects are aliases of each other: $(AB^2) = (AC) = (BC)$. Therefore, (1/)3^3 is not a good design. However, it may be useful to point out the rela- ionship between (1/3)3^3 and a Latin square design for 3 treatments (Chap. 7). oth designs have 3 factors, but in a Latin square 2 factors are nuisance var- ables used in blocking (by rows and columns). An inherent assumption is that uisance factors do not interact with treatments or with each other. If the reatments in a 3 × 3 square are designated A and the rows and columns are esignated B and C, an analogy to the (1/3)3^3 design may be made. Recall in he fractional design illustrated above that the component (BC^2) is an alias f A. Correspondingly, a component of interaction between two nuisance varia- les used as rows and columns in a Latin square would be completely confounded ith treatments; that is, if the assumption of zero interaction between two locking factors is invalid, one obtains biased estimates of the treatment ef- ects in a Latin square. An interesting but minor point is that one could onstruct all the 12 possible 3 × 3 Latin squares by using different 1/3 frac- ions from each of the 4 components of ABC interaction.

Consider 1/3 replicate of a 3^4 experiment involving 27 treatment combina- ions and an equal number of subjects. If one uses the fundamental identity, $= (ABCD^2)$, or any other component of $ABCD$ interaction, the aliases of each ain effect are components of interactions that involve at least 3 factors. here are 6 first order interactions, each of which has two orthogonal compo- ents of 2 df each. Unfortunately, only 6 of the 12 components do not have nother 2-factor component as an alias. Therefore, the design is of resolu- ion IV, and only 6 df are available for error (3 alias triplets, each involving 2 first order components of interaction) unless all first order interaction components are pooled for error (18 df in that case). Gravett (1971) has discussed a 2/3 replicate design for 3^4 having resolution V, although the amount of information obtained on first order interactions is less than that obtained on main effects.

A $(1/3)3^5$ design is of resolution V. Main effects have alias components of 4 or more factors, and first order interactions have alias components of 3 or more factors. The 80 df for the experiment may be partitioned into 10 df for main effects, 40 df for first order interactions, and 30 df for error, which consists of alias triplets of components involving 3 or more factors. Designs with 5 or more factors require a relatively large number of subjects (81 or more) for 1/3 replicate. Smaller experiments may be obtained by using smaller fractions of a replicate (Secs. 6.4.3.2, 6.4.3.3), or more efficient experiments may be obtained by blocking within 1/3 replicate.

Consider $(1/3)3^5$ in 3 blocks of 27 units each, with $I = (ABCDE^2)$. Any defining relation selected for blocking must be an alias triplet and therefore should contain at least 3 factors in each component. For example, one might select the alias triplet, $(BCD^2) = (AB^2C^2E^2) = (AD^2E^2)$. The aliases of estimable effects remain unchanged by blocking, but degrees of freedom for error are reduced from 30 to 28. For $(1/3)3^6$ in 9 blocks of 27 units each, with $I =$ $(ABCD^2E^2F^2)$, one must select 2 independent sets of aliases for defining relations. Two additional sets also are confounded with blocks automatically. For example, if the triplets $(AB^2DE^2) = (AC^2E^2F) = (BC^2D^2F)$ and $(BC^2EF^2) =$ $(AB^2D^2F) = (AC^2D^2E)$ are selected, then

$$(AB^2DE^2)(BC^2EF^2) = (AC^2DF^2) = (AB^2EF^2) = (BC^2DE^2)$$

$$(AB^2DE^2)(BC^2EF^2)^2 = (ABCDEF) = (ABC) = (DEF)$$

are automatically confounded. The design has resolution VI, as neither main effects nor first order interactions have alias components involving fewer than 4 factors. The major partitions of variation are summarized in Table 6.38.

Table 6.38. Partition of Variation for $(1/3)3^6$ in 9 Incomplete Blocks

Source of Variation	df	Aliases
Main effects	12	components involving 5 or more factors
First order interactions	60	components involving 4 or more factors
Blocks: (AB^2DE^2), (AB^2D^2F), (AB^2EF^2), (ABC)	8	components involving 3 or more factors
Error	162	
Total	242	

The analysis of data from 1/3 replicate experiments can be accomplished y "ignoring" one factor and forming summary tables of various combinations of ne remaining factors. For example, one may ignore factor D in $(1/3)3^4$, formng a 3-way (ABC) table of totals and three 2-way tables (AB, AC, BC). Sums f squares for A, B, and C follow the usual computations for main effects in eplicated factorials (Chap. 2), but sums of squares for interaction components ıst be computed as described for partial confounding in Sec. 6.4.1.2. Twoactor components involving D must be computed from their aliases that do not nvolve D.

.4.3.2. *One-Ninth Replicate Designs*. When several factors are of interest nd one wishes to examine possible nonlinear responses, even 1/3 replicate may equire too many resources. To construct 1/9 replicate designs, one must seect 2 DR to create 9 incomplete blocks and use only one block (usually the rincipal block). The 2 generalized interactions of the selected defining reations also will form the fundamental identity I. Thus, I will contain 4 omponents, and every estimable effect will have 8 aliases--2 generalized ineractions with each component of I.

For example, consider $(1/9)3^6$, with 81 subjects. If one chooses as deining relations (ABC^2D^2) and $(CDEF^2)$, then

$$(ABC^2D^2)(CDEF^2) = (ABEF^2) \qquad (ABC^2D^2)(CDEF^2)^2 = (ABCDE^2F)$$

lso are involved in I. The design has resolution IV only, because each 2actor component has one or more aliases that also involve only 2 factors. esigns with 7 or more factors have resolution V or higher for 1/9 replicate.

Most 1/9 replicates involve a sufficiently large number of subjects that fficiency may be improved by blocking within the fraction. For example, conider $(1/9)3^7$ in 9 blocks of 27 units each. If one chooses $(ABCD^2E)$ and $CD^2E^2F^2G^2)$ as defining relations for fractionation, the 2 generalized interctions also are part of the fundamental identity:

$$I = (ABCD^2E) = (CD^2E^2F^2G^2) = (ABC^2DF^2G^2) = (ABE^2FG)$$

ll aliases of main effects and first order interactions involve components ontaining at least 3 factors. Since each effect has 8 aliases, the 2 DR seected to create 9 blocks will consist of alias sets of 9 components each. lso, the 2 GI of the 2 DR will consist of alias sets of 9 components each. herefore, 36 components will be confounded with blocks. If one chooses alias ets for defining relations that involve (AB^2F^2G) and $(BCDF)$, with generalized nteractions involving $(ACDG)$ and (ABC^2D^2FG), it can be shown that no compoent of any 2-factor interaction is confounded with blocks. That is, the deign permits estimation of all main effects and first order interactions with liases involving 3 or more factors (resolution V). The 242 df may be partiioned into 8 for blocks, 14 for main effects, 84 for first order interacions, and 136 for error (which is composed of 68 unconfounded alias sets of 9 omponents, each component involving 3 or more factors).

.4.3.3. *Construction of Any Regular Fraction*. A 3^{-s} fraction of a 3^n experment is one of 3^s incomplete blocks formed by selecting s independent components of interactions (2 df each) as defining relations ($s < n$). A total of $(3^s - 1)/2$ components make up the fundamental identity I including the generalized interactions among the selected defining relations. Each estimable main effect or component has $3^s - 1$ aliases. For example, $(1/27)3^8 = 243$ units require $s = 3$ components as defining relations. Ten additional components are automatically part of the fundamental identity, and each estimable effect or component has 26 aliases.

A 3^{-s} fraction may be split into 3^u blocks of 3^{n-s-u} units each ($s + u < n$) by selecting u alias sets containing 3^s components (2 df each) per set. For $u \geq 2$, $(3^u - 2u - 1)/2$ additional alias sets will be automatically confounded with blocks. For example, $(1/27)3^8$ in 9 blocks of 27 units each $(3^{8-3-2} = 27)$ requires $s = 3$ components for fractionation (with 10 additional components in I) and $u = 2$ alias sets of 27 components each for blocking. Two additional alias sets are automatically confounded with blocks.

Tables of regular fractional plans for 3^n experiments have been provided by Connor and Zelen (1959). Webb (1971) presented plans for small irregular fractions of 3^2 and 3^3 experiments, and Hoke (1974) developed economical, nonorthogonal plans for irregular fractions with 3 or more factors.

6.5. OTHER FACTORIAL EXPERIMENTS

In most disciplines, the factorial experiments that have involved blocking and more than 2 or 3 factors have been symmetrical and restricted to no more than 3 levels per factor. However, occasionally finer resolution of nonlinear response to dose is required or symmetry is not possible because one or more factors (usually qualitative) are restricted to fewer levels or classes than the others. In the former case, one may desire symmetrical designs for factors with 4 or more levels, and in the latter case designs are required that permit different numbers of levels of the various factors.

6.5.1. Symmetrical Designs with More than Three Levels

When the subtleties of nonlinear response require more than 3 levels of each factor, one is likely to restrict the investigation to symmetry (same number of levels of each factor) and no more than 5 levels per factor. Procedures for construction of symmetrical designs for 2 and 3 levels (Secs. 6.3, 6.4) can be applied to 5- but not 4-level designs, because 4 is not a prime number.

6.5.1.1. *Prime-Level Factorials*.

The rules of modular arithmetic for 2^n and 3^n experiments in incomplete blocks may be extended for any prime number p, although for practical purposes $p > 5$ is rarely required.

In general for p^n experiments (n factors, each with p levels), any treatment combination may be symbolized by an ordered vector of factor levels, $\{x_1 x_2 \ldots x_n\}$, where each x_i may take values 0, 1, ..., $p - 1$. In working with sums of elements in vectors, one must reduce numerical results exceeding $p - 1$, modulo p (divide by p and save only the residue). There are $p^n - 1$ df among the p^n treatment combinations, and they may be partitioned into $(p^n -$

)/(p - 1) orthogonal components of p - 1 df each, conventionally symbolized $A^{a_1}B^{a_2}C^{a_3}\ldots$), each $a_i = 0, 1, \ldots, p - 1$. Each component is associated with ifferences among p groups of p^{n-1} treatment combinations specified by resiues 0, 1, ..., p - 1 from vector sums, $\sum_{i=1}^{n} a_i x_i/p$, where values of the a_i (0, , ..., p - 1 with $\sum_{i=1}^{n} a_i > 0$) are exponents of the symbols for the factors in he orthogonal component used as a defining relation. There are $p^n - 1$ possile linear functions, $\sum_{i=1}^{n} a_i x_i$, with $a_i = 0, 1, \ldots, p - 1$ and $\sum_{i=1}^{n} a_i > 0$, but nly $(p^n - 1)/(p - 1)$ orthogonal components. Therefore, to ensure uniqueness n the orthogonal partition, it is necessary to restrict the value of one of he a_i to just one of the p - 1 possible nonzero values. By convention, if $a_1 = a_2 = \ldots = a_f = 0$, then $a_{f+1} = 1$; i.e., the first factor symbol appearing in a component symbol must carry the exponent 1. (One may convert an "illegal" symbol, in which the exponent of the first factor > 1, to a "legal" symbol by aking some power of the symbol that will produce 1 as an exponent of the irst factor when each exponent of the product is reduced, modulo p.)

A p^n factorial experiment can be designed in p^s incomplete blocks of p^{n-s} nits each ($s < n$) by utilizing s independent components (p - 1 df each) as lefining relations to group the treatment combinations by residues, R_1, R_2, ..., R_s, each R taking on values 0, 1, ..., p - 1. The residue values can be ombined in p^s ways to form blocks having p^s - 1 df. Such degrees of freedom are completely confounded with $(p^s - 1)/(p - 1)$ components of interactions aving p - 1 df each. Of these components, s are the selected defining relaions, and the remaining $(p^s - ps + s - 1)/(p - 1)$ components automatically onfounded are the generalized interactions among the defining relations. Any wo defining relations have p - 1 generalized interactions, any three have $(p - 1)^2$ interactions, etc. A convenient rule for finding the generalized interactions is to find products of the symbols and powers of the symbols up to p - 1 for the defining relations and then reduce individual exponents of the results, modulo p. For example, if two components, denoted $X = (A^{a_1}B^{a_2}\ldots)$ and $Y = (A^{b_1}B^{b_2}\ldots)$, are used as defining relations to form incomplete blocks, then (XY), (XY^2), ..., (XY^{p-1}) are automatically confounded with blocks. Note that

$$(XY) = (A^{a_1+b_1}B^{a_2+b_2}\ldots) \qquad (XY^2) = (A^{a_1+2b_1}B^{a_2+2b_2}\ldots)$$

etc., and the exponents on factors A, B, etc., are to be reduced, modulo p. Since the treatment combinations assigned to the same block take one constant value, $R = 0, 1, \ldots,$ or $p - 1$, for each of the equations denoted by (XY^{λ}) ($\lambda = 1, 2, \ldots, p - 1$), the component symbolized by $[(A^{a_1}B^{a_2}\ldots)(A^{b_1}B^{b_2}\ldots)^{\lambda}]$ is confounded with block effects.

To illustrate the general procedure, consider an experiment with 2 factors, each at 5 levels. (It may be thought of as a continuation and refinement of a study that involved more factors at fewer levels in previous experiments.) Since a 5^2 experiment has only one interaction, it will be necessary to replicate and use partial confounding to obtain information about the *AB* interaction. The *AB* interaction has 16 df that may be divided into 4 orthogonal components, (AB), (AB^2), (AB^3), and (AB^4), having 4 df each. A design with balanced information on the components requires 4 replicates. Suppose it is desirable to block by litters. Then one could obtain 5 blocks of 5 animals each within 1 replicate, or 100 animals in all. Let the component (AB) be the defining relation in the first replicate. Then the 25 treatment combinations will be divided into 5 blocks according to residues ($R = 0, 1, 2, 3, 4$) from the vector sum, $(x_1 + x_2)/5$. For example, the principal block ($R = 0$) for the first replicate will contain {00, 14, 23, 32, 41}. Similarly, let (AB^2), (AB^3), and (AB^4) be the defining relations for replicates 2, 3, and 4, respectively. The corresponding vector sums are $(x_1 + 2x_2)/5$, $(x_1 + 3x_2)/5$, and $(x_1 + 4x_2)/5$. The partition of variation is summarized in Table 6.39. The error

Table 6.39. Partition of Variation for a 5^2 Experiment in 5 Blocks per Replicate

Source of Variation	df
Replicates	3
Blocks/replicates	16
A	4
B	4
AB	16
Error (*RA*+*RB*+*RAB*)	56
Total	99

is composed of 12 df each for the interactions of replicates with *A* and *B* separately and 32 df for the interaction of replicates with *A* and *B* together. The latter has only 32 df (instead of 48) because each component of *AB* interaction can be estimated in only 3 of the 4 replicates (i.e., 2 df instead of 3 for replicates in those cases). The sum of squares for *AB* interaction must be obtained by combining the results for each component computed in the 3 replicates where a particular component is not confounded with blocks.

Suppose component (AB^{λ}) is used as a defining relation in replicate $i' =$ $(\lambda = 1, 2, 3, 4)$. Let $y_{(i')j.}$ be the total of 15 responses, in the 3 repli-
ates where $i \neq i'$, to the 5 treatment combinations listed in block j of rep-
icate i'. Then, the unconfounded portion of the sum of squares for component
$AB^{\lambda})$ is

$$ss_{(AB^{i'})} = \left(\sum_{j=0}^{4} y^2_{(i')j.}/15\right) - \left(\sum_{j=0}^{4} y_{(i')j.}\right)^2/75 \qquad (6.68)$$

nd the unconfounded sum of squares for the entire AB interaction is

$$ss_{AB} = \sum_{i'=1}^{4} ss_{(AB^{i'})} \qquad (6.69)$$

Five-level experiments with 3 or more factors may be performed in a sin-
le replicate (or fractional replicate), using higher order interactions for
rror. For example, a single replicate of 5^3 in 5 blocks of 25 subjects each
equires that only 1 component of ABC interaction be confounded. Since ABC
as 16 components of 4 df each, the remaining 15 components may be used for
rror. A complete replicate of a 5^4 experiment is quite large (625 units), so
ractional replication is desirable. However, if a complete replicate is per-
ormed one can maintain a block size of 25 units without confounding first or-
er interactions by choosing 2 DR that are suitable components of 3-factor in-
eractions. For example, if (AB^2C^3) and (BCD) are selected, the 4 generalized
nteractions,

$$(AB^2C^3)(BCD) = (AB^3C^4D) \qquad (AB^2C^3)(BCD)^2 = (AB^4D^2)$$

$$(AB^2C^3)(BCD)^3 = (ACD^3) \qquad (AB^2C^3)(BCD)^4 = (ABC^2D^4)$$

re automatically confounded with blocks. The principal block would contain
reatment combinations having zero residues for $(x_1 + 2x_2 + 3x_3)/5$ and $(x_2 +$
$_3 + x_4)/5$.

A p^{-1} fraction of a p^n experiment is just one of p incomplete blocks
ormed by selecting a component of an interaction ($p - 1$ df) as a defining re-
ation. By using only a fraction, the defining relation becomes the fundamen-
al identity, $I = (A^{a_1}B^{a_2}\ldots)$, where $a_1 = 1$ and all other a_i may take values 0,
, ..., $p - 1$. Commonly, the fraction selected is the group of treatment com-
inations for which $\sum_{i=1}^{n} a_i x_i/p$ produces zero residues. Any estimable component
any component other than I) having $p - 1$ df, say $(A^{a_1}B^{a_2}\ldots)$, is an alias of
completely confounded with) the components symbolized by the products,
$A^{a_1}B^{a_2}\ldots)I^{\lambda}$ $(\lambda = 1, 2, \ldots, p - 1)$, with exponents in each product reduced,

modulo p. Thus, any estimable main effect or component has $p - 1$ aliases. For example, consider 1/5 replicate of a 5^4 experiment, with $I = (AB^2C^3D^4)$; the fraction includes treatment combinations that produce zero residue for $(x_1 + 2x_2 + 3x_3 + 4x_4)/5$. Each estimable main effect or component has 4 aliases. Consider aliases of the main effect of factor A:

$$A(AB^2C^3D^4) = (A^2B^2C^3D^4) = (A^2B^2C^3D^4)^3 = (ABC^4D^2)$$

$$A(AB^2C^3D^4)^2 = (A^3B^4CD^3) = (A^3B^4CD^3)^2 = (AB^3C^2D)$$

$$A(AB^2C^3D^4)^3 = (A^4BC^4D^2) = (A^4BC^4D^2)^4 = (AB^4CD^3)$$

$$A(AB^2C^3D^4)^4 = (B^3C^2D) = (B^3C^2D)^2 = (BC^4D^2)$$

A p^{-s} fraction of a p^n experiment is one of p^s incomplete blocks formed by selecting s independent components ($p - 1$ df each) as defining relations ($s < n$). A total of $(p^s - 1)/(p - 1)$ components form the fundamental identity I; these are the s components selected and their generalized interactions. The generalized interactions among s defining relations denoted $\{D_1, D_2, \ldots, D_s\}$ are

$$\sum_{i<i'}^{s-1} \sum^{s} \left[\sum_{\lambda=1}^{p-1} (D_i D_{i'}^{\lambda}) \right] + \sum_{i<i'<i''}^{s-2} \sum^{s-1} \sum^{s} \left[\sum_{\lambda=1}^{p-1} \sum_{\lambda'=1}^{p-1} (D_i D_{i'}^{\lambda} D_{i''}^{\lambda'}) \right] + \ldots$$

Each estimable effect has $p^s - 1$ aliases, which are the generalized interactions of the effect with each component of I and with powers of each component up to $p - 1$. For example, consider $(1/25)5^5$ with 2 DR selected, $(AB^2C^3D^4)$ and $(BCDE)$. On finding the generalized interactions, one notes that

$$I = (AB^2C^3D^4) = (BCDE) = (AB^3C^4E) = (AB^4DE^2) = (ACD^2E^3) = (ABC^2D^3E^4)$$

Each estimable effect has 24 aliases. For example, the aliases of the main effect of factor A are:

$$A(AB^2C^3D^4)^\lambda \quad A(BCDE)^\lambda \quad A(AB^3C^4E)^\lambda$$

$$A(AB^4DE^2)^\lambda \quad A(ACD^2E^3)^\lambda \quad A(ABC^2D^3E^4)^\lambda$$

where $\lambda = 1, 2, 3, 4$.

A p^{-s} fraction of a p^n experiment may be designed in p^u blocks of p^{n-s-u} units each ($s + u < n$) by selecting u alias sets of p^s components ($p - 1$ df per component) as defining relations for blocking. For $u \geq 2$, additional $[p^u - (p - 1)u - 1]/(p - 1)$ alias sets will be confounded automatically with locks. For example, consider $(1/25)5^6$ in 25 blocks of 25 units each. The undamental identity will contain 6 components of 4 df each. Two DR to establish blocking each require an alias set of 25 components. Four additional ets of 25 aliases each will be automatically confounded with blocks.

.5.1.2. *Nonprime-Level Factorials*. The rules of modular arithmetic discussed previously are not valid when the number of levels per factor is not rime. The theory of Galois fields (Kempthorne 1952; Carmichael 1956; Stewart 972) may be used to construct symmetrical factorial designs in incomplete locks when the number of levels is a prime (p) or an integral power of a rime (p^m). Because of the validity of modular rules for prime-level designs p = 2, 3, 5, 7, ...), the more complex field theory is needed only for p^m = , 8, 9, ... levels, with $p^m = 4$ being the only class of such designs that is ommon in practice.

A set of k elements, $\{u_0, u_1, \ldots, u_{k-1}\}$ is a finite field of order k if:

1. The usual commutative and distributive laws of addition and multiplication apply to the elements
2. $u_0 + u_i = u_i$ and $u_0 u_i = u_0$ $(i \geq 1)$ (u_0 acts like 0)
3. $u_1 u_i = u_i$ $(i \geq 2)$ (u_1 acts like 1 in multiplication)
4. There exists a unique u_i such that $u_i + u_{i'} = u_{i''}$
5. There exists a unique u_i such that $u_i u_{i'} = u_{i''}$ $(i' \neq 0)$

f the order of a field is prime, the elements may be represented by $\{0, 1, \ldots, p - 1\}$, in which ordinary arithmetic applies except that results are reduced, modulo p. For example, when p = 3 levels, requirement 5 implies $u_i u_1$ = u_0, u_1, u_2, which may be met uniquely by using u_i = 0, 1, 2, respectively, and $u_i u_2 = u_0, u_1, u_2$, which may be met uniquely by using u_i = 0, 2, 1, respectively [i.e., (0)(2) = 0, (2)(2) = 1, mod 3, and (1)(2) = 2]. However, if the rder of a field is nonprime, the simplified elements and modular rules do not ermit conformance with requirement 5. For example, when p = 4 levels, then $u_i u_2 = u_1$ and $u_i u_2 = u_3$ are impossible because $(i)(2)/4$ always leads to an *ven* residue for i = 0, 1, 2, 3. Also, uniqueness is not achieved for $u_i u_2$ = u_2 because both i = 1 and i = 3 produce residue 2 for $(i)(2)/4$.

Any element of a finite field of order p^m (number of levels = integral ower of a prime) satisfies an equation of degree $m' \leq m$,

$$c_0 + c_1 x + \ldots + c_{m'} x^{m'} = 0 \quad (c_{m'} \neq 0)$$

here the c_i are integral elements of the field. Let $P_m(x)$ be a polynomial of egree m, with integral coefficients not all divisible by p, and let $P_i(x)$ be ny polynomial in x with integral coefficients. Then

$$P_i(x)/P_m(x) = Q(x) + \text{remainder with degree} \leq m - 1$$

The remainder may be written $f(x) + p[q(x)]$, where

$$f(x) = b_0 + b_1 x + \dots + b_{m-1} x^{m-1}$$

with the b_i belonging to the set $\{0, 1, \dots, p - 1\}$. The function $q(x)$ is a polynomial with integral coefficients. That is, $f(x)$ is the residue of $P_i(x)$, modulo p and $P_m(x)$. Two polynomials $P_i(x)$ and $P_{i'}(x)$ are congruent (with double modulus) if they belong to the same residue class. The number of classes equals the number of functions $f(x)$, which is p^m, because each b_i takes values $\{0, 1, \dots, p - 1\}$. Therefore, the classes of residues are (in some order) the elements $\{u, u_1, \dots, u_{(p^m-1)}\}$ because classes of residues combine uniquely under addition and multiplication. However, in order that division of an arbitrary class, say C_i, by any class other than the zero class C_0 (in which all $b_i = 0$) shall lead uniquely to a third class (so that classes of residues constitute a finite field), the equation $C_i C_{i'} = C_0$ must imply $C_i = C_0$. That is possible only if p is a prime and $P_m(x)$ is irreducible, modulo p, i.e., $P_m(x)$ cannot be expressed in the form $P_1(x) + P_2(x) - pP_3(x)$, where the polynomials have integral coefficients and the degrees of $P_1(x)$ and $P_2(x)$ are less than m. The finite field formed by the p^m classes of residues is called a *Galois field*, of order p^m, or GF$[p^m]$, and contains the elements, $\{0, 1, \dots, p - 1\}$, *as well as others.*

Alternatively, Carmichael (1956) shows that any one of the m distinct roots satisfying an irreducible equation of the form,

$$x^m + c_1 x^{m-1} + \dots + c_m = 0$$

is a *primitive element* of the field, say u^*, such that $\{u^*, (u^*)^2, \dots, (u^*)^{p^m-1}\}$ all are distinct and are (in some order) the elements $\{u_1, u_2, \dots, u_{(p^m-1)}\}$, i.e., all the elements except u_0.

As an example, consider the general problem of 4-level factorial designs, i.e., $p = 2$, $m = 2$, GF$[2^2]$. Possible polynomials of degree $m = 2$, with integral coefficients not all divisible by $p = 2$, are of the form

$$P_m(x) = a_0 + a_1 x + a_2 x^2 \qquad (a_i = 0, 1, \dots)$$

If $a_0 = 0$, then $P_m(x) = x(a_1 + a_2 x)$; i.e., $P_m(x)$ is reduced. Therefore, $a_0 > 0$. For example, if $\{a_0, a_1, a_2\}$ are $\{1, 0, 1\}$, then

$$P_m(x) = 1 + x^2 = (1 + x)(1 + x) - 2x$$

is reducible (= 0, mod 2, when $x = 1$). Other sets of values of the a_i may be tried. For {1, 1, 0}, $P_m(x) = 1 + x$, which also = 0, mod 2, when $x = 1$. However, for {1, 1, 1}, $P_m(x) = 1 + x + x^2$ is irreducible (= 1, mod 2, when $x = 0$ or 1). Therefore, take $P_m(x) = 1 + x + x^2$. Possible $f(x)$ are of the form $b_0 + b_1 x$ ($b_i = 0, 1$). When b_0, b_1 take on values {0, 0}, {1, 0}, {0, 1}, and {1, 1}, then $f(x) = 0, 1, x$, and $1 + x$, respectively. That is, $\{0, 1, x, 1 + x\}$ are the elements $\{u_0, u_1, u_2, u_3\}$. Alternatively, it can be shown that the primitive element, $u^* = x$, satisfies the requirement that the quantities $\{x, x^2, x^3\}$ are (in some order) the elements $\{u_1, u_2, u_3\}$:

$$u^* = x = u_2$$

$$(u^*)^2 = x^2 = P_m(x) - (1 + x) = (1 + x) \mod [2, P_m(x)] = u_3$$

$$(u^*)^3 = x^3 = u^*(u^*)^2 = x(1 + x) = x + x^2 = P_m(x) - 1 = 1 \mod [2, P_m(x)] = u_1$$

It will be seen that u^* is especially useful in the multiplication of elements. [Note: $(u^*)^{p^m - 1} = (u^*)^3 = 1$.] Now it is possible to establish unique rules of addition and multiplication in GF[2^2]. For addition,

$$u_0 + u_i = u_i \quad (i = 0, 1, 2, 3)$$

$$u_1 + u_1 = 1 + 1 = 2 = 0 \mod 2 = u_0$$

$$u_1 + u_2 = 1 + x = u_3$$

$$u_1 + u_3 = 1 + (1 + x) = 2 + x = x \mod 2 = u_2$$

$$u_2 + u_2 = x + x = 2x = 0 \mod 2 = u_0$$

$$u_2 + u_3 = x + (1 + x) = 1 + 2x = 1 \mod 2 = u_1$$

$$u_3 + u_3 = (1 + x) + (1 + x) = 2 + 2x = 0 \mod 2 = u_0$$

For multiplication,

$$u_0 u_i = u_0 \quad (i = 0, 1, 2, 3)$$

$$u_1 u_i = u_i \quad (i = 1, 2, 3)$$

$$u_2 u_2 = u_2^2 = (u^*)^2 = u_3$$

$$u_2 u_3 = (u^*)(u^*)^2 = (u^*)^3 = 1 = u_1$$

$$u_3 u_3 = u_3^2 = [(u^*)^2]^2 = (u^*)(u^*)^3 = (u^*)(1) = (u^*) = u_2$$

The rules of addition and multiplication for 4-level designs are summarized in Table 6.40.

Table 6.40. Rules for Addition and Multiplication in GF[2^2] (4-level factorial designs)

	Addition				Multiplication			
	u_0	u_1	u_2	u_3	u_0	u_1	u_2	u_3
u_0	u_0	u_1	u_2	u_3	u_0	u_0	u_0	u_0
u_1		u_0	u_3	u_2		u_1	u_2	u_3
u_2			u_0	u_1			u_3	u_1
u_3				u_0				u_2

The treatment combinations of a $(p^m)^n$ factorial experiment (n factors, each at p^m levels) may be represented by ordered vectors of levels of the factors $\{x_1 x_2 \ldots x_n\}$, where the x_i take on values 0, 1, ..., $p^m - 1$. There are $p^{mn} - 1$ df, which may be partitioned orthogonally into $(p^{mn} - 1)/(p^m - 1)$ components of $p^m - 1$ df each. Each component, or set of degrees of freedom, is associated with contrasts among p^m groups of treatment combinations ($p^{m(n-1)}$ combinations per group) specified by the set of equations,

$$\sum_{i=1}^{n} u_{a_i} u_{x_i} = u_0 \mod p$$

$$\sum_{i=1}^{n} u_{a_i} u_{x_i} = u_1 \mod p$$

$$\ldots$$

$$\sum_{i=1}^{n} u_{a_i} u_{x_i} = u_{(p^m-1)} \qquad (6.70)$$

where the u_{a_i} and u_{x_i} are elements from GF$[p^m]$, where a_i, x_i = 0, 1, ..., p^m-1 ($\sum_{i=1}^{n} a_i > 0$). There are $p^{mn} - 1$ possible functions of the form $\sum_{i=1}^{n} u_{a_i} u_{x_i}$ ($\sum_{i=1}^{n} a_i > 0$). By restricting the first nonzero a_i to unity (one of the $p^m - 1$ possible nonzero values), the set of equations of (6.70) can be formed in only $(p^{mn} - 1)/(p^m - 1)$ ways that are equal to the number of orthogonal components, and uniqueness is ensured. By convention, the symbol $(A^{a_1}B^{a_2}\ldots)$ is used to denote a component having $p^m - 1$ df, with the first nonzero $a_i = 1$.

As an example of the orthogonal partition of components, consider a 4^2 factorial experiment (p = 2, m = 2, n = 2). The 15 df of a single replicate can be partitioned into 5 components of 3 df each:

$$A \text{ with } \sum_{i=1}^{n} u_{a_i} u_{x_i} = u_1 u_{x_1} + u_0 u_{x_2} = u_{x_1}$$

$$B \text{ with } \sum_{i=1}^{n} u_{a_i} u_{x_i} = u_0 u_{x_i} + u_1 u_{x_2} = u_{x_2}$$

$$(AB) \text{ with } \sum_{i=1}^{n} u_{a_i} u_{x_i} = u_1 u_{x_1} + u_1 u_{x_2} = (u_{x_1} + u_{x_2})$$

$$(AB^2) \text{ with } \sum_{i=1}^{n} u_{a_i} u_{x_i} = u_1 u_{x_1} + u_2 u_{x_2} = (u_{x_1} + u_2 u_{x_2})$$

$$(AB^3) \text{ with } \sum_{i=1}^{n} u_{a_i} u_{x_i} = u_1 u_{x_1} + u_3 u_{x_2} = (u_{x_1} + u_3 u_{x_2})$$

Given these formulas for vector sums, one can determine the residue class (block) to which a specific treatment combination belongs. For example, the treatment combination {23}, involving level 2 of factor A and the highest level (3) of factor B, produces residue $u_{x_1} = u_2$ when main effect A is used as a defining relation or residue $u_{x_2} = u_3$ when B is the defining relation selected. When any component of the interaction AB is selected as a defining relation, one must use the rules for addition and multiplication established for GF$[2^2]$ in Table 6.40. If (AB) is the defining relation, combination {23} produces the residue $u_{x_1} + u_{x_2} = u_2 + u_3 = u_1$. If (AB^2) is used, then

$$u_{x_1} + u_2 u_{x_2} = u_2 + u_2 u_3 = u_2 + u_1 = u_3$$

and if (AB^3) is used, then

$$u_{x_1} + u_3 u_{x_2} = u_2 + u_3 u_3 = u_2 + u_2 = u_0$$

Note that both B and (AB^2), used as defining relations, place {23} in the same block, defined by residue u_3. Assignments of treatment combinations to blocks for different defining relations are shown in Table 6.41.

Table 6.41. Possible Assignments of Treatment Combinations for 4^2 Factorial Experiment in 4 Incomplete Blocks per Replicate

DR	Blocks Defined by Residues			
	u_0	u_1	u_2	u_3
A	{00,01,02,03}	{10,11,12,13}	{20,21,22,23}	{30,31,32,33}
B	{00,10,20,30}	{01,11,21,31}	{02,12,22,32}	{03,13,23,33}
(AB)	{00,11,22,33}	{01,10,23,32}	{02,13,20,31}	{03,12,21,30}
(AB^2)	{00,13,21,32}	{03,10,22,31}	{01,12,20,33}	{02,11,23,30}
(AB^3)	{00,12,23,31}	{02,10,21,33}	{03,11,20,32}	{01,13,22,30}

In 4-level experiments with 3 or more factors, it may be desirable to divide the treatment combinations into more than 4 blocks. In general, a $(p^m)^n$ factorial plan can be constructed in p^{ms} incomplete blocks of $p^{m(n-s)}$ units each $(s < n)$ by utilizing s independent components ($p^m - 1$ df each) as defining relations to group treatment combinations according to residue values, $\{u_0, u_1, \ldots, u_{(p^m-1)}\}$, for each defining relation. Residue values for the s defining relations can be combined in p^{ms} ways to form blocks with $p^{ms} - 1$ df. These degrees of freedom are completely confounded with $(p^{ms} - 1)/(p^m - 1)$ components of $p^m - 1$ df each. Of these components, s are the selected defining relations and the others are generalized interactions of the defining relations.

For $s = 2$ DR, described by $\sum_{i=1}^{n} u_{a_i} u_{x_i}$ and $\sum_{j=1}^{n} u_{a_j} u_{x_j}$, the $p^m - 1$ components corresponding to

$$\sum_{k=1}^{n} u_{a_k} u_{x_k} = \sum_{i=1}^{n} u_{a_i} u_{x_i} + \lambda \sum_{j=1}^{n} u_{a_j} u_{x_j} \qquad (\lambda = u_1, u_2, \ldots, u_{(p^m-1)}) \tag{6.71}$$

are the generalized interactions automatically confounded with blocks. For a given λ, the result of (6.71) may correspond to an inadmissible symbol (for uniqueness), i.e., the first nonzero $a_k > 1$. A convenient rule for converting an inadmissible symbol to one that is permitted requires that $\sum_{k=1}^{n} u_{a_k} u_{x_k}$ be multiplied by a suitable value of $\lambda^* = u_1, u_2, \ldots, u_{(p^m-1)}$, such that $\lambda^* u_{a_{k'}} = u_1$, where $a_{k'}$ is the first $a_k > 1$. The "power" rule used for prime-level experiments is not appropriate, e.g., $(ABC^2)(AB^3C) = (A^2C^3)$ mod 4, but there is no integral power of 2 that will produce an exponent of 1 mod 4 on the symbol A.

For example, consider a 4^3 experiment in 16 blocks of 4 animals each ($p = n = s = 2$, $n = 3$). Let the defining relations be (ABC^2) and (AB^3C), with 3 df each. Since there will be 15 df for blocks, 3 additional components of 3 df each will be confounded automatically. The formulas for vector sums corresponding to the 2 DR are $u_1u_{x_1} + u_1u_{x_2} + u_2u_{x_3}$ and $u_1u_{x_1} + u_3u_{x_2} + u_1u_{x_3}$. Therefore, the generalized interaction produced when $\lambda = u_1$ is

$$(u_1u_{x_1} + u_1u_{x_2} + u_2u_{x_3}) + u_1(u_1u_{x_1} + u_3u_{x_2} + u_1u_{x_3})$$

$$= (u_1 + u_1u_1)u_{x_1} + (u_1 + u_1u_3)u_{x_2} + (u_2 + u_1u_1)u_{x_3}$$

$$= u_0u_{x_1} + u_2u_{x_2} + u_3u_{x_3} \qquad \text{(see Table 6.40)}$$

$$= u_2u_{x_2} + u_3u_{x_3}$$

Since u_2 is inadmissible as the leading coefficient, one must multiply the result by either $\lambda^* = u_1$ or $\lambda^* = u_3$, whichever permits $\lambda^* u_2 = u_1$. From Table 6.40 it is obvious that λ^* must be u_3, i.e., $u_3u_2 = u_1$. Thus the generalized interaction becomes

$$u_3(u_2u_{x_2} + u_3u_{x_3}) = u_1u_{x_2} + u_2u_{x_3} = (BC^2)$$

Similarly, the generalized interaction produced when $\lambda = u_2$ is

$$(u_1u_{x_1} + u_1u_{x_2} + u_2u_{x_3}) + u_2(u_1u_{x_1} + u_3u_{x_2} + u_1u_{x_3}) = u_3u_{x_1} + u_0u_{x_2} + u_0u_{x_3} = u_3u_{x_1}$$

which may be converted, using $\lambda^* = u_2$, to $u_1u_{x_1} = A$. Obviously, when a main effect is confounded the plan is not defensible unless partial confounding is used (different defining relations used in each replicate). The third generalized interaction, when $\lambda = u_3$, is

$$(u_1 u_{x_1} + u_1 u_{x_2} + u_2 u_{x_3}) + u_3(u_1 u_{x_1} + u_3 u_{x_3} + u_1 u_{x_3}) = u_2 u_{x_1} + u_3 u_{x_2} + u_1 u_{x_3}$$

which may be converted, using $\lambda^* = u_3$, to $u_1 u_{x_1} + u_2 u_{x_2} + u_3 u_{x_3} = (AB^2C^3)$.

When more than 2 DR are required (e.g., to create 64 or more blocks for 4^n experiments), higher ordered generalized interactions occur. For example, to create 64 blocks for a 4^5 experiment, one must select $s = 3$ DR, say D_1, D_2, and D_3. There will be $C(s, 2)(p^m - 1) = 3(3) = 9$ GI involving 2 DR and $C(s, 3) \cdot (p^m - 1) = (1)(3)^2 = 9$ GI involving all 3 DR--the 3 DR plus 18 GI (3 df each) are completely confounded with the 63 df among blocks. The generalized interactions may be determined by substituting the appropriate vector formulas for Ds in the following expressions:

$$(D_1 + \lambda_2 D_2) \quad (D_1 + \lambda_3 D_3) \quad (D_2 + \lambda_3 D_3) \quad (D_1 + \lambda_2 D_2 + \lambda_3 D_3)$$

where λ_2 and λ_3 take on values u_1, u_2, or u_3 independently.

Fractional replication may be desirable if the number of treatment combinations is very large or if the experimental units or procedures are quite costly or limited in some other way. A p^{-m} fraction of a $(p^m)^n$ factorial is just one of p^m incomplete blocks formed by selecting a component of an interaction ($p^m - 1$ df) as a defining relation. Using only one block forces the defining relation to be the fundamental identity, $I = (A^{a_1} B^{a_2} \ldots)$, where the a_i take values 0, 1, ..., $p^m - 1$ ($\sum_{i=1}^{n} a_i > 0$) and the first nonzero $a_i = 1$. Commonly, the fraction of treatment combinations used includes only those that produce residue u_0 for the vector sum, $\sum_{i=1}^{n} u_{a_i} u_{x_i}$, where the a_i are specified by I and x_i is the level of the ith factor in a given treatment combination ($x_i = 0, 1, \ldots, p^m - 1$). Any estimable main effect or component of $p^m - 1$ df, defined by $\sum_{j=1}^{n} u_{a_j} u_{x_j}$, has $p^m - 1$ aliases that may be determined from

$$\sum_{k=1}^{n} u_{a_k} u_{x_k} = \sum_{j=1}^{n} u_{a_j} u_{x_j} + \lambda \sum_{i=1}^{n} u_{a_i} u_{x_i} \qquad (\lambda = u_1, u_2, \ldots, u_{(p^m-1)}) \tag{6.72}$$

For example, consider 1/4 replicate of a 4^3 experiment (16 subjects) with fundamental identity $I = (AB^2C^3)$. Treatment combinations to be used are those for which $u_1 u_{x_1} + u_2 u_{x_2} + u_3 u_{x_3} = u_0$. Consider combination {013}:

$$u_1 u_0 + u_2 u_1 + u_3 u_3 = u_0 + u_2 + u_2 = u_2 + u_2 = u_0 \quad \text{(Table 6.40)}$$

Therefore, combination {013} is one of the 16 to be used. Consider the aliases of the main effect of A ($A = u_1 u_{x_1}$), for example:

$$u_1 u_{x_1} + u_1(u_1 u_{x_1} + u_2 u_{x_2} + u_3 u_{x_3}) = u_2 u_{x_2} + u_3 u_{x_3}$$

which may be converted, using $\lambda^* = u_3$, to

$$u_3(u_2 u_{x_2} + u_3 u_{x_3}) = u_1 u_{x_2} + u_2 u_{x_3} = (BC^2)$$

$u_1 u_{x_1} + u_2(u_1 u_{x_1} + u_2 u_{x_2} + u_3 u_{x_3}) = u_3 u_{x_1} + u_2 u_{x_2}$ converts to

$$u_2(u_3 u_{x_1} + u_2 u_{x_2}) = u_1 u_{x_1} + u_3 u_{x_2} = (AB^3)$$

and $u_1 u_{x_1} + u_3(u_1 u_{x_1} + u_2 u_{x_2} + u_3 u_{x_3}) = u_2 u_{x_1} + u_3 u_{x_3}$ converts to

$$u_3(u_2 u_{x_1} + u_3 u_{x_3}) = u_1 u_{x_1} + u_2 u_{x_3} = (AC^2)$$

The plan has only resolution III (Table 6.21) and should not be used if first order interactions are likely to be important. One-quarter replicates of 4^n experiments with $n \geq 4$ factors have higher resolution.

Occasionally, fractions smaller than p^{-m} are desirable. For example, a 4^5 experiment includes 1024 treatment combinations. Even 1/4 replicate may be too large for resources available, but 1/16 replicate (64 subjects) may be feasible. A p^{-ms} fraction of a $(p^m)^n$ factorial is one of p^{ms} incomplete blocks ($s < n$), and $(p^{ms} - 1)/(p^m - 1)$ components ($p^m - 1$ df each) form the fundamental identity I. In I there are s defining relations selected deliberately and their generalized interactions. Each estimable main effect or component has $p^{ms} - 1$ aliases, the generalized interactions of the $\sum_{j=1}^{n} u_{a_j} u_{x_j}$ defining the estimable effect with $\sum_{i=1}^{n} u_{a_i} u_{x_i}$ defining each component of I. For any given component of I, the aliases are determined as in (6.72). For example, in a $(1/16)4^5$ design, with $I = (AB^2C^3) = (CD^2E^3)$ plus 3 GI, every estimable effect has 15 aliases determined from 3 GI of the effect with each of the 5 components of I. Aliases of A are

$$u_1 u_{x_1} + \lambda(u_1 u_{x_1} + u_2 u_{x_2} + u_3 u_{x_3}),\ u_1 u_{x_1} + \lambda(u_1 u_{x_3} + u_2 u_{x_4} + u_3 u_{x_5}),\ \text{etc.},$$

with λ taking values u_1, u_2, and u_3 in each case.

In some cases precision can be improved by blocking within a fractional replicate. A p^{-ms} fraction of a $(p^m)^n$ factorial may be divided into p^{mu} blocks of $p^{m(n-s-u)}$ units each ($s + u < n$) by selecting u alias sets (p^{ms} components per set, $p^m - 1$ df per component) as defining relations for allocating treatment combinations to blocks. For $u \geq 2$, $[p^{mu} - (p^m - 1)u - 1]/(p^m - 1)$ additional alias sets (the generalized interactions of the u selected sets) will be automatically confounded with blocks. For example, $(1/16)4^5$ in 4 blocks requires 2 alias sets of 5 components each as defining relations. An additional 3 sets of 5 components each will be confounded automatically with blocks.

Another procedure for constructing nonprime-level designs for symmetrical factorial experiments is known as the "method of pseudofactors" and often is simpler to apply than the theory of Galois fields. It has the advantage of permitting more flexibility in block size, because one may form p^s incomplete blocks per replicate for any $s < n$ instead of the p^{ms} blocks required by GF$[p^m]$. For example, GF$[2^2]$ for 4-level experiments permits the number of blocks per replicate to be only $4^s = 4, 16, 64, \ldots$, but the pseudofactor method allows $2^s = 2, 4, 8, 16, 32, 64, \ldots,$. The pseudofactor method requires correspondence to be established between combinations of m pseudofactors, each having p levels, and the p^m levels of each factor in a $(p^m)^n$ factorial experiment. For example, consider a 4^2 factorial with factors denoted A and B. Let pseudofactors U and V (having only 2 levels each) stand for A and W and Z stand for B. Let the levels of A, i.e., 0, 1, 2, 3, correspond to combinations of levels of U and V, 00, 01, 10, 11. Establish similar correspondence between the 4 levels of B and the 4 combinations of levels of W and Z. The 3 df among the levels of A correspond to the same number of degrees of freedom among the combinations of U and V, and the latter may be partitioned into 3 orthogonal contrasts:

$$q_U = (-1)y_{00.} + (-1)y_{01.} + (+1)y_{10.} + (+1)y_{11.} \tag{6.73}$$

$$q_{UV} = (+1)y_{00.} + (-1)y_{01.} + (-1)y_{10.} + (+1)y_{11.} \tag{6.74}$$

$$q_V = (-1)y_{00.} + (+1)y_{01.} + (-1)y_{10.} + (+1)y_{11.} \tag{6.75}$$

One may rename the orthogonal partitions, $q_{A'}$, $q_{A''}$, and $q_{A'''}$, and express them in terms of responses to the 4 levels of A instead of the 4 combinations of U and V. These components of main effect A have no direct practical meaning; they are merely a device for constructing incomplete block designs. However, if the levels of A are quantitative and equally spaced, polynomial interpretations (L = linear, Q = quadratic, C = cubic) may be made as follows:

$$q_{A_L} = 2q_{A'} + q_{A'''} \qquad q_{A_Q} = q_{A''} \qquad q_{A_C} = 2q_{A'''} - q_{A'}$$

Note that the coefficients c_{ik} of any one of (6.73) to (6.75) are products of the coefficients of the other two equations, i.e., $c_{i1}c_{i2} = c_{i3}$, $c_{i1}c_{i3} = c_{i2}$, and $c_{i2}c_{i3} = c_{i1}$. Therefore, the generalized interactions of A', A'', and A''' are

$$A'A'' = A''' \qquad A'A''' = A'' \qquad A''A''' = A' \tag{6.76}$$

The forms $A'A'$, $A''A''$, and $A'''A'''$ each correspond to the grand total of responses and are taken as unity in generalized interactions. Similar correspondence applies for B', B'', B''' and pseudoeffects W, WZ, and Z.

Consider application of the pseudofactor method to construction of 2 incomplete blocks for a 4^2 factorial in which blocking is to be based on litters of pigs. Obviously, one must use partial confounding if the AB interaction is to be estimated in its entirety. Suppose the component $(A'''B'')$ is selected as the defining contrast. Since the A''' contrast may be represented by $-a_0 + a_1 - a_2 + a_3$ and the B'' contrast by $+b_0 - b_1 - b_2 + b_3$, the $(A'''B'')$ component is the product,

$$\{+a_0b_1 + a_0b_2 + a_1b_0 + a_1b_3 + a_2b_1 + a_2b_2 + a_3b_0 + a_3b_3\}$$

$$- \{a_0b_0 + a_0b_3 + a_1b_1 + a_1b_2 + a_2b_0 + a_2b_3 + a_3b_1 + a_3b_2\}$$

the last 8 combinations forming the principal block and the first 8 forming the alternate block. Alternatively, one may construct the blocks by noting that $(A'''B'') = (VWZ)$ and using the pseudovector sum, $(x_2 + x_3 + x_4) \bmod 2$. For example, treatment combination $a_1b_2 = u_0v_1w_1z_0 = \{0110\}$ is a pseudovector. Then $(x_2 + x_3 + x_4)/2$ produces $(1 + 1 + 0)/2$ with residue 0, indicating that the combination a_1b_2 belongs in the principal block. Similarly, $a_1b_3 = u_0v_1w_1z_1 = \{0111\}$ produces residue 1 and belongs in the alternate block.

Of course, the pseudofactor method may be extended for 4^n experiments with 3 or more factors or for experiments with more than 2 incomplete blocks. Consider a 4^3 factorial in 4 blocks of 16 units each. Let $A = ST$, $B = UV$, and $C = WZ$ to establish pseudofactors. Also, let $A' = S$, $A'' = ST$, $A''' = T$, etc. Suppose one selects $(STUW) = (A''B'C')$ and $(SUVZ) = (A'B''C''')$ as defining contrasts. The generalized interaction may be determined from either of

$$(STUW)(SUVZ) = S^2TU^2VWZ = TVWZ = A'''B'''C''$$

$$(A''B'C')(A'B''C''') = A'''B'''C''$$

The principal block will contain treatment combinations that produce zero residues for both $(x_1 + x_2 + x_3 + x_5)/2$ and $(x_1 + x_3 + x_4 + x_6)/2$. The 63 df in one replicate comprise 3 for blocks; 9 for main effects; 27 for first order interactions; and 24 for error, which is composed of unconfounded components of the interaction *ABC*.

The pseudofactor method also may be used to plan fractional replication. For example, one may use only the principal block from the 4^3 experiment in 4 blocks (discussed above) as 1/4 replicate. John (1970) developed the design $(3/16)4^3$ for an experiment with only 12 units, which still permits estimation of main effects. Consider a half-replicate of 4^4, with pseudofactor correspondence $A = QR$, $B = ST$, $C = UV$, and $D = WZ$. Let $QRSTUVWZ$, or $(A''B''C''D'')$, be the defining contrast, and use treatment combinations having zero residue for the pseudofactor vector sum, $\sum_{i=1}^{8} x_i/2$. Each main effect has 3 aliases, the generalized interactions of its components (e.g., A', A'', A''') with the defining contrast. For example, the aliases that involve the main effect of A are

$$A'(A''B''C''D'') = (A'''B''C''D'')$$

$$A''(A''B''C''D'') = (B''C''D'')$$

$$A'''(A''B''C''D'') = (A'B''C''D'')$$

The plan has resolution IV (Table 6.21) because $(A''B'')$ has alias $(C''D'')$, i.e., some of the first order interaction components are confounded. A 1/4 of the 4^4 factorial may be obtained by using 2 DC, say

$$(A''B'C'D') \text{ and } (A'B''C'''D''')$$

with generalized interaction $(A'''B'''C''D'')$ and then using only one of the 4 blocks formed. Each component of a main effect has 3 aliases; e.g., for A',

$$A'(A''B'C'D') = (A'''B'C'D')$$

$$A'(A'B''C'''D''') = (B''C'''D''')$$

$$A'(A'''B'''C''D'') = (A''B'''C''D'')$$

and similarly for A'' and A'''. Since B'' has alias $(C'''D''')$, the plan has only resolution III. Analysis of results from fractional plans of 4^n experiments is parallel to the procedures discussed for fractional replicates of 3^n experiments in Sec. 6.4.

The pseudofactor method may be used to construct a class of symmetrical designs that cannot be developed directly from Galois fields--designs of the $(p_1p_2)^n$ class, where p_1 and p_2 are different prime numbers. The only practical example involves $p_1 = 2$, $p_2 = 3$, or 6-level symmetrical factorials (6^n). One can utilize 2 pseudofactors, one having 2 levels and the other 3, for each of the actual factors involved. Construction follows procedures described for asymmetrical factorial designs.

6.5.2. Asymmetrical Designs

Symmetrical factorial designs are the rule for experiments in which all the factors are quantitative with number of levels easily manipulated. However, some experiments involve one or more qualitative factors for which "level" is merely a convenient label for "class," the numbers of which are constrained by nature, resources available, etc. For example, in some experiments in which sex of animal may influence response, sex should be treated as a factor with 2 levels (3 in cases involving males, females, and neutered males). If the other factors are quantitative and one wishes to examine nonlinear response, 3 or more levels should be used for those factors (circumstances may demand a plan for a 2×3^2 experiment instead of 3^3, for example).

6.5.2.1. *Replicated Designs*. Asymmetrical factorial designs of practical interest may be divided into two classes: (1) $(p^m)^{n_1} \times (p^{m'})^{n_2}$ designs, where n_1 factors have p^m levels each and n_2 factors have $p^{m'}$ levels each (p = prime, $m \neq m'$); and (2) $(p)^{n_1} \times (p')^{n_2}$ designs, where n_1 factors have p levels each and n_2 factors have p' levels each (p, p' = prime, $p \neq p'$). For practical purposes the first class of designs may be restricted to $2^{n_1} \times 4^{n_2}$. For example, a $2^3 \times 4^2$ factorial involves factors A, B, and C at 2 levels each and factors D and E at 4 levels each. Since the number of levels of D and E is not prime, one must use pseudofactors, say $D = UV$ and $E = WZ$, and construct the design as if it were a 2^7 factorial (see Raktoe 1970; Raktoe and Federer 1972 for an alternative method). If one desires 4 incomplete blocks of 32 units each, let the defining contrasts be $(ABUW)$ and $(ACVZ)$ (or $ABD'E'$ and $ACD'''E'''$), for example [(6.73)-(6.75)]. The generalized interaction, $(ABUW)(ACVZ) = BCUVWZ = BCD''E''$, also is confounded with blocks.

Most of the $(p)^{n_1} \times (p')^{n_2}$ designs of interest are $2^{n_1} \times 3^{n_2}$ although $2^{n_1} \times 5^{n_2}$ and $3^{n_1} \times 5^{n_2}$ may be useful in a few cases. Possible numbers of incomplete blocks per replicate are mathematical integral factors of the number of treatment combinations. For example, a 2×3^2 design has 18 treatment combinations (which may be divided into 2, 3, 6, or 9 blocks per replicate) and a $2^2 \times 3^2$ design has 36 combinations (which may be divided into 2, 3, 4, 6, 9, 12, or 18 blocks per replicate). If the desired number of blocks is a power of p, say p^s, the effects selected for confounding with blocks should come only from interactions among the n_1 factors that have p levels each. Similarly, if the desired number of blocks is $(p')^s$, one should select interactions among the n_2 factors having p' levels each. When the desired number of blocks is not p^s or $(p')^s$ but $p^s(p')^{s'}$, such as 6 or 12, one must select s effects involving only the n_1 factors having p levels and s' effects involving only the

n_2 factors having p' levels. Do not select an interaction involving factors from both groups. Consider, for example, a $2^2 \times 3^2$ experiment in 6 blocks of 6 units each (A, B with 2 levels; C, D with 3 levels). Suppose one attempts to confound the AC interaction, which involves a factor from each group. The contrast coefficients of main effect response totals for A and the linear (L) and quadratic (Q) effects of C are

$$A\text{: } (+1, -1) \quad C_L\text{: } (+1, 0, -1) \quad C_Q\text{: } (+1, -2, +1)$$

and the coefficients for AC combination totals are

$$AC_L\text{: } (+1, 0, -1, -1, 0, +1) \text{ and } AC_Q\text{: } (+1, -2, +1, -1, +2, -1)$$

If one forms 6 blocks based on the 6 combinations of coefficients for AC_L and AC_Q, {(+1, +1), (0, -2), (-1, +1), (-1, -1), (0, +2), (+1, -1)}, then

$$A = (\text{blocks } 1, 2, 3) - (\text{blocks } 4, 5, 6)$$

$$C_L = (\text{blocks } 1, 4) - (\text{blocks } 3, 6)$$

$$C_Q = (\text{blocks } 1, 3, 5) - (\text{blocks } 2, 4, 6)$$

That is, if 2 df for AC interaction are deliberately confounded with blocks, the remaining 3 df among blocks are automatically confounded with the main effects of A and C. See Das and Rao (1967) for proper use of polynomial components in the construction of incomplete blocks.

White and Hultquist (1965) generalized Galois field theory for constructing designs for mixed factorials of the $(p)^{n_1} \times (p')^{n_2}$ class, and another alternative to the pseudofactor method has been developed by Raktoe (1969, 1970) and Raktoe and Federer (1972). Worthley and Banerjee (1974) devised the first completely generalized method of construction for arbitrary numbers of levels of the various factors involved. Analysis of data from these designs has been discussed by Kempthorne (1952), Cochran and Cox (1957), and Kirk (1968). See Winer (1962) or Das and Rao (1967) for numerical examples.

6.5.2.2. *Fractional Replication*. When practical considerations dictate fractionally replicated designs, the factors to be studied usually are quantitative so that symmetrical experiments may be planned. However, when one or more factors are qualitative and have a fixed number of classes, it may be necessary to construct fractional replicates for $2^{n_1} \times 3^{n_2}$ or $2^{n_1} \times 4^{n_2}$ experiments. The construction problem is basically the same as in the construction of incomplete blocks, but fractional designs rarely are practical unless the number of factors having a common number of levels is sufficient to use interactions among those factors only as the fundamental identity. A large number of plans of the $2^{n_1} \times 3^{n_2}$ class have been published by Connor and Young (1961). Addelman (1963) discussed the general problem of constructing complex fractional designs. The plans are orthogonal or nonorthogonal. The orthogonal plans usually provide more precise estimates but often require more subects or experimental units to obtain the estimates.

Addelman (1962) showed that orthogonal plans of resolution III (Table .21) for $2^{n_1} \times 3^{n_2}$ experiments could be obtained from 2^n fractions of resolu-ion III by replacing n_2 sets of three 2-level factors by n_2 4-level factors nd then "collapsing" each 4-level factor to a 3-level factor by corresponding evels (0, 1, 2, 3) to (0, 1, 2, 1). The procedure is appropriate only when he number of treatment combinations in the fraction is some power of 2. Mar-olin (1968) extended the method for fractions in which the number of combina-ions is a multiple of 8 but not a power of 2 (e.g., 24, 40, 48, ...). Plans f resolution III require negligible first order interactions for validity nd therefore are not widely useful. Margolin (1969b) discussed the construc-ion of orthogonal plans of resolution IV that permit estimation of main ef-ects, assuming only that interactions involving 3 or more factors are negli-ible. If one wishes to be able to estimate first order interactions, orthog-nal plans of resolution V may be obtained by selecting appropriate subsets of he effects of a 2^n factorial and replacing sets of three 2-level factor des-gnations by 4-level factors. These $2^{n_1} \times 4^{n_2}$ plans can be converted to $2^{n_1} \times 3^{n_2}$ plans by changing level 3 of a 4-level factor to level 1 of a 3-level fac-or.

Nonorthogonal fractional plans for $2^{n_1} \times 3^{n_2}$ experiments have been dis-ussed by Margolin (1969c). Webb (1971) has catalogued fractional plans for $2^{n_1} \times 3$ $(n_1 \leq 7)$, $2^{n_1} \times 3^2$ $(n_1 \leq 5)$, and $2^{n_1} \times 3^3$ $(n_1 \leq 2)$, all of which uti-ize no more than 20 treatment combinations and subjects. These plans permit he estimation of all main effects and none, some, or all first order interac-ions. Draper and Stoneman (1968a) presented a method for fitting response urfaces (see Chap. 9) for $2^{n_1} \times 3^{n_2}$ and $2^{n_1} \times 4^{n_2}$ fractional designs when in-teractions involving the 2-level factors are assumed to be negligible.

XERCISES

onfactorial Experiments

.1. An experiment was designed to study the effects of 21 different diets on nonprotein nitrogen in the colostrum of rats. Twenty-one blocks (block = 5 rats from the same litter) were used to obtain complete balance with $\lambda = 1$ comparison of any 2 treatments in the same block. Results (mg/100 mL) are shown with treatment numbers in parentheses.

1		2		3		4		5		6		7	
74	(1)	68	(1)	85	(1)	74	(1)	74	(1)	102	(2)	86	(2)
92	(2)	80	(6)	86	(10)	89	(14)	83	(18)	108	(6)	71	(7)
69	(3)	83	(7)	66	(11)	72	(15)	72	(19)	132	(10)	55	(11)
79	(4)	89	(8)	78	(12)	82	(16)	82	(20)	127	(14)	76	(15)
65	(5)	97	(9)	66	(13)	81	(17)	85	(21)	120	(18)	56	(19)
379		417		381		398		396		589		344	

8		9		10		11		12		13		14	
95	(2)	101	(2)	76	(3)	73	(3)	71	(3)	80	(3)	85	(4)
87	(8)	100	(9)	84	(6)	89	(7)	71	(8)	100	(9)	81	(6)
79	(12)	66	(13)	65	(13)	85	(12)	52	(11)	108	(10)	59	(11)
82	(16)	83	(17)	73	(15)	92	(14)	61	(17)	102	(16)	93	(16)
69	(20)	91	(21)	76	(20)	90	(21)	80	(18)	84	(19)	84	(21)
412		441		374		429		335		474		402	

15		16		17		18		19		20		21	
70	(4)	96	(4)	82	(4)	75	(5)	77	(5)	80	(5)	86	(5)
78	(7)	116	(8)	99	(9)	96	(6)	101	(7)	112	(8)	92	(9)
86	(10)	79	(13)	85	(12)	84	(12)	80	(13)	93	(10)	64	(11)
69	(17)	111	(14)	84	(15)	74	(17)	98	(16)	84	(15)	105	(14)
78	(20)	81	(19)	98	(18)	62	(19)	104	(18)	106	(21)	90	(20)
381		483		448		391		460		475		437	

(a) Compute intrablock estimates of treatment means.
(b) Which treatments differ significantly?
(c) Compare combined estimates of treatment means (recovering interblock information) with the results in (a).
(d) Estimate the efficiency of the combined estimates relative to intrablock estimates, i.e., the efficiency of recovering interblock information.

6.2. An experiment with lesser weasels will involve prey selection in a Y-shaped run. There are t = 10 species of prey to be compared in pairs (k = 2) in the legs of the run. For each trial, rankits (normal order scores) will be recorded for the species selected first and second (automatic). Nine weasels are available for the trials. Construct a completely balanced incomplete block plan for the experiment.

6.3. An experiment involving 6 treatments is to be performed on male pigs. Assume that blocking by litter is desirable and one cannot expect to obtain 6 male pigs in every litter.
(a) How many pigs are required for the smallest incomplete block experiment that will permit complete balance?
(b) Complete balance for this experiment is possible with any incomplete block size ($1 < k < 6$). Which of the possible block sizes will not permit minimum size of experiment?
(c) Which of the designs that permit minimum size is likely to be the most economical and sensitive? Why?
(d) If the design selected in (c) is used and mean square error is expected to be approximately 0.01, how large is the expected standard error of an adjusted treatment mean?

6.4. Construct a partially balanced incomplete block design with 2 associate classes for 9 replications of 20 treatments in blocks (litters) of 6 animals. (Hint: First establish possible combinations of λ_1 and λ_2 and derive the association structures $\{p^1_{hh'}\}$ and $\{p^2_{hh'}\}$ for each combination. Then eliminate impossible constructions by studying the algebra of the necessary conditions. Finally, construct the remaining design by trial and error arrangements.)

Two-Level Factorial Experiments

6.5. A colleague at another institution sends you a copy of the design and data from an experiment involving 4 replicates of 4 factors (A, B, C, D) present or absent in all possible combinations, applied to rats. Treatments assigned to the first litter in each replicate were designated $\{(1), d, ab, ac, bc, abd, acd, bcd\}$. Describe briefly how you can make sense from the data without further details about the design. Include a list of sources of variation and degrees of freedom.

6.6. A new research student who has designed a 2^3 factorial experiment plans to use the following 2 incomplete blocks: $\{(1), a, ac, abc\}$ and $\{b, c, ab, bc\}$. Is the design valid? Why?

6.7. A 2^3 experiment on mental patients was designed in 2 incomplete blocks per replicate. The factors were drug A (present, absent), drug B (present, absent), and C = psychotherapy (present, absent). Treatment combinations were administered to groups of 5 patients, each group being an experimental unit. The response recorded for the unit was mean "improvement," as scored subjectively by a panel of experts. To obtain a sufficient number of patients, 4 units were taken from each of 4 hospitals, 2 hospitals making a complete replicate (Winer 1962).

Replicate 1				Replicate 2			
Hospital 1		Hospital 2		Hospital 3		Hospital 4	
a	6	(1)	2	a	14	(1)	3
b	10	ab	4	b	15	ab	6
c	6	ac	15	c	9	ac	25
abc	8	bc	18	abc	12	bc	22
	30		39		50		56

(a) Analyze the data by Yates's method.
(b) If any 2-factor interaction is significant, test the significance of each factor in the presence and in the absence of the other.

6.8. A nutritional experiment on dairy cows is to be conducted in a herd divided into 2 breeding groups based on differential selection for milk production. The experiment will involve 5 factors, each at 2 levels (present, absent) not replicated.
(a) Using the breeding groups as incomplete blocks, construct a design for the experiment showing proper allocation of treatment combinations to blocks.
(b) List sources of variation and degrees of freedom for analysis of the data.

6.9. A physiological experiment on pigs will involve 6 factors, each at 2 levels. Eight litters of pigs are available.
(a) Using litters as blocks, construct a design that will permit estimation of all main effects and 2-factor interactions.

(b) List sources of variation and degrees of freedom for analysis of the data.

6.10. A large scale trial on the effect of semen diluents on conception rate in the artificial insemination of dairy cows is to be performed as a 2^4 factorial. Factor A is buffer (citrate or phosphate) and factors B, C, and D are sulfanilamide, streptomycin, and penicillin (each present or absent). The experimental unit is one-half a semen sample collected from one bull; the entire semen sample from one bull represents a block of size 2. Eight bulls are to be sampled 4 times each; i.e., there will be 4 replicates.

(a) Construct a design that will permit estimation of all effects with 3/4 information on main effects and 1/2 information on first order interactions.

(b) List sources of variation and degrees of freedom for analysis of the data.

6.11. An exploratory experiment on dogs involved 1/4 replicate of a 2^6 factorial in 2 blocks (8 dogs of each sex): block 1 = {(1), *acd*, *bde*, *abce*, *abdf*, *bcf*, *aef*, *cdef*} and block 2 = {*acf*, *bcd*, *ce*, *abcdef*, *ab*, *df*, *ade*, *bef*}. All interactions involving 3 or more factors were thought to be negligible. An estimate of error ($ms_E \simeq 12$) was available from previous similar work (30 df). Yates's analysis produced the following contrast sums of squares (ss_{q_k}): $A = 88.3$, $B = 23.8$, $C = 9.4$, $D = 18.6$, $AB = 58.4$, $AC = 31.0$, $AD = 14.5$, $BC = 91.4$, $BD = 42.2$, $CD = 13.6$, $ABC = 73.5$, $ABD = 8.7$, $ACD = 28.1$, and $ABCD = 30.4$.

(a) What conclusions should be drawn from these results?

(b) What would you recommend as the next step in the investigation?

6.12. A chemotherapy trial will involve 9 factors (present or absent) in combinations. Not more than 70 suitable patients will be available during the time allotted for the study. The patients used should be grouped by age into 4 blocks. Design the experiment so that all main effects and first order interactions are estimable.

Three-Level Factorial Experiments

6.13. A physiological experiment on cows will involve 3 kinds of injections (B, C, D), each at 3 levels (coded 0, 1, 2, where 0 = saline), and factor A, the mode of injection (let 0 = intravenous, 1 = intramuscular, 2 = subcutaneous), which is thought not likely to interact with the other factors. Cows to be injected should be blocked by age (2 yr, 3 yr, mature). Although $3^4 = 81$ cows, fewer than 50 are available for this study. Construct 1/3 replicate design in 3 blocks (9 cows each) that will have resolution V with reference to estimation of 2-factor interactions among B, C, and D. Find the aliases of main effects and estimable first order interactions. List sources of variation and degrees of freedom.

6.14. A food science experiment will involve studying shelf life of a product treated with various combinations of levels (coded 0, 1, 2) of 5 different emulsifiers and stabilizers. For convenience, the experiment will be arranged in 9 blocks according to time at which the experiment is carried out. Construct a suitable design such that not more than one 3-factor interaction is confounded with blocks.

6.15. A computer simulation is planned for evaluation of different selection criteria and mating schemes for maximizing genetic improvement in beef cattle. Factors are A = mating system (random, positive assortative mating, parent-offspring inbreeding), B = heritability of yearling weight (0.2, 0.4, 0.6), C = number (out of 4 bulls saved) of young bulls (0, 2, 4), D = percentage of heifers saved (20, 50, 80), E = selection criteria for young animals (estimated breeding value, yearling weight ratio, index of weaning weight and average daily gain), and F = selection criteria for older animals (estimated breeding value, average progeny yearling weight ratio, index of progeny average weaning weight and average daily gain). To reduce computer time and cost it is desirable to run only one-third of the 3^6 = 729 combinations and obtain resolution VI, i.e., to be able to estimate main effects and first order interactions unconfounded with 3-factor interactions. Describe the method for constructing a proper design.

6.16. A swine nutrition experiment is to be performed using 5 dietary factors at 3 levels each. It is thought unlikely that 2 factors will interact with each other. Litters should be used as blocks; only about 10 to 12 litters are available for this study. Construct 1/3 replicate design for the 3^5 experiment in 9 blocks (litters) of 9 pigs each so that no more than one first order interaction is confounded with blocks. List aliases of main effects and 2-factor interactions.

6.17. A medical trial is to involve 4 factors, each at 3 levels. Practical considerations dictate that the study should utilize sequential fractions of 1/9 replicate each (9 patients) over a relatively long period of time. The project director is adamant that combination 1212 be run in the first fraction because of prior knowledge and special interest. Design the first 1/9 replicate in such a way that 1212 is included and all main effects are estimable with resolution III.

6.18. An investigation was planned to study the effects of 7 factors, each at 3 levels, on the adrenal weight (as percent of body weight) of female mice. Mice used were "salvaged" from a 3 × 3 diallelic crossbreeding experiment involving 3 strains. The 9 diallelic groups (3 purebred, 3 crossbred, 3 reciprocally crossbred) were used as blocks with reference to the adrenal investigation. To reduce the 3^7 experiment (2187 treatment combinations) to manageable proportions and a size compatible with the number of mice available, only 1/9 replicate was used. Components selected for the fundamental identity were $(ABCD^2E)$ and $(CD^2E^2F^2G^2)$. Defining relations selected for blocking were (AB^2F^2G) and $(BCDF)$. Show that no component of any 2-factor interaction is confounded, i.e., the plan has resolution V.

Other Factorial Experiments

6.19. Four antihelminthics for sheep are to be studied at 4 doses each, but fewer than 40 animals are available for this study. Construct $(1/8)4^4$ design for 32 animals so that main effects are estimable with resolution III. List 2-factor aliases of main effects.

6.20. The storage of sour cream to retain maximum viscosity is a problem in dairy manufacturing. It is desirable to examine 6 stabilizers at 5 levels each, but 5^6 = 15,625 units. Show how to construct $(1/25)5^6$ in 25 storage blocks of 25 units each so that not more than 2 first order interactions are confounded with blocks. Show how to find aliases of estimable effects.

6.21. A complex computer simulation of genetic improvement of animal populations involved 4 levels each of population size, selection intensity, environmental variation, and linkage. Because of cost, complexity, and the exploratory nature of the study, only 1/16 replicate was used. Use the theory of Galois fields to construct the design. List 2-factor aliases of main effects.

6.22. Crosses are to be made in all combinations of male (A) and female (B) parental genotypes representing 8 inbred lines of rats ($8^2 = 64$ crosses) in 8 blocks of cages. Construct a proper design for the allocation of crosses to blocks, using the theory of Galois fields.

6.23. A swine feeding trial will involve both gilts and barrows (sex = A) in 2 types of housing (B) with 3 dietary factors (C, D, E), each at 3 levels. Pigs available were sired by 6 boars. Using the sire groups as blocks, show how to construct one replicate of a $2^2 3^3$ factorial in 6 incomplete blocks of 18 pigs each (9 male, 9 female).

CHAPTER 7

Double Blocking: Latin Squares and Other Designs

"Double blocking" describes the process used to minimize experimental error in the designs of this chapter, but the equivalent expression, "two-way elimination of heterogeneity," has been used more frequently. Either expression means that two nuisance variables or criteria are used to stratify the experimental units or subjects into homogeneous groups instead of the single criterion employed in the designs of Chaps. 5 and 6. The chief purpose is to eliminate more of the nuisance variation from experimental error. For example, subjects often could be stratified according to body weight and age, for traits dependent on weight and age, to achieve smaller standard errors than could be achieved by blocking on age or weight alone. For double blocking in experiments in which the units are animals or humans, the subjects themselves often serve as one blocking criterion and time is the other, i.e., nonrandom repeated measurements are obtained from each subject assigned to a sequence of treatments. Such cases are considered in the discussion of changeover designs in the first section of Chap. 8. The basic structure of changeover designs is the same as that of designs not involving repeated measurement and is discussed in this chapter. More advanced material is in a text by Denes and Keedwell (1974) that is devoted solely to Latin squares and their applications from a mathematical point of view.

7.1. TWO RESTRICTIONS ON RANDOMIZATION; ASSUMPTIONS

For complete and incomplete block designs (Chaps. 5, 6), stratification of subjects into homogeneous blocks is followed by random assignment of subjects to treatments *separately* for each block, the separateness constituting a restriction on the randomization process as initially conceived for completely randomized experiments. Designs utilizing double blocking impose a second restriction on the randomization process, which can be illustrated by discussing the most common type of design, the Latin square. The name derives from the traditional use of Latin letters in an old mathematical puzzle. As the structure of the puzzle came to be used in experimentation, the Latin letters were used to designate the treatments imposed on agronomic plots arranged in a square to permit two-way elimination of natural soil fertility gradients across the land available.

A B C

C A B

B C A

The portrayal of the design as a square is convenient even when the nuisance variables are nonspatial factors, such as age and race of subjects.

To permit unbiased estimation of treatment effects, each treatment must occur exactly once in each row (e.g., age group) and once in each column (e.g., racial group). Additionally (as shown in Sec. 6.4.3.1), to avoid bias, rows and columns (e.g., age and race) must not interact in their effects on the variable of interest. Randomization consists of giving each possible configuration of treatments, in such squares of a given size, equal chance of being selected; i.e., a random square is to be chosen.

First, a *standard* (reduced) square must be defined. Any square in which the treatment symbols $\{A, B, C, \ldots\}$ occur exactly once in each row and once in each column and are so ordered in the first row and first column is a standard Latin square. Only one standard square exists when the number of treatments t is 2 or 3, but many such squares exist when the number of treatments is large. By permuting (randomly ordering) the rows and columns of a standard square one can obtain $t!(t - 1)!$ other Latin squares. Recall that $t! = t(t - 1)(t - 2)\ldots(1)$ and $(t - 1)! = t!/t$. The possible numbers of standard squares and all squares derivable from them for $2 \leq t \leq 6$ are given in Table 7.1. Obviously, no one has published complete tables of Latin squares from

Table 7.1. Possible Numbers of Latin Squares Involving t Treatments

Number of Treatments (t):	2	3	4	5	6
Number of standard squares	1	1	4	56	9,408
Number derivable from each standard	2	12	144	2,880	86,400
Total number of possible squares	2	12	576	155,280	812,851,200

which an experimenter might make a random selection. Therefore, a random square must be obtained by beginning with a standard square and performing certain permutations of rows and columns to give each of the possible squares an equal chance of being chosen. For larger squares, one may begin with the unique standard square in which successive rows are cyclic versions of the first row. Standard squares of size t = 2, 3, and 4 and a cyclic standard for t = 5 are shown in Table 7.2.

For t = 2, 3, or 4 the randomization process proceeds as follows: (1) begin with the unique standard square (t = 2 or 3) or a randomly selected standard (t = 4), (2) randomly permute (reorder) all rows except the first, and (3) randomly permute all columns of the result.

For $t \geq 5$ treatments it is not convenient to seek a random standard square because so many exist. Therefore, a procedure that permits approximately equal chance of selection for any square is: (1) randomly assign Latin symbols $\{A, B, C, \ldots\}$ to the actual treatments, (2) begin with the cyclic standard square of size t, and (3) randomly permute all rows and then all columns of the result.

Addelman (1975) has suggested additional considerations for cases in which systematic trends among experimental units are likely to occur (e.g., a diagonal trend when rows and columns represent ordered age and weight classes of subjects). Diagonal squares (same treatment in a line connecting opposite corners) have been shown to cause experimental error to be underestimated; and completely balanced or "Knut Vik" squares (Kempthorne 1952) cause overestima-

Table 7.2. All Possible Standard Latin Squares for t=2,3, or 4 Treatments; Cyclic Standard for t=5

A B *B A*	*A B C* *B C A* *C A B*	*A B C D E* *B C D E A* *C D E A B* *D E A B C* *E A B C D*

A B C D *B C D A* *C D A B* *D A B C*	*A B C D* *B A D C* *C D A B* *D C B A*	*A B C D* *B A D C* *C D B A* *D C A B*	*A B C D* *B D A C* *C A D B* *D C B A*

ion in such cases. Addelman's recommendations may be summarized as follows:

1. For 5 × 5 or larger squares, rerandomize if a "systematic" square is irst obtained (fully diagonal, diagonal in 2 × 2 or 3 × 3 subsquares, etc.).
2. For 4 × 4 squares, rerandomize if for the square first obtained, the umber of different treatment letters in *each* 2 × 2 corner subsquare plus the umber of different letters in the four corners of *each* 3 × 3 subsquare is ess than 6 or more than 7.
3. Avoid using 3 × 3 or 2 × 2 squares if systematic trends in units are ikely.

Note that for the changeover designs of Sec. 8.1.3 in which rows and colmns of a Latin square are subjects and time periods, respectively, the ranlomization problem is easier because the subjects can be randomly assigned to ows, i.e., to a particular sequence of treatments. If it is believed that he results in one period of the sequence will not affect results in subseuent periods, random assignment of subjects to treatment sequences in any roper Latin square should be adequate insurance against bias. However, when rovision must be made for residual effects of treatments in subsequent periods (Sec. 8.1.4), one must ensure that each treatment follows each of the others an equal number of times. For example, see the right-hand standard square of size t = 4 in Table 7.2.

In nonchangeover designs, additional insurance against bias, as well as replication for better precision and power, can be provided by utilizing a balanced set of squares. In a balanced set of t squares for t treatments, each treatment occurs exactly once in conjunction with each combination of row and column classes somewhere in the set of squares. A balanced set of squares is equivalent to a full replicate of a 3-factor experiment in which treatments, rows, and columns are the factors. A single square is equivalent to a $1/t$ fractional replicate for a t^3 factorial experiment (Sec. 6.4). If a balanced set of squares is used, interactions that may occur among rows, columns, and treatments inflate experimental error but do not cause bias in the estimates of treatment effects. However, if a single square is used, a component of the interaction of rows with columns (i.e., the 2 nuisance variables) is completely confounded with treatment effects. A balanced set of squares may be obtained by using a randomly selected square and $t - 1$ other squares obtained from it by cyclic permutation of the columns (or rows). For example, for $t = 3$ treatments, suppose the randomly selected square has rows (*BAC*), (*ACB*), and (*CBA*). Then a balanced set that includes that square is as shown in Table 7.3.

Table 7.3. Example of Balanced Set of Latin Squares for t=3 Treatments

B A C	*A C B*	*C B A*
A C B	*C B A*	*B A C*
C B A	*B A C*	*A C B*

7.2. ANALYSIS OF DATA FROM LATIN SQUARES

The model for analysis of nonfactorial data from a single Latin square may be written

$$Y_{ijk} = \mu + \rho_i + \gamma_j + \tau_k + E_{(ijk)} \qquad (i, j, k = 1, 2, \ldots, t; n = t^2) \quad (7.1)$$

where Y_{ijk} is an observable variable; μ = population mean; ρ_i = the effect of the ith row (ith class of the first nuisance variable); γ_j = effect of the jth column (jth class of the second nuisance variable); τ_k = effect of kth treatment; and $E_{(ijk)}$ is the residual effect, including random experimental error and various components of any interactions that exist. In contrast to the factorial models of Chap. 2, the subscript indices on Y here are partially redundant. That is, any two of the three subscripts will serve to identify uniquely an experimental unit or cell of the square because there are t^3 combinations of i, j, and k but only t^2 observations, as in a fractional replicate of a t^3 factorial experiment. Since there are but $t^2 - 1$ df, and $t - 1$ are accounted for by each of the variables (rows, columns, and treatments), only $(t^2 - 1) - 3(t - 1) = (t - 1)(t - 2)$ remain for residual error. For t = 2, 3, 4, and 5, the corresponding degrees of freedom for error are ν_E = 0, 2, 6, and 12. Therefore, an experiment with a single square of fewer than 5 treatments often does not provide enough information for adequate power and precision.

The model for analysis of nonfactorial data from a set of m Latin squares may be written

$$Y_{ijkl} = \mu + \theta_i + \rho_{(i)j} + \gamma_{(i)k} + \tau_l + E_{(ijkl)}$$

$$(i = 1, 2, \ldots, m;\ j, k, l = 1, 2, \ldots, t;\ n = mt^2) \quad (7.2)$$

nere θ_i = the effect of the ith square, $\rho_{(i)j}$ = the effect of the jth row ithin the ith square, $\gamma_{(i)k}$ = the effect of the kth column within the ith quare, and the other symbols are as defined for (7.1). When the squares used xactly constitute a balanced set, then also $i = 1, 2, \ldots, t$, i.e., $m = t$ quares. When the row and column classifications (e.g., age and race) are onsidered to be uniformly defined across all squares, degrees of freedom for rror can be conserved by not nesting row and column effects within squares.

.2.1. Estimation of Parameters

As in completely randomized designs, the parameters of chief interest are he set of treatment effects $\{\tau_k\}$ or treatment means $\{\mu + \tau_k\}$. The unbiased stimators of means are, as usual, the set of sample treatment means, $\{\overline{Y}_{..k}\}$ or a single square or $\{\overline{Y}_{...l}\}$ for m squares. To avoid mathematical discrep-ncies of notation caused by redundancy of indexing, let indices for row and olumn numbers (and squares if necessary) uniquely identify each observation r cell of a square. Then a sample treatment mean from a single square may be ymbolized by

$$\overline{y}_{..k} = \sum_{i=1}^{t} \sum_{j=1}^{t} \delta_{ij}^{k} y_{ijk}/t \tag{7.3}$$

here $\delta_{ij}^{k} = 1$ if the kth treatment was given to the unit in the ith row and th column or $\delta_{ij}^{k} = 0$ otherwise. Row, column, and overall totals may be des-gnated $y_{i..}$, $y_{.j.}$, and $y_{...}$, where subscript k is ignored in the summing rocess. For data from m squares, the equivalent notation for a treatment ean is

$$\overline{y}_{...l} = \sum_{i=1}^{m} \left(\sum_{j=1}^{t} \sum_{k=1}^{t} \delta_{jk}^{l} y_{ijkl} \right)/(mt) \tag{7.4}$$

quare, row, column, and overall totals may be designated $y_{i...}$, $y_{ij..}$, $y_{i.k.}$, nd $y_{....}$, where subscript l is ignored in the summing process.

As in the block designs (Chaps. 5, 6) one cannot make proper probability tatements about differences among the true treatment means or the reliability f estimates of them unless the effects of nuisance variables are statistical-y eliminated to obtain a proper estimate of the experimental error variance. iven the definitions of treatment means in (7.3) and (7.4) (or the corre-ponding treatment totals) and the defined symbols of other totals, the compu-ations for an analysis of variance are relatively straightforward, as de-cribed for completely randomized designs. Computational formulas for the ums of squares for data from a single square or from m squares are given in able 7.4.

The partition of variance for a single square is shown in Table 7.5. As sual, the experimental error variance is estimated by the mean square for er-or (ms_E). The standard error of a treatment mean is $\sqrt{ms_E/t}$ for a single quare or $\sqrt{(ms_E/mt)}$ for m squares. The analysis of variance for m squares is

Table 7.4. Sums of Squares for Analysis of Variance of 1 or m Latin Squares for t Treatments*

	1 Square	m Squares
Squares	...	$ss_S = \sum_{i=1}^{m} y_{i...}^2/t^2 - y_{....}^2/n$
Rows	$ss_R = (\sum_{i=1}^{t} y_{i..}^2/t) - (y_{...}^2/n)$	$ss_{R/S} = (\sum_{i=1}^{m} \sum_{j=1}^{t} y_{ij..}^2/t) - (\sum_{i=1}^{m} y_{i...}^2/t^2)$
Columns	$ss_C = (\sum_{j=1}^{t} y_{.j.}^2/t) - (y_{...}^2/n)$	$ss_{C/S} = (\sum_{i=1}^{m} \sum_{k=1}^{t} y_{i.k.}^2/t) - (\sum_{i=1}^{m} y_{i...}^2/t^2)$
Treatments	$ss_T = (\sum_{k=1}^{t} y_{..k}^2/t) - (y_{...}^2/n)$	$ss_T = (\sum_{\ell=1}^{t} y_{...\ell}^2/mt) - (y_{....}^2/n)$
Error	$ss_E = ss_y - ss_R - ss_C - ss_T$	$ss_E = ss_y - ss_S - ss_{R/S} - ss_{C/S} - ss_T$
Total	$ss_y = (\sum_{i=1}^{t} \sum_{j=1}^{t} y_{ijk}^2) - (y_{...}^2/n)$	$ss_y = (\sum_{i=1}^{m} \sum_{j=1}^{t} \sum_{k=1}^{t} y_{ijk\ell}^2) - (y_{....}^2/n)$
	$n = t^2$	$n = mt^2$

*Ignore the treatment subscript (k for one square, ℓ otherwise) when computing sums for effects other than treatments.

Table 7.5. Analysis of Variance for t Treatments in Single Latin Square

Source of Variation	df	ss	ms	$E[MS]$
Rows	$t-1$	ss_R	ms_R	$\sigma^2 + t\Sigma\rho_i^2/(t-1)$
Columns	$t-1$	ss_C	ms_C	$\sigma^2 + t\Sigma\gamma_j^2/(t-1)$
Treatments	$t-1$	ss_T	ms_T	$\sigma^2 + t\Sigma\tau_k^2/(t-1)$
Error	$(t-1)(t-2)$	ss_E	ms_E	σ^2
Total	t^2-1	ss_y		

a straightforward extension of that for a single square. Squares have $m - 1$ df. Rows and columns each have $m(t - 1)$ df if nested ($t - 1$ if not), and error has $(t - 1)(mt - m - 1)$ df if rows and columns are nested in squares or $(t - 1)(mt + m - 3)$ otherwise. A word of caution: expected mean squares have been derived under the assumption that the 2 nuisance variables (rows, col-

mns) do not interact with each other or with treatments. If in fact interac-
ions exist in a case with a single square, certain components of interaction
re completely confounded with error, perhaps inflating it badly; and other
omponents of interaction are completely confounded with the effects of rows,
olumns, or treatments. Specifically, if rows interact with columns (the 2
uisance variables), estimates of treatment effects and the treatment mean
quare are biased (Sec. 6.4.3.1). As mentioned previously, the use of a bal-
nced set of squares minimizes the possibility of bias of treatment effects.

.2.2. Testing Hypotheses

If the assumption of negligible interactions is correct, an unbiased test
f the hypothesis that treatment effects do not differ, $H{:}\tau_k = 0$, may be per-
ormed by comparing the test statistic,

$$f = ms_T/ms_E \tag{7.5}$$

ith $f_{\alpha,t-1,\nu_E}$ (App. A.5), where $\nu_E = (t - 1)(t - 2)$ for a single square, $(t -$
$)(mt - m - 1)$ for m squares with nested rows and columns, and $(t - 1)(mt + m$
$3)$ for m squares without nesting. As usual the hypothesis is rejected if f
f_α. Specific contrasts should be designed and evaluated when possible.
therwise postdata comparisons may be made (see Chap. 2). The various tech-
iques make use of the standard error of a treatment mean, $\sqrt{ms_E/t}$ [or $\sqrt{(ms_E/mt)}$
or m squares], or of the difference between two means, $\sqrt{2ms_E/t}$ [or $\sqrt{(2ms_E/mt)}$].

A partial check on the validity of having assumed additivity of the ef-
ects of rows, columns, and treatments (i.e., negligible interactions) in a
ingle square was proposed by Tukey (1955) as an extension of his test of non-
dditivity (1949) in randomized block designs. Let the fitted values (obserations minus residuals) be defined by

$$\hat{y}_{ijk} = \hat{\mu} + \hat{\rho}_i + \hat{\gamma}_j + \hat{\tau}_k \tag{7.6}$$

here $\hat{\mu} = \bar{y}_{...}$, $\hat{\rho}_i = (\bar{y}_{i..} - \bar{y}_{...})$, $\hat{\gamma}_j = (\bar{y}_{.j.} - \bar{y}_{...})$, and $\hat{\tau}_k = (\bar{y}_{..k} - \bar{y}_{...})$
s defined previously. Let x_{ijk} be the square of a fitted value $(\hat{y}^2_{ijk})$, or
or computational convenience let $x_{ijk} = c_1(\hat{y}_{ijk} - c_2)^2$, where c_1 and c_2 are
rbitrary constants. For example, c_1 could be 10, 1, 0.1, or 0.01 and c_2
ould be $\bar{y}_{...}$, the overall mean. In any case the sum of squares for nonaddi-
ivity has 1 df and is computed from

$$ss_{\mathrm{NA}} = \Big[\sum_{i=1}^{t}\sum_{j=1}^{t}(x_{ijk}e_{ijk})\Big]^2/ss_{E(x)} \tag{7.7}$$

here $e_{ijk} = y_{ijk} - \hat{y}_{ijk}$, a residual effect, and $ss_{E(x)}$ is the error sum of
quares derived from an analysis of variance on the x_{ijk} in the manner of Table
.5. Then, the hypothesis of additivity is rejected if the test statistic,

$$f = ss_{\mathrm{NA}}/[(ss_{E(y)} - ss_{\mathrm{NA}})/(\nu_E - 1)] > f_{\alpha,1,\nu_E-1} \tag{7.8}$$

where $ss_{E(y)}$ is the error sum of squares from analysis of variance of the original data (y_{ijk}) and $\nu_E = (t - 1)(t - 2)$ for a single square. As for other relatively insensitive tests of this kind, it may be advisable to reject the hypothesis at $\alpha = 0.25$ or some other value of α greater than the usual 0.05. Rejection implies that the nuisance variables may be interacting with each other or with treatments to such degree that the interactions should be examined in a replicated 3-factor experiment. Rojas (1973) has shown that the test of nonadditivity is equivalent to the test of zero regression slope in an analysis of covariance of y_{ijk} and x_{ijk}.

In experiments with m squares, the test for nonadditivity may be performed by adding the estimated effects of different squares, $\hat{\theta}_i = \bar{y}_{i...} - \bar{y}$, to the definition of fitted values [see (7.6)] and adjusting subscripts to fit the multiple-square model [see (7.2)]. Alternatively, if several squares with the same arrangement of treatments constitute random replication of experimental units (rather than groupings of units that differ from square to square) and if the row and column classifications are uniformly defined across all squares, the data could be pooled to permit direct computation of a sum of squares for combined interactions. Suppose the data from m squares are pooled so that each of the t^2 cells of a $t \times t$ square has m observations. Let the data be represented by y_{ijkl}, where i, j, and k index row, column, and treatment as in a single square and l indexes observations within a cell. The sums of squares and degrees of freedom among cells, rows, columns, and treatments and new computations for interactions and error are given in Table 7.6. The test of nonadditivity may be made by comparing the test statistic,

$$f = ms_I/ms_E \tag{7.9}$$

with the critical value, $f_{\alpha,(t-1)(t-2),t^2(m-1)}$.

In some experiments the experimental unit representing each cell of a square is sampled so that multiple determinations lead to additional random variation (e.g., multiple laboratory determinations on blood drawn from each subject). Such effects are nested within subject effects (experimental error) and may be termed subsampling error (as in Chap. 2).

When treatments consist of combinations of the levels or classes of 2 or more factors, the models must be expanded so that the effects of treatment combinations are partitioned into main effects and interactions (Sec. 7.6).

7.2.3. Missing Cells

In even the best-designed experiments one or more observations may be missing because an animal died; a laboratory sample was accidentally destroyed, lost, or contaminated; or a gross error was discovered too late for rectification. The computations for statistical analysis may be facilitated by estimating the missing values and placing the estimates with the observed data for analysis.

One procedure for estimating missing values involves minimizing the residual sum of squares as for complete block designs (also see Haseman and Gaylor 1973). In cases involving a single square, the proper sum of squares for treatments may be obtained from

$$ss_T = (ss'_y - ss'_R - ss'_C) - ss_E \tag{7.10}$$

Table 7.6. Sums of Squares and Degrees of Freedom when Data from m Identical Latin Squares are Pooled

Source of Variation	df	ss
[Cells]	$[t^2-1]$	$ss_{cells}=(\sum_{i=1}^{t}\sum_{j=1}^{t} y_{ij..}^2/m)-(y_{....}^2/m)$
Rows	$t-1$	$ss_R=(\sum_{i=1}^{t} y_{i...}^2/mt)-(y_{....}^2/m)$
Columns	$t-1$	$ss_C=(\sum_{j=1}^{t} y_{.j..}^2/mt)-(y_{....}^2/m)$
Treatments	$t-1$	$ss_T=(\sum_{k=1}^{t} y_{..k.}^2/mt)-(y_{....}^2/m)$
Interactions	$(t-1)(t-2)$	$ss_I=ss_{cells}-ss_R-ss_C-ss_T$
Error (within cells)	$t^2(m-1)$	$ss_E=ss_y-ss_{cells}$
Total	mt^2-1	$ss_y=(\sum_{i=1}^{t}\sum_{j=1}^{t}\sum_{\ell=1}^{m} y_{ijk\ell}^2)-(y_{....}^2/m)$

where ss'_y, ss'_R, and ss'_C are sums of squares for total data, rows, and columns computed from observed values only in the manner of a one-way analysis of variance with unequal replication, and ss_E is the residual sum of squares obtained when the complete set of values (observed plus estimated) is analyzed as in Table 7.5. Degrees of freedom for error should be reduced by the number of missing values estimated. When only one cell is missing, the minimization of residual sum of squares leads to a formula for estimating the missing value,

$$\hat{y}_{ijk} = [t(y'_{i..} + y'_{.j.} + y'_{..k}) - 2y'_{...}]/[(t - 1)(t - 2)] \tag{7.11}$$

where $y'_{i..}$, $y'_{.j.}$, and $y'_{..k}$ are totals of $t - 1$ observations each of row, column, and treatment involved; and $y'_{...}$ is the total of all observations $(t^2 - 1)$. In such cases, an alternative to (7.10) for adjusted treatment sum of squares is

$$ss_T = [(\sum_{k=1}^{t} y_{..k}^2/t) - (y_{...}^2/t^2)]$$

$$- \{[y'_{...} - y'_{i..} - y'_{.j.} - (t - 1)y'_{..k}]^2/[(t - 1)^2(t - 2)^2]\} \tag{7.12}$$

where the first difference is the usual treatment sum of squares with the estimated value included. For computation of the standard error of the difference between two treatment means, one of which includes the estimated value, the usual $\sqrt{2ms_E/t}$ should be replaced by $\sqrt{(2ms_E/t) + (ms_E/\nu_E)}$, where ν_E is the number of degrees of freedom for the error mean square. When several cells are missing, a computation of "exact" standard errors of differences is very tedious. However, upper and lower limits may be computed to indicate appropriate reliability. For two means having m_1 and m_2 estimated values, respectively, the lower limit of the standard error (se) of the difference is

$$se_{min} = \sqrt{ms_E[1/(t - m_1) + 1/(t - m_2)]} \tag{7.13}$$

and the upper limit is

$$se_{max} = \sqrt{2ms_E/(t - m_1 - m_2)} \tag{7.14}$$

For practical purposes one could compute an approximate standard error from

$$se_{approx} = \sqrt{[(se_{min})^2 + (se_{max})^2]/2} \tag{7.15}$$

Alternative methods of estimating values for missing cells are (1) covariance technique (Chap. 3), (2) iterative procedure, and (3) general application of general least squares estimation by computer (Chap. 4). To use the iterative procedure, guess arbitrary but reasonable values for all but one of the missing values and estimate the latter as in (7.11). Next, one of the arbitrary values is removed and the missing value again is estimated. The process continues until all arbitrary values have been replaced by estimates. Then, another cycle or two of removing earlier estimates one at a time and replacing with current estimates should stabilize the values sufficiently for practical purposes. Adjustments in treatment sum of squares, residual degrees of freedom, and standard errors of treatment differences apply as before.

If an entire row or column (but not both) is missing, one of the following alternatives may be useful for analysis: (1) ignore one of the nuisance variables entirely and analyze the data as if the experiment had been performed in a randomized complete block design, (2) analyze as an incomplete Latin square (Sec. 7.8), or (3) use a generalized least squares computer program.

7.3. ESTIMATION OF NUMBER OF SQUARES REQUIRED

Since small Latin squares have few degrees of freedom for experimental error (Sec. 7.2), two or more squares must be used to obtain reasonably precise estimates of treatment means and sufficiently sensitive tests of hypotheses.

One can estimate the power of the test provided by a single square of specified size to determine if additional squares will be needed. For the hypothesis, $H:\tau_k = 0$ (no treatment effects), the appropriate abscissa scale for App. Figs. A.15.17-A.15.21 is

$$\phi = \sqrt{\sum_{k=1}^{t} (\tau_{dk}/\sigma)^2} \tag{7.16}$$

where the array of treatment effects one wishes to detect $\{\tau_{dk}\}$ must be specified such that $\Sigma\tau_{dk} = 0$. One may estimate σ from a prior similar experiment performed as a Latin square. Or one may make an educated guess of the potential range of response for subjects treated alike and sharing similar values or classes with respect to the 2 nuisance variables and convert that to an estimate of σ by using App. A.14 to find a proper division of the range. Usually a divisor of 5 provides a reasonable approximation. Selection of a proper power curve depends on $\nu_1 = t - 1$ df among t treatments, the probability of Type I error permitted (α), and $\nu_2 = (t - 1)(t - 2)$ df for experimental error. If the intersection of the curve for ν_2 with ϕ indicates an unacceptably high probability of Type II error (β), additional squares should be considered. As for simpler designs, it is often difficult to specify the array of treatment effects $\{\tau_{dk}\}$ one would like to detect, especially if the number of treatments is more than 2 or 3. Alternatively, one may select orthogonal contrasts to be performed and use the abscissa scale,

$$\phi = (\Delta_d/\sigma)\sqrt{(t/8)\left(\sum_{i=1}^{t}|c_{ik}|\right)^2 / \sum_{i=1}^{t} c_{ik}^2} \tag{7.17}$$

where Δ_d is the mean contrast difference one wishes to detect and the c_{ik} are coefficients of treatment totals in the kth contrast. (In such cases use $\nu_1 = 1$ and App. A.15.17 or A.15.18.)

If additional squares are needed, the abscissa scale for the overall test of treatments becomes

$$\phi = \sqrt{m \sum_{l=1}^{t} (\tau_{dl}/\sigma)^2} \tag{7.18}$$

for m squares, with $\nu_1 = t - 1$ and $\nu_2 = (t - 1)(mt - m - 1)$ if rows and columns are nested in squares or $(t - 1)(mt + m - 3)$ if not. Alternatively, for orthogonal contrasts

$$\phi = (\Delta_d/\sigma)\sqrt{(mt/8)\left(\sum_{i=1}^{t}|c_{ik}|\right)^2 / \sum_{i=1}^{t} c_{ik}^2} \tag{7.19}$$

with $\nu_1 = 1$ and ν_2 as above. In either case, one should take a trial value for the number of squares m and compute ϕ and ν_2. If the intersection of ϕ and ν_2 indicates unacceptably high β, revise m upward and try again. Keep in mind the additional advantages of minimizing bias by using a balanced set (or sets) of t squares.

7.4. EFFICIENCY OF DOUBLE BLOCKING

The concept of relative efficiency of two designs is discussed in Sec. 5.6. Recall that the efficiency of design 1 relative to design 2 (with means in each case estimated from the same number of observations) is

$$E_{1:2} = [(\nu_1 + 1)/(\nu_1 + 3)][(\nu_2 + 3)/(\nu_2 + 1)](\hat{\sigma}_2^2/\hat{\sigma}_1^2)100\% \quad (7.20)$$

where the νs and $\hat{\sigma}^2$s are residual degrees of freedom and estimated variances. The obvious standard of comparison for block designs is the completely randomized design. For Latin squares, three possibilities arise: (1) Latin square (LS) versus completely randomized design (*CRD*); (2) Latin square versus randomized complete block design, in which the blocks are rows (first nuisance variable) of the Latin square (*RCBD*:*R*); and (3) Latin square versus randomized complete block design, in which the blocks are columns (second nuisance variable) of the Latin square (*RCBD*:*C*). As before, the pertinent variances (all to be estimated post facto with data from a Latin square design) are those expected to occur under the various forms of blocking (or nonblocking) in the absence of treatment effects, i.e., as in a randomization trial to study variation without treatments.

For estimation of the full effect on efficiency of using both nuisance variables, let the Latin square be design 1 and the hypothetical completely randomized experiment be design 2. Then $\hat{\sigma}_1^2 = ms_E$, from the Latin square analysis (Table 7.5). The degrees of freedom are those that would be associated with experimental error in a randomization trial with double blocking but no treatments. Therefore, $\nu_1 = \nu_T + \nu_E$. For the hypothetical completely randomized experiment, all the variation encountered in the Latin square except the portion caused by treatments must be designated as experimental error but the size of the experiment must be the same. That is achieved by multiplying ms_E from the Latin square by the degrees of freedom for residual and treatments combined and adding the result to the sums of squares caused by the nuisance variables. For a single Latin square, then,

$$\hat{\sigma}_2^2 = [ss_R + ss_C + (t - 1)^2 ms_E]/(t^2 - 1) \quad (7.21)$$

with $\nu_2 = t^2 - 1$, the total degrees of freedom without blocking or treatments. For m squares with nested rows and columns the comparable estimate is

$$\hat{\sigma}_2^2 = [ss_S + ss_{R/S} + ss_{C/S} + (\nu_T + \nu_E)ms_E]/(mt^2 - 1) \quad (7.22)$$

with $\nu_2 = mt^2 - 1$. Of course ν_2 is smaller if rows, columns, or both are not nested.

The values of the adjustments for degrees of freedom for $t > 5$ in the efficiency calculation [(7.20)] are sufficiently near unity that they may be ignored, so $E_{1:2} = (\hat{\sigma}_2^2/\hat{\sigma}_1^2)100\%$. For squares having $3 \leq t \leq 5$ treatments the adjustment may be ignored if one uses $m > 5$ squares. Values of the adjustment for $t \leq 5$, $m \leq 5$, and with rows and columns nested are shown in Table 7.7.

Perhaps more informative than comparison with a completely randomized design is the efficiency obtained by adding a second nuisance variable to one that could have been used in a RCBD. Let the Latin square be design 1 and a hypothetical block design using the row nuisance variable as blocks be design

Table 7.7. Adjustments for Degrees of Freedom $[(\nu_1+1)/(\nu_1+3)][(\nu_2+3)/(\nu_2+1)]$ for Efficiency of m Latin Squares (nested rows and columns) Relative to Completely Randomized Design [see (7.20)]

m	t=2*	3	4	5
1	0.750	0.873	0.938	0.966
2	.750	.909	.961	.981
3	.778	.931	.972	.986
4	.804	.944	.978	.990
5	0.825	0.954	0.982	0.992

*For t=2 and m>5, compute $(4m^2+6m+2)/(4m^2+12m)$.

2, *RCBD*:*R*. Then $\hat{\sigma}_1^2 = ms_E$ from the Latin square analysis. The corresponding degrees of freedom include those for error and treatments as before; $\nu_1 = \nu_T + \nu_E$. However, $\hat{\sigma}_2^2$ must now be computed as if the column effects (second nuisance variable) had been included in the error of *RCBD*:*R*. Therefore,

$$\hat{\sigma}_2^2 = [ss_C + (\nu_T + \nu_E)ms_E]/(\nu_C + \nu_T + \nu_E) \tag{7.23}$$

with $\nu_2 = \nu_C + \nu_T + \nu_E$ df for columns, treatments, and error of the Latin squares. For experiments with m squares, ss_C may be interpreted as $ss_{C/S}$, where columns are nested within squares. The remaining sums of squares, $ss_S + ss_{R/S}$, represent the sum of squares attributable to blocks in *RCBD*:*R*.

Adjustments for degrees of freedom in (7.20) are sufficiently near unity for $t > 3$ treatments that they may be ignored. For $t = 3$, the adjustments for m = 1, 2, 3, 4, 5 squares are 0.918, 0.944, 0.958, 0.966, 0.972 and may be ignored for $m > 5$. For $t = 2$ and m squares, compute the adjustment from $(2m^2 + 5m + 3)/(2m^2 + 7m + 3)$.

When design 1 is a Latin square and design 2 is a hypothetical block design using the column nuisance variable as blocks (*RCBD*:*C*), then $\hat{\sigma}_1^2 = ms_E$, $\nu_1 = \nu_T + \nu_E$, and

$$\hat{\sigma}_2^2 = [ss_R + (\nu_T + \nu_E)ms_E]/(\nu_R + \nu_T + \nu_E) \tag{7.24}$$

with $\nu_2 = \nu_R + \nu_T + \nu_E$. For experiments with m squares, ss_R must be interpreted as $ss_{R/S}$, where rows are nested within squares. Comments about the adjustments for degrees of freedom for *RCBD*:*R* apply here as well.

7.5. NUMERICAL EXAMPLE IN LATIN SQUARE

Activity of follicle-stimulating hormone (FSH) in cows was measured in bioassay by the ovarian weight (mg) of immature rats. Two nuisance variables known to influence ovarian weight of the rats were genetic constitution and body weight. It was believed that the relation of ovarian weight to body weight was independent of genetic differences so that a Latin square design would be valid. Two randomly selected 4 × 4 squares were used with rows = litters of rats and columns = body weight classes. Because the range of body weights was consistent from litter to litter, the experimenter decided to consider differences in body weight across both squares (rather than nested within squares) to conserve degrees of freedom for experimental error. The selected squares and responses are shown in Table 7.8.

Table 7.8. Ovarian Weight (mg) of Immature Rats (FSH bioassay for cows) in 2 Latin Squares with 4 Treatments

	Square 1						Square 2				
Litter	1	2	3	4	Total	Litter	1	2	3	4	Total
1	D_{44}	C_{39}	B_{52}	A_{73}	208	5	B_{51}	C_{63}	A_{74}	C_{82}	270
2	B_{26}	A_{45}	D_{49}	C_{58}	178	6	D_{62}	A_{74}	C_{75}	B_{79}	290
3	C_{67}	D_{71}	A_{81}	B_{76}	295	7	A_{71}	D_{67}	B_{60}	C_{74}	272
4	A_{77}	B_{74}	C_{88}	D_{100}	339	8	C_{49}	B_{47}	D_{58}	A_{68}	222
Total	214	229	270	307	1020		233	251	267	303	1054

Treatment Totals: 563(*A*), 465(*B*), 513(*C*), 533(*D*)

Variation between squares and among litters within squares may be combined and computed as variation among the 8 litters. Data from both squares may be pooled to compute variation among the 4 body weight classes. Sums of squares for litters, body weight classes, and treatments are, respectively,

$$ss_L = (208^2 + 178^2 + \dots + 222^2)/4 - (1020 + 1054)^2/32 = 4869$$

$$ss_B = [(214 + 233)^2 + \dots + (307 + 303)^2]/8 - (1020 + 1054)^2/32 = 1914$$

$$ss_T = (563^2 + \dots + 533^2)/8 - (1020 + 1054)^2/32 = 635$$

Subtracting those results from the total sum of squares, $ss_y = 7710$, one obtains the sum of squares for error, $ss_E = 292$. The analysis of variance is

Table 7.9. Analysis of Variance of Latin Square Data (see Table 7.8)

Source of Variation	df	*ss*	*ms*
Litters	7	4869	696
Body weight classes	3	1914	638
Treatments	3	635	212
Error	18	292	16

shown in Table 7.9. The treatment means, 70.4 (A), 58.1 (B), 64.1 (C), and 66.6 (D) each have standard error equal to $\sqrt{16/8} = 1.4$. The hypothesis of zero treatment effects may be rejected with very high confidence because $f = 212/16 = 13.25$ far exceeds $f_{0.001,3,18} = 8.49$. Such a result is often expected in bioassay work and is secondary to estimation of potency.

The efficiency of the Latin square design relative to a CRD may be estimated if one uses the information in Table 7.9. Degrees of freedom for the two designs are $\nu_1 = \nu_T + \nu_E = 3 + 18 = 21$ and $\nu_2 = mt^2 - 1 = 2(4^2) - 1 = 31$, and the adjustment factor for efficiency is

$$[(\nu_1 + 1)/(\nu_1 + 3)][(\nu_2 + 3)/(\nu_2 + 1)] = (22/24)(34/32) = 0.974$$

Estimated variances for the two designs are $\hat{\sigma}_1^2 = ms_E = 16$ and

$$\hat{\sigma}_2^2 = (7ms_L + 3ms_B + 21ms_E)/31 = [4869 + 1914 + 21(16)]/31 = 230$$

The relative efficiency of double blocking is 0.974(230/16)100% = 1400%, which is very large indeed. From the mean squares for litters and body weight (Table 7.9), it is evident that the 2 nuisance variables are approximately equal in importance.

Suppose that in a future experiment of similar type and design, one desires 90% power to detect a mean difference of $\Delta_d = 4.2$ mg for an orthogonal contrast between two treatment means at a 99% significance level. The number of squares required (m) may be estimated by referring to (7.19) and Fig. A.15.17. One may use $\sqrt{ms_E} = \sqrt{16} = 4$ as an estimate of σ. For $t = 4$ treatments,

$$\phi = \sqrt{mt/4}(\Delta_d/\hat{\sigma}) = \sqrt{m}(4.2/4)$$

If body weight classes are to be combined over all squares as before, the degrees of freedom for error will be

$$\nu_2 = (n - 1) - \nu_L - \nu_B - \nu_T = (mt^2 - 1) - (mt - 1) - (t - 1) - (t - 1)$$

$= (t - 1)(mt - 2) = 3(3m - 2)$

As a first trial, use $m = 8$ squares. Then $\phi = 3.0$ and $\nu_2 = 66$. Figure A.15.17 indicates $\beta \simeq 0.06$ or 94% power. For $m = 7$, one obtains $\phi = 2.8$, $\nu_2 = 57$, and $\beta \simeq 0.10$ or 90% power. Therefore 7 squares of 16 animals each is the minimal requirement, or one could use 2 balanced sets of 4 squares of 4×4.

An alternative design, utilizing both nuisance variables but not restricting the number of litters to multiples of 4, is an RCBD with covariance analysis in which litters are blocks and body weight is the covariate.

7.6. FACTORIAL EXPERIMENTS IN LATIN SQUARES

The Latin square design can easily accommodate small factorial experiments, i.e., those having relatively few treatment combinations. (See Sec. 7.8.4 for large factorials.) The randomization procedure discussed in Sec. 7.1 is applicable with the treatment combinations labeled A, B, C, etc. Analysis of data from such experiments is a relatively straightforward extension of the analysis in Secs. 7.2 and 7.5, with factorial partition of variation (described in Chap. 2). A general model for a 2-factor experiment in a Latin square is

$$Y_{ijkl} = \mu + \rho_i + \gamma_j + \alpha_k + \beta_l + (\alpha\beta)_{kl} + E_{(ijkl)} \quad (7.25)$$

where $i = j = 1, 2, \ldots, t = ab$; $k = 1, 2, \ldots, a$ levels of factor A; and $l = 1, 2, \ldots, b$ levels of factor B. As in (7.1), ρ_i and γ_j are row and column effects of first and second nuisance variables and $E_{(ijkl)}$ is experimental error. Main effects and interaction of the 2 factors, symbolized by α_k, β_l, and $(\alpha\beta)_{kl}$, may be considered partitions of the treatment effects, symbolized τ_k in the nonfactorial model. Just as (7.25) is a special case of (7.1), so the factorial model for 2 or more squares is a special case of (7.2). Sums of squares for main effects and interaction of the 2 factors are orthogonal decompositions of the overall treatment sum of squares (Table 7.4) and may be computed in accordance with the principles demonstrated in Chap. 2.

To illustrate factorial experiments in Latin squares, consider the following example. A study was made of the influence of preliminary treatments on the lethal dose of mercupurin in rabbits. The Latin square design was selected to eliminate daily variation (rows) and to permit study of the main effect of time between injections (columns), given the belief that the effects of time between injections would not interact with effects of days or preliminary treatments. Treatments used were A = control, B = ammonium chloride-low dose (NL) plus phenobarbital-low dose (PL), C = N-high dose (NH) plus (PL), D = (NL) plus P-high dose (PH), and E = (NH) plus (PH). Thus, the experiment involves a 2^2 factorial plus control. The response variable shown in Table 7.10 is Y = (log mL injected for threshold toxic dose of mercupurin/log body weight in kg).

Total sum of squares for the data in Table 7.10 is $ss_y = 1.211774$, sum of squares for rows (days) is

$$ss_R = [(6.048^2 + 5.786^2 + \ldots + 4.933^2)/5] - (27.277^2/25) = 0.172473$$

Table 7.10. Example of Factorial Plus Control in Latin Square Design (lethal dose of mercupurin in rabbits)

Day	Minutes between Injections					Total
	0.5	1.0	1.5	2.0	2.5	
1	*A*	*C*	*D*	*E*	*B*	
	1.576	1.161	1.231	1.032	1.048	6.048
2	*C*	*E*	*B*	*A*	*D*	
	1.432	1.168	1.220	1.031	0.935	5.786
3	*B*	*D*	*E*	*C*	*A*	
	1.394	1.266	0.580	0.934	0.925	5.099
4	*E*	*B*	*A*	*D*	*C*	
	1.199	1.240	1.168	1.139	0.665	5.411
5	*D*	*A*	*C*	*B*	*E*	
	0.856	1.182	1.023	0.889	0.983	4.933
Total	6.457	6.017	5.222	5.025	4.556	27.277
Treatment	*A*	*B*	*C*	*D*	*E*	
Total	5.882	5.791	5.215	5.427	4.962	
Mean	1.176	1.158	1.043	1.085	0.992	

sum of squares for columns (minutes between injections) is

$$ss_C = [(6.457^2 + 6.017^2 + \dots + 4.556^2)/5] - (27.277^2/25) = 0.473447$$

sum of squares for treatments is

$$ss_T = [(5.882^2 + 5.791^2 + \dots + 4.962^2)/5] - (27.277^2/25) = 0.119331$$

and sum of squares for experimental error is

$$ss_E = 1.211774 - (0.172473 + 0.473447 + 0.119331) = 0.446523$$

The sum of squares of treatments may be partitioned into sums of squares for orthogonal contrasts: (1) control versus others, (2) main effect of ammonium chloride (treatments *B*, *D* versus *C*, *E*), (3) main effect of phenobarbital treatments (*B*, *C* versus *D*, *E*), and (4) interaction of ammonium chloride and phenobarbital (*B*, *E* versus *C*, *D*). Each contrast sum of squares takes the form $q_k^2/r\Sigma c_{ik}^2$; or sums of squares for the 3 factorial effects may be computed as for the general case (Chap. 2):

$$ss_1 = [4(5.882) - (5.791 + \ldots + 4.962)]^2/[5(20)] = 0.045497$$

$$ss_2 = ss_A = [(5.791 + 5.427) - (5.215 + 4.962)]^2/[5(4)] = 0.054184$$

$$ss_3 = ss_B = [(5.791 + 5.215) - (5.427 + 4.962)]^2/[5(4)] = 0.019034$$

$$ss_4 = ss_{AB} = [(5.791 + 4.962) - (5.215 + 5.427)]^2/[5(4)] = 0.000616$$

Note that the sum of these four results exactly equals ss_T. A summary of the analysis of variance is shown in Table 7.11. None of the f ratios is signifi-

Table 7.11. Analysis of Variance of Factorial Plus Control in Latin Square Design (see Table 7.10)

Source of Variation	df	ss	ms	f Ratios
Days	4	0.17247	0.0431	1.159
Minutes between injections	4	.47345	.1184	3.183
Treatments	4	.11933	.0298	0.801
Control vs. others	1	.04550	.0455	1.223
Ammonium chloride dose	1	.05418	.0542	1.457
Phenobarbital dose	1	.01903	.0190	0.511
Interaction	1	.00062	.0006	0.016
Experimental error	12	0.44652	0.0372	

cant at a 95% confidence level; there is suggestion of the effects of minutes between injections ($\alpha \simeq 0.05$ to 0.06) and hints that days may differ, the preliminary treatments may enhance toxicity of mercupurin, and the high dose of ammonium chloride was more effective than the low dose. A larger experiment that would provide more degrees of freedom for error or a different design that would permit separation of interaction of minutes and treatments from error might provide more convincing evidence.

7.7. TRIPLE BLOCKING: GRAECO-LATIN SQUARES

Graeco-Latin (or Euler) squares make use of 3 separate classifications or nuisance variables for blocking to reduce experimental error. The 3 nuisance variables constitute 3 restrictions on randomization; each of t classes of a nuisance variable occurs exactly once in conjunction with each of t treatments and exactly once in conjunction with each of t classes of each of the other 2 blocking variables. The mathematics of the designs date back to Euler's conjecture in 1782 that such combinations were impossible for t = 2, 6, 10, 14, etc. In 1959 Euler's conjecture was proven wrong for $t \geq 10$ by Bose and associates at the University of North Carolina. Because of complex restrictions on combinations and numbers of classes of nuisance variables, the designs have been used only infrequently. The use of several variables also leads to increased doubt that interactions are negligible, i.e., doubt about the validity of assuming that treatments are not confounded with interactions of blocking factors.

The nature of the designs may be illustrated for t = 4 treatments. Let the 3 blocking factors be represented by rows, columns, and Greek letters. Let the treatments (or treatment combinations) be symbolized by Latin letters. One must ensure that each Latin and each Greek letter occurs exactly once in each row, in each column, and with each letter from the other alphabet. The usual method of construction of the designs is to superimpose two orthogonal Latin squares (Fisher and Yates 1963) as shown below.

$$\begin{matrix} A & B & C & D \\ B & A & D & C \\ C & D & A & B \\ D & C & B & A \end{matrix} \quad + \quad \begin{matrix} \alpha & \beta & \gamma & \delta \\ \delta & \gamma & \beta & \alpha \\ \beta & \alpha & \delta & \gamma \\ \gamma & \delta & \alpha & \beta \end{matrix} \quad = \quad \begin{matrix} A(\alpha) & B(\beta) & C(\gamma) & D(\delta) \\ B(\delta) & A(\gamma) & D(\beta) & C(\alpha) \\ C(\beta) & D(\alpha) & A(\delta) & B(\gamma) \\ D(\gamma) & C(\delta) & B(\alpha) & A(\beta) \end{matrix}$$

Randomization proceeds by permuting rows, then columns. Levels or classes of the third blocking factor are randomly assigned to Greek letters and levels or classes of treatments are randomly assigned to Latin letters.

The analysis of data from Graeco-Latin squares proceeds in essentially the same fashion as for Latin squares (Tables 7.4, 7.5) except that an additional sum of squares with $t - 1$ df is removed from experimental error to account for the third (Greek) nuisance variable. Degrees of freedom for error in a single square are $(t - 1)(t - 3)$. Therefore, squares with t = 3, 4, 5, 7, 8, 9 treatments have ν_E = 0, 3, 8, 24, 35, 48 df for error. Obviously, the smaller squares usually require replication to ensure adequate precision.

Hypersquares that make use of 4 or more nuisance variables for blocking may be constructed, but the exercise is rarely practical. Some other exotic derivatives of Latin square designs, such as Latin-, Graeco-Latin-, and hypercubes; plaid- and half-plaid-Latin squares; magic- and supermagic-Latin squares, have been discussed by Federer (1955).

7.8. CONFOUNDING IN INCOMPLETE BLOCKS

Suppose double blocking is desirable but block size is restricted for practical reasons or the number of treatments is relatively large so that assembly of incomplete blocks of homogeneous subjects could permit reduction of experimental error. Designs (double blocking, small block size) are available that can fulfill both requirements.

For nonfactorial experiments, Youden squares are the double-blocking counterpart of the balanced incomplete block designs of Sec. 6.2.1. Since a Youden square is an incomplete Latin square it is also related to designs discussed in the first sections of this chapter. Partially balanced incomplete Latin squares (Sec. 7.8.2) are counterparts of the partially balanced incomplete block designs (Sec. 6.2.2); balanced, semibalanced, and partially balanced lattice squares (Sec. 7.8.3) are counterparts of the multidimensional balanced and partially balanced lattices (Sec. 6.2.3). Double blocking designs for factorial experiments in incomplete blocks involve double confounding in Latin and quasi-Latin squares (Sec. 7.8.4), which are counterparts of some of the replicated symmetrical designs for factorial experiments in incomplete blocks (Secs. 6.3, 6.4). See Classification of Experimental Designs following Chap. 9.

7.8.1. Youden Squares

The term "Youden square" is a misnomer to a degree because the designs are rectangles. The name stems from extensive work done by Youden (1937, 1940) to adapt certain cyclic Latin squares to experimental needs by deleting some of the columns. The designs also can be constructed by rearranging balanced incomplete block designs that have exactly as many blocks (b) as treatments (t). Since $n = bk = tr$ for such designs, the number of subjects per block (k) equals the number of replications of the treatments (r). The advantage of the Youden square over a Latin square is that the number of replications need not equal the number of treatments. Therefore, an experiment that involves 2 nuisance variables, one of which is restricted to a certain number of classes (not equal to the number of treatments), may fit a Youden square design. For example, consider an animal nutrition experiment with 7 treatments performed at 3 locations (a nuisance variable restricted to 3 classes). The other nuisance variable is initial weight of the animals. An appropriate design, shown in Table 7.12, utilizes columns 1, 2, and 4 of the cyclic standard 7 × 7 Latin square. The same experiment could be performed as a complete block design (locations as blocks) with covariance analysis (x = initial weight).

Table 7.12. Youden Square Design for 7 Treatments

	Location		
Blocks	1	2	3
1	1	2	4
2	2	3	5
3	3	4	6
4	4	5	7
5	5	6	1
6	6	7	2
7	7	1	3

Youden squares possess all the properties of balanced incomplete block (BIB) designs (Sec. 6.2.1) and also balance with respect to columns, each of which is a distinct replication. Therefore, treatment means must be adjusted for block (row) differences as in the BIB designs, but column differences are eliminated automatically from treatment comparisons. See Table 6.1 for the general analysis of variance. The only change is that a sum of squares for c columns,

$$ss_C = (\sum_{j=1}^{c} y_{\cdot j \cdot}^2 / t) - (y_{\cdot\cdot\cdot}^2 / n) \quad (7.26)$$

where $y_{\cdot j \cdot}$ is a column total, also is removed from intrablock error [see (6.16)]. Analysis can be simplified because $t = b$. See Cochran and Cox (1957) for detailed analyses of simple Youden squares and some extensions. They also published several design plans for Youden squares containing from 3 to 11 treatments and others for larger numbers of treatments. Also, see Winer (1962) or Kirk (1968) for numerical examples. DeGray (1970) introduced sequential Youden squares that permit step-wise evaluation of treatment effects and subsequent decisions to continue or terminate the experiment.

7.8.2. Partially Balanced Incomplete Latin Squares

Partially balanced incomplete Latin squares serve the same purpose as Youden squares (two-way elimination of heterogeneity with incomplete blocks for nonfactorial experiments) but are less restrictive because only partial balance of treatments with blocks is achieved. Some are constructed by omitting the last column of a Latin square (rows = incomplete blocks, columns = replications); others are merely two-way layouts of blocks and replications for partially balanced incomplete block designs (Sec. 6.2.2) that have number of replications some low multiple of block size ($r = mk$, $m = 1, 2, 3$) or vice versa ($k = mr$, $m = 1, 2, 3$). An example of a design for 15 treatments with block size of $k = 4$ and $r = 4$ replications (columns 1, 3, 4, 12 of a 15 × 15 cyclic Latin square) is shown in Table 7.13. An index to partially balanced designs with two-way elimination of heterogeneity has been given by Cochran and Cox (1957). Bose et al. (1954) provided design plans and details of the statistical analysis. See Sec. 6.2.2.2 for the general analysis of partially balanced designs with 2 associate classes.

Mandel (1954) developed "generalized chain block designs" that are the double blocking analogue of simple chain block designs (Sec. 6.2.2.4). Treatments must be adjusted for both blocking variables (columns as well as rows). An example, a design for 8 treatments, is shown in Table 7.14. Note that treatments 3 and 4 link the columns 1 and 2, 5 and 6 link columns 2 and 3, 7 and 8 link columns 3 and 4, and 1 and 2 link columns 4 and 1. Similarly, treatments 3 and 7 link rows 1 and 2, 2 and 6 link rows 2 and 3, 4 and 8 link rows 3 and 4, and 1 and 5 link rows 4 and 1. A single repetition of this design provides only 2 df for error, so 2 or more repetitions are needed in most practical situations. Mandel provided designs for $t = 12, 16, 18, 20, 24$, and 30 treatments with instructions for analysis. Also see generalized Youden designs (Kiefer 1958; Ruiz and Seiden 1974).

7.8.3. Lattice Squares

Lattice squares are special cases of balanced and partially balanced two-dimensional lattices that permit two-way elimination of heterogeneity (i.e., 2 blocking variables). As for other lattices, their utility is restricted by the fact that the number of treatments (t) must be the square of the block size (k), which is restricted to prime numbers or powers of primes. Complete balance requires $r = k + 1$ replications to permit each treatment to occur λ times with each of the other treatments in the same row and in the same column. Table 7.15 contains a balanced design for 9 treatments. Note that each square is a replicate. Fewer replicates can be used [$r = (k + 1)/2$], but only semibalance is achieved because each treatment occurs λ times with each of the

Table 7.13. Partially Balanced Incomplete Latin Square for 15 Treatments

Blocks	Replication 1	2	3	4
1	1	3	4	12
2	2	4	5	13
3	3	5	6	14
4	4	6	7	15
5	5	7	8	1
6	6	8	9	2
7	7	9	10	3
8	8	10	11	4
9	9	11	12	5
10	10	12	13	6
11	11	13	14	7
12	12	14	15	8
13	13	15	1	9
14	14	1	2	10
15	15	2	3	11

Table 7.14. Generalized Chain Block Design for 8 Treatments

Rows	Columns 1	2	3	4
1	1	3	5	7
2	3	6	7	2
3	2	4	6	8
4	4	5	8	1

other treatments either in the same row or the same column but not both (see replicates I and III of Table 7.15). For these designs the standard error of the difference between two treatments that occur in the same row is not the same as the standard error for treatments in the same column, although the two standard errors may be quite similar in magnitude. For the larger designs (t

Table 7.15. Balanced Lattice Square Design for 9 Treatments

I	II	III	IV
1 2 3	1 4 7	1 6 8	1 9 5
4 5 6	2 5 8	9 2 4	6 2 7
7 8 9	3 6 9	5 7 3	8 4 3

49), partial balance permits even fewer replicates to be used. In that case third standard error is necessary for differences between treatments that do ot occur together in any row or column. Table 7.16 provides a summary of the

Table 7.16. Index of Lattice Square Designs for t Treatments in Blocks of Size k

		Number of Replicates Required		
t	k	Complete Balance	Semibalance	Partial Balance
9	3	4	2	...
16	4	5	...	...
25	5	6	3	...
49	7	8	4	3
64	8	9	...	3 or 4
81	9	10	5	3 or 4
121	11	12	6	3, 4, or 5
169	13	14	7	3, 4, 5, or 6

tructure of useful designs. See Cochran and Cox (1957) for specific plans. etails of analysis have been given by them and by Kempthorne (1952); a numercal example has been given by Winer (1962).

.8.4. Double Confounding in Latin and Quasi-Latin Squares

The incomplete block designs for factorial experiments in Secs. 6.3 and .4 permit only 1 blocking variable. Factorial experiments in complete Latin quares utilize 2 blocking variables but often are too large to permit homogeeous units in both rows and columns, thus restricting the intended efficiency f double blocking and jeopardizing the unbiasedness of estimates of treatment eans because of potential row-column interaction. Too, for practical reasons ome nuisance variables are restricted to fewer classes than the number of treatments. Double confounding permits reduction in block size for both nuisance variables (rows, columns) but does not eliminate completely the possibility of biased treatments because of row-column interaction. For example, consider a dairy nutrition experiment involving 4 production groups (rows based on prior records or first month of production), 4 stages of lactation, and a 2^3 factorial (8 treatment combinations). The experiment could be performed in 4 replicates of 2 incomplete blocks each (Table 6.16), with production groups as replicates and blocks as 2 stages of lactation (instead of 4); but that would be inefficient in removing variation among cows in different months of production from experimental error. An 8×8 Latin square would not be appropriate because only 4 production groups are involved, although the number of such classes might be extended from 4 to 8. Two double confounding schemes fit the requirements of the experiment. The first involves 2 (or multiples of 2) 4×4 Latin squares for which the 8 treatment combinations are assigned, 4 to each square, using complete confounding of the *ABC* interaction with squares. Treatment combinations 000, 110, 101, and 011 produce zero residues for the vector sum $(x_1 + x_2 + x_3)/2$ and are assigned to the first Latin square. The other 4 combinations (100, 010, 001, and 111) produce residue $R = 1$ and are assigned to the second square (see design I in Table 7.17). The

Table 7.17. Double Confounding for 2^3 Factorial Experiment in Latin or Quasi-Latin Squares

Design I:

Square 1 (R=0 for *ABC*)				Square 2 (R=1 for *ABC*)			
000	011	101	110	010	100	111	001
101	110	000	011	111	001	100	010
011	000	110	101	100	010	001	111
110	101	011	000	001	111	010	100

Design II:

		Square 1				Square 2					
		ABC		*AC*		*AB*		*BC*			
		R=1	R=0	R=1	R=0	R=1	R=0	R=1	R=0		
AB	R=1	001	000	111	110	001	100	111	010	R=1	*ABC*
	R=0	100	101	010	011	000	101	011	110	R=0	
BC	R=1	111	011	000	100	111	010	000	101	R=1	*AC*
	R=0	010	110	101	001	110	011	100	001	R=0	

second suitable design involves partial confounding of the 4 interactions (*AB*, *AC*, *BC*, *ABC*) with rows and columns in each square (see Table 6.16 for assignment scheme). In this design, each treatment combination occurs in both squares but in only half of the rows or columns instead of in each row and each column as in an ordinary Latin square (see design II in Table 7.17). Yates (1937) termed such designs "quasi-Latin squares." In either design the rows and columns should be permuted randomly before experimental units are treated. The analysis of data proceeds as in Table 7.4 for Latin squares except that the sum of squares for treatments is subdivided into its orthogonal factorial components (see analysis of data in Table 6.18).

A number of other factorial arrangements are suited to double confounding. An 8 × 8 quasi-Latin square will accommodate a 2^4 factorial in 4 replicates, with *ABCD* completely confounded and each 3-factor interaction confounded with 2 rows. A square of the same size also can be used for a 2^5 or 2^6 factorial by confounding some or most of the higher order interactions. For 3-level symmetrical factorial experiments, 3^3 and 3^4 plans can be constructed in one or two 9 × 9 quasi-Latin squares (Cochran and Cox 1957) so that main effects and 2-factor interactions are estimable. Kempthorne (1952) also discusses these designs.

EXERCISES

7.1. A pilot experiment was designed to study the effects of dose of releasing hormone TRH (thyroid-stimulating hormone) on peak serum prolactin response in Holstein heifers 6-8 mo of age. Recognized nuisance variables were genetic breeding groups to which the heifers belonged and body weight of the heifers. A Latin square design was used with rows = body weight classes [(1) heavy, (2) medium, (3) light], and columns = breeding groups [(1) bulls with highest producing daughters, (2) random matings to available bulls, (3) "worst" bulls available]. Treatment doses were A = 50 μg, B = 100 μg, C = 150 μg. Transformed data follow.

	Breeding Groups		
Weight	1	2	3
1	57.1(*A*)	94.8(*C*)	89.2(*B*)
2	144.8(*C*)	73.4(*B*)	37.1(*A*)
3	116.2(*B*)	75.2(*A*)	79.5(*C*)

(a) Approximately how much confidence can one have that the dose differences used are sufficient to elicit different true mean response?
(b) Estimate treatment means and their standard errors.
(c) Test for nonadditivity and make suggestions for design of the next experiment.
(d) How many squares should be used in the next experiment to have at least 0.9 power to detect an array of dose effects {-20, 0, +20} at a 99% confidence level?

7.2. A comparative assay of selenium preparations is to be made with respect to their toxicity to rats. Treatments A, B, and C will be 3 doses of a standard preparation; D, E, and F will be 3 doses of a test preparation. Nuisance variables are the litters of rats and the days on which tests will be made, since only 6 animals can be tested each day. To ensure adequate precision for such quantal responses several Latin squares will be needed.
(a) Construct a random 6 × 6 Latin square and derive from it a balanced set of squares.
(b) For an orthogonal contrast, (A, B, C vs. D, E, F), what magnitude of Δ/σ should one expect to detect with 90% power at a 95% significance level?

7.3. An experiment was designed to determine the effects of 3 diets on liver cholesterol in rats (A = control, B = control + vegetable fat, C = control + animal fat). Body weight of the rats and the litters from which they came were used to form a balanced set of Latin squares. Consider litters nested in squares; rows (weight) are not nested.

Weight	Litter 1	2	3
H	1.60(*B*)	1.97(*A*)	2.07(*C*)
M	1.83(*C*)	1.71(*B*)	1.56(*A*)
L	1.44(*A*)	1.84(*C*)	1.72(*B*)
	4	5	6
H	1.71(*A*)	2.02(*C*)	1.85(*B*)
M	1.63(*B*)	1.75(*A*)	2.06(*C*)
L	1.70(*C*)	1.59(*B*)	1.68(*A*)
	7	8	9
H	2.09(*C*)	1.83(*B*)	1.98(*A*)
M	1.63(*A*)	1.91(*C*)	1.83(*B*)
L	1.67(*B*)	1.63(*A*)	2.00(*C*)

(a) Does either experimental diet increase liver cholesterol significantly?
(b) Estimate the efficiency of the Latin square design compared to a completely randomized design. In terms of efficiency, is body weight more important than litter as a nuisance variable?
(c) Is an experiment of this size adequate for 0.9 power to detect treatment effects {-0.04, -0.02, +0.06} at a 95% significance level in a future experiment? How many squares should be used?

7.4. Response to dose (*A*, *B*, *C*, *D*) of an injected treatment in 32 Holstein heifers was studied in a Latin square design with rows = genetic groups from which the heifers came, columns = method of injection [(1) intravenous, (2) intramuscular, (3) subcutaneous I, (4) subcutaneous II].

	Square I				Square II			
	1	2	3	4	1	2	3	4
1:	26.3(*D*)	23.7(*A*)	19.0(*B*)	27.9(*C*)	26.0(*B*)	25.0(*A*)	22.3(*D*)	27.9(*C*)
2:	22.5(*B*)	27.7(*C*)	19.6(*D*)	22.3(*A*)	27.7(*C*)	24.0(*D*)	18.3(*A*)	20.9(*B*)
3:	23.7(*A*)	24.2(*B*)	26.6(*C*)	24.1(*D*)	26.1(*D*)	24.2(*B*)	25.3(*C*)	19.2(*A*)
4:	29.0(*C*)	24.2(*D*)	20.1(*A*)	21.7(*B*)	24.9(*A*)	27.8(*C*)	21.1(*B*)	25.8(*D*)

(a) Assume doses are relatively spaced 0, 1, 5, 10 in order *A*, *B*, *C*, *D*. Test the shape of response to dose via orthogonal polynomial contrasts.
(b) Is mean response significantly different for different methods of injection? Perform contrasts 1, 2 vs. 3, 4; 1 vs. 2; and 3 vs. 4.

(c) With respect to efficiency, would you recommend dropping genetic groups as a blocking variable in future experiments of this type?

7.5. For an experiment to involve 7 treatments, 6 or 7 rats from each of 7 litters may be used. Initial weight of the rats and litter differences are considered to be nuisance variables unlikely to interact with each other or with treatments. An approximately linear relation of initial weight and primary response is expected. For each of the following designs list sources of variation and degrees of freedom, and discuss the relative merits of the designs for this experiment: (1) Latin square, (2) randomized complete blocks with covariance, (3) balanced incomplete blocks of 3 rats, and (4) balanced incomplete blocks of 2 rats.

7.6. The assay of digitalislike principles in 12 cardiac substances was performed by determining the just-fatal dose in cats (Chen et al. 1942). The 12 treatments were A = α-antiarin, B = β-antiarin, C = bufotalin, D = calotoxin, E = calotropin, F = convallatoxin, G = coumingine HCl, H = cymarin, I = emicymarin, J = ouabain (the standard), K = periplocymarin, and L = uscharin. It was not practical for one technician to test more than 4 cats per day, and it was thought that the general level of tolerance or response from technician to technician might change from day to day. A 12 × 12 Latin square design was used with columns = days and with 6 rows A.M. and 6 rows P.M. Among each set of 6 rows, the first two contained data from cats observed on any given day by technician 1, the next two by technician 2, and the last two by technician 3. It was known that heart weight would affect tolerance, so it was included as a covariate. In the table below the first record for each cat is x = (100)(log heart weight in g) and the second is y = (1000)(log dose in μg). Compute the analysis of covariance and use Scheffé's procedure to test all pairs of adjusted treatment means.

	Day												
	1	2	3	4	5	6	7	8	9	10	11	12	Total
							A.M.						
1	*I*	*J*	*B*	*L*	*H*	*G*	*F*	*K*	*D*	*E*	*A*	*C*	
	95	104	89	91	85	92	88	87	83	86	88	90	1078
	525	273	315	557	189	228	54	473	165	254	193	358	3584
	K	*G*	*J*	*H*	*I*	*B*	*L*	*C*	*E*	*F*	*D*	*A*	
	108	97	83	100	100	112	99	86	97	76	98	93	1149
	737	345	195	425	444	350	557	209	335	22	605	237	4461
2	*B*	*L*	*G*	*C*	*D*	*J*	*K*	*E*	*H*	*A*	*F*	*I*	
	89	92	92	103	103	104	99	79	93	93	86	96	1129
	293	427	371	413	515	307	446	301	266	368	198	515	4420
	E	*D*	*F*	*G*	*J*	*K*	*A*	*L*	*C*	*I*	*B*	*H*	
	94	102	100	92	85	100	91	100	91	89	97	81	1122
	299	437	411	400	173	661	250	449	573	316	347	228	4544
3	*C*	*K*	*A*	*B*	*F*	*L*	*I*	*D*	*G*	*H*	*J*	*E*	
	91	93	98	113	106	89	95	86	92	87	85	95	1130
	601	398	400	502	385	443	329	394	444	377	307	247	4827
	F	*H*	*K*	*E*	*G*	*C*	*D*	*B*	*A*	*L*	*I*	*J*	
	73	94	87	95	82	85	98	83	87	81	81	89	1035
	35	355	384	451	378	444	394	211	218	442	473	239	4024

	Day												
	1	2	3	4	5	6	7	8	9	10	11	12	Total
							P.M.						
1	*J*	*C*	*E*	*K*	*A*	*I*	*H*	*F*	*B*	*G*	*L*	*D*	
	92	87	97	86	98	107	89	81	96	98	95	88	1114
	376	540	350	512	501	674	256	126	253	336	373	132	4429
	D	*F*	*I*	*A*	*L*	*E*	*C*	*G*	*J*	*B*	*H*	*K*	
	93	93	91	90	90	105	86	108	98	90	88	96	1128
	313	284	348	326	537	501	523	402	199	270	387	473	4563
2	*A*	*B*	*C*	*D*	*E*	*F*	*G*	*H*	*I*	*J*	*K*	*L*	
	81	85	94	94	100	102	99	88	91	96	86	85	1101
	309	294	446	336	349	283	322	377	477	305	650	580	4728
	H	*E*	*L*	*J*	*C*	*A*	*B*	*I*	*K*	*D*	*G*	*F*	
	97	96	94	97	100	100	86	91	98	82	83	100	1124
	261	419	625	368	426	460	211	348	716	289	402	181	4706
3	*G*	*I*	*D*	*F*	*K*	*H*	*J*	*A*	*L*	*C*	*E*	*B*	
	89	101	94	106	93	93	95	91	93	99	85	76	1115
	363	606	651	360	453	336	185	205	437	632	337	167	4732
	L	*A*	*H*	*I*	*B*	*D*	*E*	*J*	*F*	*K*	*C*	*G*	
	94	104	96	103	100	100	98	93	87	98	88	85	1146
	521	387	326	692	461	369	348	313	139	439	447	398	4840
Total	1096	1148	1115	1170	1142	1189	1123	1073	1106	1075	1060	1074	13371
	4633	4765	4822	5342	4811	5056	3875	3808	4222	4050	4719	3755	53858

Treatment Total

A	*B*	*C*	*D*	*E*	*F*	*G*	*H*	*I*	*J*	*K*	*L*
1114	1116	1100	1121	1127	1098	1109	1091	1140	1121	1131	1103
3854	3674	5612	4600	4191	2478	4389	3783	5747	3240	6342	5948

7.7. An assay of 3 preparations of streptomycin (differing in assumed potencies) against a standard, at 2 doses each, was performed in an 8 × 8 Latin square (modified from Brownlee et al. 1948). Let the first factor be preparations and the second be dose. The 8 treatment combinations (a_1b_1, a_1b_2, a_2b_1, ..., a_4b_2) were labeled *A*, *B*, *C*, ..., *H* for purposes of randomizing the Latin square (a_4 is the standard). A plate containing 64 cavities in 8 rows of 8 was used. Position on the plate was expected to affect response, so the rows and columns of the plate are the rows and columns of the Latin square. Each cavity was filled with agar and inoculated with *Bacillus subtilis*. The response variable was diameter of zone of inhibition of bacterial growth. Data are expressed in units of 0.1 mm - 100.

	Column								
Row	1	2	3	4	5	6	7	8	Total
1	9(*A*)	44(*D*)	11(*G*)	49(*H*)	9(*E*)	46(*F*)	11(*C*)	52(*B*)	231
2	42(*D*)	5(*A*)	40(*F*)	7(*E*)	45(*H*)	12(*G*)	46(*B*)	11(*C*)	208
3	5(*G*)	42(*F*)	2(*A*)	37(*B*)	6(*C*)	45(*D*)	7(*E*)	51(*H*)	195
4	35(*B*)	3(*C*)	37(*H*)	6(*G*)	39(*F*)	5(*E*)	45(*D*)	9(*A*)	179
5	7(*C*)	35(*B*)	0(*E*)	36(*F*)	6(*G*)	40(*H*)	4(*A*)	44(*D*)	172
6	32(*F*)	0(*G*)	35(*D*)	1(*C*)	35(*B*)	1(*A*)	42(*H*)	7(*E*)	153
7	0(*E*)	36(*H*)	2(*C*)	36(*D*)	2(*A*)	39(*B*)	6(*G*)	41(*F*)	162
8	36(*H*)	0(*E*)	35(*B*)	2(*A*)	37(*D*)	5(*C*)	40(*F*)	8(*G*)	163
Total	166	165	162	179	179	193	201	223	1463

Treatment totals:	Preparation				
Dose	1	2	3	4 (std.)	Total
1	34	46	35	54	169
2	314	328	316	336	1294
Total	348	374	351	390	1463

(a) Complete an analysis of variance, including partition of the factorial treatment combinations into preparations, dose, and interaction (parallelism of response to dose).
(b) Estimate the efficiency of the Latin square design relative to a completely randomized design.

7.8. Sixteen cows will be used for a pilot experiment to involve 4 treatments. Nuisance variables that should be accounted for are genetic groups (4), previous production levels of the cows, and number of days "fresh" at initiation of the experiment (stage of lactation).
(a) Construct a Graeco-Latin square design for the experiment and show the form of analysis, including sources of variation and degrees of freedom.
(b) Compare alternative designs: (1) Latin square with previous production as a covariate, (2) Latin square with days "fresh" as a covariate, and (3) randomized complete block design with 2 covariates.

7.9. Suppose the experiment in 7.8 involved only 3 genetic groups but 4 stage-of-lactation groups, and you decide to utilize prior production as a covariate.
(a) Construct a Youden square for an experiment with 12 cows.
(b) Compute the minimum efficiency of the design relative to a complete Latin square.

7.10. Nine drugs are to be clinically tested in 3 cooperating hospitals. Hospital variation and age of patient are to be considered nuisance variables.

(a) If the patients are grouped by ages ≤ 55, 56-70, and > 70, how many patients will be required for a completely balanced lattice square design? How many for semibalance?

(b) What is the minimum efficiency of the balanced design?

7.11. An animal scientist who intends to use pigs as experimental animals knows that 2 nuisance variables, weight and age, are likely to be responsible for a large proportion of variation encountered among responses of the 64 pigs available. The researcher is planning to perform a 2^3 factorial experiment in which the animals must be sacrificed to obtain the data. For each of the following designs list sources of variation and degrees of freedom and discuss the relative merits and possible limitations of the designs for this experiment: (1) double confounding in Latin squares, (2) 8×8 Latin square, (3) randomized complete blocks (blocks based on age and weight jointly), and (4) incomplete blocks (blocks based on age and weight jointly).

CHAPTER 8

Repeated Measurement and Split-Plot Designs

A significant proportion of studies in which the experimental units are animals or humans involve nonrandom repeated measurement of the units. The reasons for performing experiments in this manner are many and varied. First, one may attempt to conserve resources and reduce experimental error by using each subject as a block with several treatments applied sequentially to each subject. Large animals are expensive, and numbers of humans often are limited. Expensive facilities or extensive conditioning and training of subjects may be involved. Changeover designs (Sec. 8.1) minimize the number of subjects and expense involved while maintaining reasonable sensitivity of the experiment. They are particularly effective in reducing experimental error if the major source of random variation is the subjects themselves, and standard blocking or analysis of covariance is not sufficient for reduction of experimental error. Changeover designs remove differences among subjects from residual error, thereby improving the precision of estimates and the power of tests of hypotheses. Each subject acts as its own control in an extension of the before-after pairing discussed in Chap. 1. When such designs are used it is often necessary to remove known nuisance trends over time or space. For example, the normal decline in milk production of dairy cows over the months of single lactation must be accounted for if the cows are exposed to a sequence of experimental diets or other treatments in weekly or monthly treatment periods.

Second, repeated measurement is necessary if the residual effects of treatments in a time sequence are likely to be important either as a confounding nuisance (e.g., lingering effect of a drug) or as effects of major interest (e.g., learning and behavior). Some of the changeover designs are balanced in such a way that residual effects may be separated from direct effects of treatments.

Third, many experiments involve measurement of interesting trends in individual responses to treatment (Sec. 8.2). For example, complex physiological responses to injections of hormones or hormonal-releasing factors may be studied by successively sampling the blood for several minutes, hours, or even days following treatment. In a few cases space instead of time is the repeated factor, as in study of responses in different sections of the gut. Most experiments are performed in block or split-plot designs, where each subject makes a complete or incomplete block. Some are performed within the context of a changeover design, with macro trend being nuisance and micro trend being of major interest.

8.1. CHANGEOVER DESIGNS: SEQUENCES OF TREATMENTS

These designs share the attribute of sequential application of 2 or more treatments to each unit. Complete block designs (subject = block) have (1) randomized sequence with no nuisance trend anticipated or (2) balanced sequence with nuisance trend likely to be a significant factor but uniform for all subjects. All other designs assume the necessity of removing (accounting for) nuisance trend, not necessarily uniform for all subjects, in time or space.

8.1.1. Removal of Nuisance Trend

When a sequence of observations on a subject is expected to show trend even under standardized conditions and the trend itself is well known, it is merely a nuisance to be accounted for. The decline in milk production of cows over successive weeks or months of lactation is a classic example. The other major trend problem relates to residual effects of treatments. These may be physical, as in radioactive decay; physiological, as in the action of nutrients or hormones slowly metabolized; or psychological, as in learning and behavior studies. In some cases the residual effects are of major interest; but if they are merely a nuisance, the problem often may be avoided by spacing treatments sufficiently to effectively dissipate the lingering influence. In other studies, such as those involving learning, the time required for dissipation is much too long for practical purposes, so designs must be selected that permit separation of the direct and residual effects of treatments. Choice of proper design, then, depends on expectation of the existence of nuisance trend, severity of trend, uniformity or diversity of trend for subjects involved, and anticipation of the nature of residual effects of treatments.

8.1.2. Simple (complete block) Crossover Design

The simplest case of crossover design involves random assignment of treatments in sequence to each subject (block). It should be used only when one is confident a priori that no nuisance trend is likely to be involved, i.e., a subject's responses over time or space would be expected to be relatively uniform or only randomly dispersed if no changes in treatment were made. Also, if the repeated factor is time, one should be willing to assume that the applications of treatments will be spaced sufficiently that the effects of one treatment will not carry over into subsequent treatment periods. As an example of time as the repeated factor, consider a study of the effect of vasopressor drugs on arterial pressure in dogs. The drugs may be given to each dog in random order, with the pressure being allowed to return to normal after each treatment. Space as a repeated factor may be illustrated by an experiment to compare the effects of different methods of treatment of plasma on clotting time of plasma. For each subject the treatments may be assigned randomly to an equal number of samples of plasma from the subject. In some cases the treatments may represent factorial combinations, as when A = method of injection, B = dose of drug given.

Random order experiments with repeated measurement may be described by the standard model for randomized complete block experiments [see (5.1)],

$$Y_{ij} = \mu + \tau_i + D_j + (\tau D)_{ij} + E_{(ij)} \tag{8.1}$$

if one ignores the restricted randomization effect. The τ_i still represent treatment effects. The main distinction is that the δ_j of (5.1) represent fixed effects of blocks of homogeneous subjects in the standard design of

Chap. 5, but the D_j of (8.1) represent random effects of individual subjects in experiments with repeated measurement. Likewise, the $E_{(ij)}$ no longer represent variation *among* homogeneous subjects treated alike but measure variation *within* subjects treated alike, i.e., as if each subject is measured repeatedly over time or space without changing treatments. The assumption that block-treatment interaction $(\tau D)_{ij}$ is zero or trivial is still important, for substantial interaction will badly inflate the estimate of experimental error variance with consequent decrease in precision of estimated treatment means. However, if the effects of subjects are random, interaction does not bias the test of treatment effects (for a parallel case see Chap. 2 with mixed model). An additional assumption (ignored as unimportant or unnecessary in the standard block experiments of Chap. 5) is that the correlations between responses in any two periods (treatments) are uniform, i.e., responses in two adjacent periods are no more (or no less) correlated than responses in first and last (or any other) periods. This assumption may be perfectly valid for most experiments in random order designs but is less likely to be correct for designs in which trends must be accounted for.

Consider a random order experiment with 4 trained subjects (Scott and Chen 1944). Each subject was given a 200 mg oral dose of each treatment: (1) placebo, (2) 1-ethyl theobromine, and (3) caffeine, in random order. Trained response was rate of finger tapping 2 hours after each treatment was given. The data in Table 8.1 are rounded averages of 3 trials by each subject on each treatment. Significant increases in tapping rate were obtained by use of ei-

Table 8.1. Analysis of Complete Block Crossover Design with Treatments (stimulants) Administered to Subjects in Random Order

	Subject Number					
Treatment	1	2	3	4	Total	Mean
1	451	496	455	446	1848	462
2	466	523	474	453	1916	479
3	460	511	481	472	1924	481
Total	1377	1530	1410	1371	5688	

Source of Variation	df	*ss*	*ms*	*f*	$f_{0.05,2,6}$
Treatments	2	872	436	7.93	5.14
Blocks	3	5478	1826		
Error	6	332	55		

Dunnett's tests against placebo (one-sided): (2)t_D=3.24 (P<0.05); (3)t_D=3.63 (P<0.01).

ther stimulant. The efficiency of this design (RCBD) relative to a completely randomized design (CRD), in which 4 different subjects would be assigned to each treatment, may be of interest (Sec. 5.6). For the block design, $\hat{\sigma}^2_{\text{RCBD}} = ms_E = 55$ and $\nu_{\text{RCBD}} = \nu_T + \nu_E = 8$. For the hypothetical CRD,

$$\hat{\sigma}^2_{\text{CRD}} = [(3)(1826) + (8)(55)]/11 = 538$$

and $\nu_{\text{CRD}} = n - 1 = 11$. The efficiency of blocking is

$$E_{\text{RCBD:CRD}} = (9/11)(14/12)(538/55)100\% = 934\%$$

A completely randomized design would require approximately (9.34)(4) = 37 different subjects assigned to each treatment group to provide the same precision as the changeover design with only 4 subjects in the entire experiment. In general, estimation of number of subjects required to obtain given power for the changeover design proceeds as indicated in Sec. 5.5 except that the relevant estimate of σ pertains to prior information from a similar changeover experiment or to information on the variability of a subject from time to time under uniform conditions.

Experiments in which nuisance trends in time (or space) are anticipated require designs in which the treatment order is balanced rather than random. However, the simple complete block design should be selected over more complex designs only if no residual effects of treatments are expected and it is likely that all subjects would have a similar trend if repeated measurements were taken but no treatments were imposed. For example, in an experiment on milk production of cows fed different diets, it may be reasonable to assume uniform trend if all the cows are genetically similar, are mature, and produce similar amounts of milk under standard conditions. Because of increased likelihood of diverse trends over longer periods of time, the simple design is rarely used in experiments with more than 4 treatments. For 2 treatments, *A* and *B*, the only treatment sequences possible are *AB* and *BA*. For 3 treatments, two sets of $t = 3$ balanced cyclic sequences are possible, {*ABC*, *BCA*, *CAB*} and {*ACB*, *CBA*, *BAC*}, but normally only one set is used. For 4 treatments, six sets of $t = 4$ balanced cyclic sequences are possible, e.g., {*ABCD*, *BCDA*, *CDAB*, *DABC*}. Note that within any set a particular treatment always follows the same treatment in the sequence. Therefore it is imperative that these designs not be used if the treatments are likely to have residual effects or interfere with each other in any way. Randomization proceeds by randomly allocating r subjects equally to the t treatment sequences in the set of sequences selected for the experiment. Treatments are balanced with respect to period effects because each treatment is applied the same number of times in each period.

In contrast to random order designs, balanced simple crossover designs cannot be adequately described by the standard RCBD model [(8.1)] because the nuisance trend (period effects) must be accounted for. An appropriate model is

$$Y_{ijk} = \mu + D_i + \rho_j + \tau_k + E_{(ijk)}$$

$$(i = 1, 2, \ldots, r;\ j = 1, 2, \ldots, t;\ k = 1, 2, \ldots, t;\ n = tr) \quad (8.2)$$

Interactions among effects of subjects (D_i), periods (ρ_j), and treatments (τ_k) are assumed to be zero or trivial. For simplicity, interactions and randomization error are not listed in the model. The error term $E_{(ijk)}$ reflects random variation of repeated information within subjects as if treatments were not imposed and the nuisance trend (period effects) did not exist. Sums of squares for subjects (D = blocks), periods (P), and treatments (T) are similar computationally to those for Latin square designs (Table 7.4):

$$ss_D = \left(\sum_{i=1}^{r} y_{i..}^2/t\right) - (y_{...}^2/n) \tag{8.3}$$

$$ss_P = \left(\sum_{j=1}^{t} y_{.j.}^2/r\right) - (y_{...}^2/n) \tag{8.4}$$

$$ss_T = \left(\sum_{k=1}^{t} y_{..k}^2/r\right) - (y_{...}^2/n) \tag{8.5}$$

The subscript indices are understood to be redundant so that a block total is $y_{i..} = \sum_{j=1}^{t} y_{ijk}$, a period total is $y_{.j.} = \sum_{i=1}^{r} y_{ijk}$, the grand total is $y_{...} = \sum_{i=1}^{r} \sum_{j=1}^{t} y_{ijk}$, and a treatment total is

$$y_{..k} = \sum_{i=1}^{r} \sum_{j=1}^{t} \delta_{ij}^{k} y_{ijk} \tag{8.6}$$

where $\delta_{ij}^{k} = 1$ if the kth treatment was given to the ith subject in the jth period or $\delta_{ij}^{k} = 0$ otherwise. Error sum of squares ss_E may be obtained as residual from the total sum of squares ss_y:

$$ss_y = \sum_{i=1}^{r} \sum_{j=1}^{t} y_{ijk}^2 - (y_{...}^2/n) \tag{8.7}$$

$$ss_E = ss_y - ss_D - ss_P - ss_T \tag{8.8}$$

The analysis of variance is summarized in Table 8.2.

As an example of a simple crossover experiment with nuisance trend, consider a study on the effect of thyrotropin-releasing hormone (TRH) on milk production of Holstein cows (kg/day). Let A = 50 μg TRH injected twice daily for 8 days and B = 500 μL saline injected twice daily for 8 days. Twenty cows received sequence AB; another 20 received BA. In each case a rest of two days between treatments was specified to dissipate any short-term residual effects

Table 8.2. Analysis of Variance for Complete Block Changeover Experiment with Nuisance Trend (period effects)

Source of Variation	df	*ss*	*ms*	*E*[*MS*]
Subjects	$r-1$	ss_D	ms_D	$\sigma^2+t\sigma_D^2$
Periods	$t-1$	ss_P	ms_P	$\sigma^2+r\sum_{j=1}^{t}\rho_j^2/(t-1)$
Treatments	$t-1$	ss_T	ms_T	$\sigma^2+r\sum_{k=1}^{t}\tau_k^2/(t-1)$
Error	$(r-2)(t-1)$	ss_E	ms_E	σ^2

of treatments. Since the periods were quite short, relatively uniform trend in production was expected. The data are shown in Table 8.3 and the analysis of variance is given in Table 8.4.

Treatment means and standard errors are 20.8 ± 0.38 for TRH and 20.2 ± 0.38 for saline. Evidence is not strong ($f = 1.08 < f_{0.25,1,38} = 1.36$) for a significant effect of TRH. However it is clear that the crossover design was effective in reducing experimental error by removing variation among cows, and it appears that the postulated nuisance trend (period effect) was substantial. Computation of the efficiency of the design (CROSS) relative to a hypothetical completely randomized design (CRD) requires $\hat{\sigma}^2_{\text{CROSS}} = 5.8959$ and, ignoring period effects,

$$\hat{\sigma}^2_{\text{CRD}} = [1903.9255 + 40(5.8959)]/79 = 27.0856$$

Therefore, $E_{\text{CROSS:CRD}} = (27.0856/5.8959)100\% = 459\%$. That is, attainment of equivalent sensitivity by randomly assigning cows to treatments without repeated measurement would require $(4.59)(20) \simeq 92$ cows per treatment. Within the context of changeover designs, one may be interested in the efficiency of the crossover design with period effects accounted for relative to a random order design that would ignore trend:

$$\hat{\sigma}^2_1 = 5.8959 \quad \text{(balanced sequence)}$$

$$\hat{\sigma}^2_2 = [20.4020 + 39(5.8959)]/40 = 6.2586 \quad \text{(random order)}$$

$$E_{1:2} = (6.2586/5.8959)100\% = 106\%$$

The gain in efficiency is small, but the only costs were extra care in treating animals in proper order, a bit of extra computation, and 1 df.

Table 8.3. Crossover Experiment on Effect of Thyrotropin-Releasing Hormone (A=TRH, B=saline) on Milk Production (kg/day)

Cow	1(A)	2(B)	Total	Cow	1(B)	2(A)	Total
1	25.1	19.7	44.8	21	21.1	21.2	42.3
2	18.6	16.2	34.8	22	25.7	26.3	52.0
3	29.1	26.6	55.7	23	17.7	14.2	31.9
4	14.2	10.8	25.0	24	30.4	30.2	60.6
5	30.6	31.2	61.8	25	16.7	16.9	33.6
6	18.5	12.5	31.0	26	19.2	17.7	36.9
7	28.3	32.7	61.0	27	27.4	24.8	52.2
8	26.8	19.7	46.5	28	20.3	22.7	43.0
9	23.8	19.2	43.0	29	16.2	15.0	31.2
10	25.7	20.1	45.8	30	26.2	26.8	53.0
11	20.5	25.2	45.7	31	12.8	13.8	26.6
12	18.4	24.4	42.8	32	21.1	22.2	43.3
13	20.4	19.6	40.0	33	20.2	19.9	40.1
14	23.7	15.2	38.9	34	17.2	16.4	33.6
15	16.8	23.9	40.7	35	18.7	14.7	33.4
16	18.9	12.5	31.4	36	16.8	18.8	35.6
17	13.2	13.2	26.4	37	26.9	24.6	51.5
18	19.8	20.5	40.3	38	12.6	13.0	25.6
19	17.7	16.0	33.7	39	24.0	25.3	49.3
20	15.9	15.3	31.2	40	22.7	20.5	43.2
Total	426.0	394.5			413.9	405.0	

Period 1, 839.9; period 2, 799.5;

A, 831.0; B, 808.4; $y_{...}$=1639.4

Table 8.4. Analysis of Variance for Effect of TRH on Milk Production (see Table 8.3)

Source of Variation	df	*ss*	*ms*	*f*
Cows	39	1903.9255	48.8186	7.28
Periods	1	20.4020	20.4020	3.46
Treatments	1	6.3845	6.3845	1.08
Error	38	224.0435	5.8959	

8.1.3. Latin Square Crossover Design

The Latin square design is intended for use in experiments for which a nuisance trend is anticipated, as for the designs with balanced sequences of treatments in the previous section. The Latin squares have the advantage of permitting more complete removal of nuisance trend when the trend is unlikely to be uniform for all subjects. For example, dairy cows of different breeds or different genetic producing ability are likely to have different rates of decline in milk production. Animal-period interaction can be minimized by using a set of Latin squares with each square made up of animals expected to have similar trends. Avoid the interaction if possible because it is completely confounded with treatment effects (Sec. 6.4). Some experiments are performed in a single Latin square, which differs from the simple crossover only in the fact that the number of subjects must be exactly equal to the number of treatments.

Randomization may be accomplished by selecting a random square or set of squares (Sec. 7.1); however in crossover designs the rows of the squares (subjects) can be assigned randomly to the various sequences of treatments. If no residual effects of treatments are expected, random assignment of subjects to cyclic sequences within each square should provide adequate insurance against bias.

An appropriate model for a single square is essentially the same as the model for a simple crossover [(8.2)],

$$Y_{ijk} = \mu + D_i + \rho_j + \tau_k + E_{(ijk)} \qquad (i,\ j,\ k = 1,\ 2,\ \ldots,\ t;\ n = t^2) \tag{8.9}$$

the only distinction being that there are exactly t subjects (D_i) instead of the more flexible r for an experiment involving sequences of t treatments (τ_k) across t periods (ρ_j). By changing r to t in (8.3)-(8.8) and Table 8.2, one may adapt the simple crossover analysis to the analysis for a single Latin square.

When two or more Latin squares are used for a crossover experiment the model is very similar to the one used for noncrossover experiments in Latin squares [(7.2)]:

$$Y_{ijkl} = \mu + \theta_i + D_{(i)j} + \rho_{(i)k} + \tau_l + E_{(ijkl)}$$
$$(i = 1,\ 2,\ \ldots,\ m;\ j,\ k,\ l = 1,\ 2,\ \ldots,\ t;\ n = mt^2) \tag{8.10}$$

Here the effects of squares are denoted θ_i as before but the row effects are those of random subjects nested within squares ($D_{(i)j}$), and the columns represent the fixed effects of periods nested within squares ($\rho_{(i)k}$). The nesting of periods is necessary to accomplish more complete elimination of differences in average nuisance trend from square to square despite the fact that precisely the same periods may be represented in each square. The nesting also allows the use of subjects that are initially at different points in the trend (e.g., stages of lactation of cows) by grouping those at similar stage in the same square and permits different starting times for different squares if necessary. As for noncrossover designs, it is convenient to denote a treatment mean by

$$\bar{y}_{...l} = \sum_{i=1}^{m} \left(\sum_{j=1}^{t} \sum_{k=1}^{t} \delta^{l}_{jk} y_{ijkl} \right) / (mt) \tag{8.11}$$

where $\delta^{l}_{jk} = 1$ if the lth treatment was given to the jth subject in the kth period or $\delta^{l}_{jk} = 0$ otherwise. Totals for square, subject (within square), period (within square), and all data may be designated $y_{i...}$, $y_{ij..}$, $y_{i.k.}$, and $y_{...}$, respectively, where it is understood that the treatment subscript l is ignored in the summing process because of the redundancy of indexing that occurs in Latin square models. Note that the effects of interaction between squares and treatments may be separated from residual error if necessary. Winer (1962, pp. 538-77) has discussed many other variations of this basic model for crossover experiments.

Computational formulas for sums of squares corresponding to the effects in the model (8.10) are listed in Table 7.4 for m squares. The analysis of variance is summarized in Table 8.5. Johnson and Grizzle (1971) have discussed the analysis of ordinal (ranked) data in crossover experiments, and Koch (1972) developed a nonparametric analysis for 2-period changeover experiments in which an assumption of normal distribution of errors is likely to be invalid.

Table 8.5. Analysis of Variance for Latin Square Changeover Experiment with Nuisance Trend (period effects) and m Squares

Source of Variation	df	*ss*	*ms*	$E[MS]$
Squares	$m-1$	ss_S	ms_S	$\sigma^2 + t^2\Sigma\theta_i^2/(m-1)$
Subjects/squares (rows)	$m(t-1)$	$ss_{D/S}$	ms_D	$\sigma^2 + t\sigma_D^2$
Periods/squares (cols.)	$m(t-1)$	$ss_{P/S}$	ms_P	$\sigma^2 + t\Sigma\rho_{(i)k}^2/[m(t-1)]$
Treatments	$t-1$	ss_T	ms_T	$\sigma^2 + mt\Sigma\tau_l^2/(t-1)$
Error	$(t-1)(mt-m-1)$	ss_E	ms_E	σ^2

As an example of a crossover experiment in Latin squares, consider a study of rate of gain of lambs fed 8 experimental diets (A, B, C, ..., H). The 16 lambs available were divided into 2 relatively homogeneous groups (squares) by weight and age. Each lamb was fed a particular diet for a week, then was given another diet for another week, etc., until 8 weeks had elapsed. The natural trend in rate of growth was accounted for by period effects. Gains (kg/day) are shown in Table 8.6.

The analysis of variance is presented in Table 8.7. The mean square for diets is very large compared with error, indicating high significance of differences among diets. Tukey's test of all pairs of means (Chap. 2) requires

Table 8.6. Latin Square Crossover Experiment on Gains of Lambs Fed 8 Experimental Diets

	Period								
Lamb	1	2	3	4	5	6	7	8	Total
1	0.394(*G*)	0.337(*C*)	0.350(*H*)	0.368(*E*)	0.309(*B*)	0.388(*A*)	0.310(*F*)	0.347(*D*)	2.803
2	0.201(*F*)	0.255(*B*)	0.374(*G*)	0.255(*D*)	0.333(*A*)	0.333(*H*)	0.383(*E*)	0.348(*C*)	2.482
3	0.253(*B*)	0.223(*F*)	0.311(*C*)	0.350(*H*)	0.392(*E*)	0.292(*D*)	0.363(*A*)	0.446(*G*)	2.640
4	0.350(*C*)	0.426(*G*)	0.306(*D*)	0.396(*A*)	0.306(*F*)	0.435(*E*)	0.369(*B*)	0.451(*H*)	3.039
5	0.362(*E*)	0.334(*A*)	0.274(*F*)	0.363(*C*)	0.408(*H*)	0.485(*G*)	0.357(*D*)	0.379(*B*)	2.962
6	0.366(*H*)	0.296(*D*)	0.383(*A*)	0.268(*F*)	0.382(*C*)	0.359(*B*)	0.460(*G*)	0.450(*E*)	2.964
7	0.345(*A*)	0.370(*E*)	0.311(*B*)	0.444(*G*)	0.341(*D*)	0.401(*C*)	0.416(*H*)	0.332(*F*)	2.960
8	0.232(*D*)	0.339(*H*)	0.353(*E*)	0.295(*B*)	0.431(*G*)	0.291(*F*)	0.379(*C*)	0.377(*A*)	2.697
Total	2.503	2.580	2.662	2.739	2.902	2.984	3.037	3.140	22.547
9	0.181(*F*)	0.281(*C*)	0.299(*B*)	0.297(*A*)	0.367(*H*)	0.424(*G*)	0.393(*E*)	0.311(*D*)	2.483
10	0.215(*B*)	0.336(*G*)	0.214(*F*)	0.328(*E*)	0.279(*D*)	0.358(*C*)	0.359(*A*)	0.393(*H*)	2.482
11	0.262(*E*)	0.257(*B*)	0.322(*A*)	0.342(*H*)	0.404(*G*)	0.293(*F*)	0.318(*D*)	0.381(*C*)	2.579
12	0.271(*A*)	0.163(*F*)	0.302(*E*)	0.241(*D*)	0.309(*C*)	0.281(*B*)	0.349(*H*)	0.424(*G*)	2.340
13	0.277(*H*)	0.271(*E*)	0.169(*D*)	0.271(*C*)	0.299(*B*)	0.313(*A*)	0.384(*G*)	0.250(*F*)	2.164
14	0.294(*G*)	0.200(*D*)	0.249(*C*)	0.219(*B*)	0.278(*A*)	0.309(*H*)	0.270(*F*)	0.344(*E*)	2.163
15	0.240(*C*)	0.268(*H*)	0.327(*G*)	0.221(*F*)	0.331(*E*)	0.246(*D*)	0.254(*B*)	0.348(*A*)	2.235
16	0.183(*D*)	0.266(*A*)	0.307(*H*)	0.361(*G*)	0.247(*F*)	0.379(*E*)	0.348(*C*)	0.308(*B*)	2.399
Total	1.923	2.042	2.119	2.280	2.444	2.603	2.675	2.759	18.845
Tr. Total	5.373(*A*)	4.522(*B*)	5.308(*C*)	4.373(*D*)	5.723(*E*)	4.044(*F*)	6.424(*G*)	5.625(*H*)	41.392
Mean	0.336	0.283	0.332	0.273	0.358	0.253	0.402	0.352	

Table 8.7. Analysis of Gains of Lambs Fed 8 Experimental Diets (see Table 8.6)

Source of Variation	df	*ss*	*ms*
Squares	1	0.107069	0.107069
Lambs/squares	14	.055387	.003956
Periods/squares	14	.131706	.009408
Diets	7	.279279	.039897
Error	91	0.016083	0.000177

computation of the standard error of a mean, $\sqrt{(ms_E/mt)} = \sqrt{0.000177/[(2)(8)]} = 0.0033$, and critical value $q_{\alpha,8,91}$ (App. A.8). For $\alpha = 0.05$, the critical value 4.402 may be multiplied by the standard error to provide the minimum significant difference: (0.0033)(4.402) = 0.0145. All comparisons are significant except *A* versus *C*, *B* versus *D*, and *E* versus *H*.

Estimates of random variation are $\hat{\sigma}^2 = 0.000177$ (error within lambs) and $\hat{\sigma}^2_D = (0.003956 - 0.000177)/8 = 0.000472$ for variation among lambs within blocks based on weight and age. The efficiency of the crossover design relative to an RCBD design with blocks based on weight and age and only one feeding period requires

$$\hat{\sigma}^2_{\text{RCBD}} = [14(0.003956) + 112(0.000177)]/126 = 0.000597$$

to give $E_{\text{CROSS:RCBD}} = (0.000597/0.000177)100\% = 337\%$. If the lambs were not blocked but were assigned completely at random, then

$$\hat{\sigma}^2_{\text{CRD}} = [(1)(0.107069) + 14(0.003956) + 112(0.000177)]/127 = 0.001435$$

$$E_{\text{CROSS:CRD}} = (0.001435/0.000177)100\% = 811\%$$

That is, to obtain precision equal to the crossover design with 16 lambs, approximately (3.37)(16) = 54 lambs are required in an ordinary block design, or (8.11)(16) = 130 lambs in a completely randomized design.

8.1.4. Estimation of Residual Effects of Treatments

Experiments in which a sequence of treatments is applied to each subject may require analysis of residual effects of treatments if (1) a rest period between treatments of sufficient length to dissipate residual effects is not practical or (2) the residual effects are of interest per se (e.g., learning). In the first case residual effects usually are estimated primarily to make adjustments to obtain unbiased estimates of the direct effects of treatments (effects during the period of application). In some studies it is the permanent effect (cumulative response per period) that is of interest. For example, if a dairy cow is given a sequence of 3 treatments during 3 periods of one month each, one may be interested in the cumulative effect that would result if a particular treatment were applied over all periods. The permanent effect is approximated by adding estimates for direct and residual effects. Designs for experiments with major emphasis on residual or permanent effects are discussed in Sec. 8.1.6.

Certain of the Latin square crossover designs may be used to adjust for residual effects when the direct effects of treatments are of major interest; then each treatment is preceded by each of the other treatments an equal number of times. Ordinarily it is reasonable to assume that the residual effects will have dissipated or become trivial by the second period after application of a treatment. For example, let the direct effects of treatments A, B, C, etc., be designated τ_A, τ_B, τ_C, etc., and the residual effects be designated τ'_A, τ'_B, τ'_C, etc. If an animal receives treatments in the order B, A, C, the treatment effects on response in the 3 periods are τ_B, $(\tau_A + \tau'_B)$, and $(\tau_C + \tau'_A)$, respectively. When the number of treatments is even, a single square that is balanced for residual effects may be found. However, more than one square may be required to obtain adequate sensitivity. When the number of treatments is odd, an even number of squares is required for balance. Grizzle (1965) has discussed conditions under which crossover designs for only 2 treatments may compare unfavorably with completely randomized designs. Succinctly, if treatment-subject interaction is likely or if one expects rather unequal residual effects for the 2 treatments, the crossover design should not be used. Designs balanced for first residuals of $t = 2, 3, 4, 5$, or 6 treatments are shown in Table 8.8, with rows = subjects and columns = periods. Berenblut and Webb (1974) have shown that such designs are optimal for minimization of the generalized variance of estimates of parameters in the presence of first order autocorrelated errors that exist routinely in crossover experiments.

The model for an analysis including residual effects in m squares is the same as (8.10) plus a term $\tau'_{l'}$ for the residual effect of the treatment l' given in the previous period. However, the separation of direct and residual effects requires additional computations that are not difficult to perform but difficult to symbolize. The form of analysis was developed by Williams (1949, 1950). As in the case without residuals, totals for square, subject within square, period within square, and all data may be designated $y_{i...}$, $y_{ij..}$, $y_{i.k.}$, and $y_{....}$, respectively, where it is understood that the treatment subscript l is ignored in the summing process because of redundancy of indexing. The total for the first period across all squares $y_{..1.}$ also will be needed. Other totals will be required for each treatment. As before, an ordinary treatment total of mt observations may be designated

$$y_{...l} = \sum_{i=1}^{m} \left(\sum_{j=1}^{t} \sum_{k=1}^{t} \delta^{l}_{jk} y_{ijkl} \right) \quad (8.12)$$

where $\delta^{l}_{jk} = 1$ if the lth treatment was given to the jth subject in the kth period or $\delta^{l}_{jk} = 0$ otherwise. Also, a "residual" total ($y^{r}_{...l}$) and a "final" to

Table 8.8. Crossover Designs Balanced for Residual Effects of t=2, 3, 4, 5, or 6 Treatments (rows=subjects)

Periods	
1	2
A	*B*
B	*A*

Periods			
1	2	3	4
A	*B*	*C*	*D*
B	*D*	*A*	*C*
C	*A*	*D*	*B*
D	*C*	*B*	*A*

Periods					
1	2	3	4	5	6
A	*C*	*B*	*E*	*F*	*D*
B	*D*	*C*	*F*	*A*	*E*
C	*E*	*D*	*A*	*B*	*F*
D	*F*	*E*	*B*	*C*	*A*
E	*A*	*F*	*C*	*D*	*B*
F	*B*	*A*	*D*	*E*	*C*

Periods					
1	2	3	1	2	3
A	*B*	*C*	*A*	*C*	*B*
B	*C*	*A*	*C*	*B*	*A*
C	*A*	*B*	*B*	*A*	*C*

Periods									
1	2	3	4	5	1	2	3	4	5
A	*B*	*D*	*E*	*C*	*A*	*C*	*B*	*E*	*D*
B	*C*	*E*	*A*	*D*	*B*	*D*	*C*	*A*	*E*
C	*D*	*A*	*B*	*E*	*C*	*E*	*D*	*B*	*A*
D	*E*	*B*	*C*	*A*	*D*	*A*	*E*	*C*	*B*
E	*A*	*C*	*D*	*B*	*E*	*B*	*A*	*D*	*C*

al ($y^f_{...l}$) must be computed for each treatment. The residual total is

$$y^r_{...l} = \sum_{i=1}^{m} \left(\sum_{j=1}^{t} \sum_{k=2}^{t} \delta^l_{j,k-1} y_{ijkl} \right) \tag{8.13}$$

here $\delta^l_{j,k-1}$ = 1 if the lth treatment was given to the jth subject in the preeding period (k - 1) or $\delta^l_{j,k-1}$ = 0 otherwise. It is the total of $m(t - 1)$ bservations (t - 1 in each of m squares); i.e., it includes observations in a eriod immediately following application of the treatment in question; obserations in the first period are never included. The final total for each reatment is

$$y^f_{...l} = \sum_{i=1}^{m} \left(\sum_{k=1}^{t} \delta^l_{jt} y_{ijkl} \right) \tag{8.14}$$

here δ^l_{jt} = 1 if the lth treatment was applied to jth subject in the final peiod (period t) or δ^l_{jt} = 0 otherwise. It is the total of mt observations and ncludes all observations for one row (subject) in each square. To clarify

Table 8.9. Computation of Residual and Final Totals for Treatments in 2 Squares for 3 x 3 Crossover Experiment

Subjs. /Sq.	Square (i=1) Periods/Square k=1	k=2	k=3	Subjs. /Sq.	Square (i=2) Periods/Square k=1	k=2	k=3
j=1	A $y_{1111}=9$	B $y_{1122}=6$	C $y_{1133}=6$	j=1	A $y_{2111}=7$	C $y_{2123}=5$	B $y_{2132}=$
j=2	B $y_{1212}=7$	C $y_{1223}=6$	A $y_{1231}=7$	j=2	C $y_{2213}=6$	B $y_{2222}=4$	A $y_{2231}=$
j=3	C $y_{1313}=8$	A $y_{1321}=8$	B $y_{1332}=4$	j=3	B $y_{2312}=5$	A $y_{2321}=6$	C $y_{2333}=$

Residual total: (A) $y^r_{...A}=y_{1122}+y_{1332}+y_{2123}+y_{2333}=6+4+5+5=20$

(B) $y^r_{...B}=y_{1133}+y_{1223}+y_{2231}+y_{2321}=6+6+5+6=23$

(C) $y^r_{...C}=y_{1231}+y_{1321}+y_{2132}+y_{2222}=7+8+2+4=21$

Final total: (A) $y^f_{...A}=y_{1212}+y_{1223}+y_{1231}+y_{2213}+y_{2222}+y_{2231}=7+6+7+6+4+5=35$

(B) $y^f_{...B}=y_{1313}+y_{1321}+y_{1332}+y_{2111}+y_{2123}+y_{2132}=8+8+4+7+5+2=34$

(C) $y^f_{...C}=y_{1111}+y_{1122}+y_{1133}+y_{2312}+y_{2321}+y_{2333}=9+6+6+5+6+5=37$

Treatments: $y_{...A}=42$, $y_{...B}=28$, $y_{...C}=36$;

Period 1: $y_{..1.}=9+7+8+7+6+5=42$; Overall: $y_{....}=106$

(8.13) and (8.14), observations to be summed are indicated in Table 8.9 for a example with 3 treatments in 2 squares.

To see which parameters are estimated by the various totals, first consider the residual total for treatment A:

$$y^r_{...A} = 4\hat{\mu} + 2\hat{\theta}_1 + 2\hat{\theta}_2 + \hat{D}_{(1)1} + \hat{D}_{(1)3} + \hat{D}_{(2)1} + \hat{D}_{(2)3} + \hat{\rho}_{(1)2} + \hat{\rho}_{(1)3} + \hat{\rho}_{(2)2} + \hat{\rho}_{(2)3} + 2\hat{\tau}_B + 2\hat{\tau}_C + 4\hat{\tau}'_A$$

where $\hat{\tau}'_A$ is the estimated residual effect of A. Similarly, the final total for A is

$$y^f_{...A} = 6\hat{\mu} + 3\hat{\theta}_1 + 3\hat{\theta}_2 + 3\hat{D}_{(1)2} + 3\hat{D}_{(2)2} + \sum_{ik}\sum\hat{\rho}_{(i)k} + 2(\hat{\tau}_A + \hat{\tau}_B + \hat{\tau}_C) + 2(\hat{\tau}'_B + \hat{\tau}'_C)$$

The ordinary treatment total for A is

$$y_{...A} = 6\hat{\mu} + 3\hat{\theta}_1 + 3\hat{\theta}_2 + \sum_{ij}\sum \hat{D}_{(i)j} + \sum_{ik}\sum \hat{\rho}_{(i)k} + 6\hat{\tau}_A + 2(\hat{\tau}'_B + \hat{\tau}'_C)$$

the total for the first period is

$$y_{..1.} = 6\hat{\mu} + 3\hat{\theta}_1 + 3\hat{\theta}_2 + \sum_{ij}\sum \hat{D}_{(i)j} + 3\hat{\rho}_{(1)1} + 3\hat{\rho}_{(2)1} + 2(\hat{\tau}_A + \hat{\tau}_B + \hat{\tau}_C)$$

and the overall total is

$$y_{....} = 18\hat{\mu} + 9\hat{\theta}_1 + 9\hat{\theta}_2 + 3\sum_{ij}\sum \hat{D}_{(i)j} + 3\sum_{ik}\sum \hat{\rho}_{(i)k} + 6(\hat{\tau}_A + \hat{\tau}_B + \hat{\tau}_C) + 4(\hat{\tau}'_A + \hat{\tau}'_B + \hat{\tau}'_C)$$

Williams (1949) determined that the combination

$$q_l = (t^2 - t - 1)y_{...l} + ty^r_{...l} + y^f_{...l} + y_{..1.} - ty_{....} \quad (8.15)$$

would eliminate all estimated parameters except the direct effects. For the data of Table 8.9, $(t^2 - t - 1) = 5$, so

$$q_A = 5y_{...A} + 3y^r_{...A} + y^f_{...A} + y_{..1.} - 3y_{....} = 5(42) + 3(20) + 35 + 42 - 3(106) = +29$$

$$q_B = 5(28) + 3(23) + 34 + 42 - 3(106) = -33$$

$$q_C = 5(36) + 3(21) + 37 + 42 - 3(106) = +4$$

Note that $q_A + q_B + q_C = 0$, indicating correct computation. The quantity q_A actually measures $16[\hat{\tau}_A - (\hat{\tau}_B + \hat{\tau}_C)/2]$. Similarly, q_B measures $16[\hat{\tau}_B - (\hat{\tau}_A + \hat{\tau}_C)/2]$, so

$$q_A - q_B = 16[(3/2)\hat{\tau}_A - (3/2)\hat{\tau}_B] = 24(\hat{\tau}_A - \hat{\tau}_B)$$

Therefore, $(q_A - q_B)/24 = 2.58$ and is an estimate of $\tau_A - \tau_B$, the difference in direct effects of treatments A and B. Note that the difference in unadjusted treatment means is $(y_{...A} - y_{...B})/6 = (42 - 28)/6 = 2.33$. In general, for a treatment in m squares,

$$\hat{\tau}_A - \hat{\tau}_B = (q_A - q_B)/[mt(t - 2)(t + 1)] \quad (8.16)$$

with similar computations holding for comparisons of C with A or B. The method does not permit a unique estimate of each direct effect. However, if the treatment parameters are defined such that $\Sigma\tau = 0$, an estimated adjusted treatment mean (direct effect) is

$$\hat{\mu} + \hat{\tau}_l = \bar{y} + q_l/[mt(t - 2)(t + 1)] \quad (8.17)$$

The standard error of the difference between two direct effects is

$$s_{\hat{\tau}_A - \hat{\tau}_B} = \sqrt{(2ms_E/mt)[(t^2 - t - 1)/(t^2 - t - 2)]} \tag{8.18}$$

In similar fashion, Williams (1949) determined that the combination

$$q'_l = t^2 y^r_{...l} + t(y_{...l} + y^f_{...l} + y_{..1.}) - (t + 2)y_{....} \tag{8.19}$$

would eliminate all estimated parameters except the residual effects of treatments. For the data of Table 8.9,

$$q'_A = 9(20) + 3(42 + 35 + 42) - 5(106) = +7$$

$$q'_B = 9(23) + 3(28 + 34 + 42) - 5(106) = -11$$

$$q'_C = 9(21) + 3(36 + 37 + 42) - 5(106) = +4$$

Note that $q'_A + q'_B + q'_C = 0$, indicating correct computation. As for direct effects, q'_A measures only a function of the three residual effects, $16[\hat{\tau}'_A - (\hat{\tau}'_B + \hat{\tau}'_C)/2]$, rather than $\hat{\tau}'_A$ alone. Then, using a similar function for q'_B, one obtains

$$q'_A - q'_B = 24(\hat{\tau}'_A - \hat{\tau}'_B) \quad \text{or} \quad \hat{\tau}'_A - \hat{\tau}'_B = [7 - (-11)]/24 = 0.75$$

the difference in residual effects of treatments A and B. In general, for t treatments in m squares, the difference in residual effects and its standard error may be computed from

$$\hat{\tau}'_A - \hat{\tau}'_B = (q'_A - q'_B)/[mt(t - 2)(t + 1)] \qquad s_{\hat{\tau}_A - \hat{\tau}_B} = \sqrt{(2ms_E/mt)[t^2/(t^2 - t - 2)]} \tag{8.20}$$

Examination of (8.18) and (8.20) should make it apparent that the precision of estimating residual effects is poorer than that for direct effects. Let R be the ratio of the standard error of the difference of two residual effects to the standard error of the difference of two direct effects. Then, for t = 2, 3, 4, 5, or 6 treatments, R = 2, 1.34, 1.21, 1.15, or 1.11, respectively. If residual effects are of major interest, it is apparent that the standard Latin square crossover is relatively inefficient when only 2 treatments are used. See Sec. 8.1.6 for some alternative designs.

Sums of squares for total (y), squares (S), subjects within squares (D/S), and periods (P/S) follow customary computations:

$$ss_y = \sum_{i=1}^{m}\sum_{j=1}^{t}\sum_{k=1}^{t} y^2_{ijkl} - \left(\sum_{i=1}^{m}\sum_{j=1}^{t}\sum_{k=1}^{t} y_{ijkl}\right)^2/(mt^2) \tag{8.21}$$

$$ss_S = \sum_{i=1}^{m} (y^2_{i...}/t^2) - \left(\sum_{i=1}^{m}\sum_{j=1}^{t}\sum_{k=1}^{t} y_{ijkl}\right)^2/(mt^2) \tag{8.22}$$

$$ss_{D/S} = \sum_{i=1}^{m} \sum_{j=1}^{t} (y_{ij..}^2/t) - \sum_{i=1}^{m} (y_{i...}^2/t^2) \quad (8.23)$$

$$ss_{P/S} = \sum_{i=1}^{m} \sum_{k=1}^{t} (y_{i.k.}^2/t) - \sum_{i=1}^{m} (y_{i...}^2/t^2) \quad (8.24)$$

If the nuisance trends of the subjects over time are likely to be relatively uniform over all squares, one could conserve degrees of freedom for error by computing the sum of squares for periods across all squares (t - 1 df) instead of within squares [$m(t - 1)$ df]:

$$ss_P = \sum_{k=1}^{t} (y_{..k.}^2/mt) - (\sum_{i=1}^{m} \sum_{j=1}^{t} \sum_{k=1}^{t} y_{ijkl})^2/(mt^2) \quad (8.25)$$

Sums of squares for direct (T) and residual (T') effects of treatments are not completely orthogonal. The adjusted and unadjusted forms have the relationship:

$$ss_{T(\text{unadj})} + ss_{T'(\text{adj})} = ss_{T(\text{adj})} + ss_{T'(\text{unadj})}$$

For practical purposes, one computes only the first three of those:

$$ss_{T(\text{unadj})} = \sum_{l=1}^{t} (y_{...l}^2/mt) - (\sum_{i=1}^{m} \sum_{j=1}^{t} \sum_{k=1}^{t} y_{ijkl})^2/(mt^2) \quad (8.26)$$

$$ss_{T'(\text{adj})} = \sum_{l=1}^{t} (q_l')^2/[mt^3(t - 2)(t + 1)] \quad (8.27)$$

$$ss_{T(\text{adj})} = \sum_{l=1}^{t} (q_l)^2/[mt(t^2 - t - 1)(t - 2)(t + 1)] \quad (8.28)$$

Finally, the sum of squares for error may be obtained by difference:

$$ss_E = ss_y - ss_S - ss_{D/S} - ss_{P/S} - ss_{T(\text{unadj})} - ss_{T'(\text{adj})} \quad (8.29)$$

The analysis of variance is summarized in Table 8.10.

Analysis of the artificial data of Table 8.9 is given in Table 8.11. The standard error of the difference between any two direct effects [(8.18)] is $\sqrt{[2(0.285)/6](5/4)} = 0.34$, and the standard error of the difference between any two residual effects [(8.20)] is $\sqrt{[2(0.285)/6](9/4)} = 0.46$. Note that the estimates of direct effect differences all exceed three standard errors [$(\hat{\tau}_A - \hat{\tau}_B) = 2.58$, $(\hat{\tau}_A - \hat{\tau}_C) = 1.04$, $(\hat{\tau}_B - \hat{\tau}_C) = -1.54$], whereas none of the estimates of residual effect differences is as large as two standard errors [$(\hat{\tau}_A' - \hat{\tau}_B') = 0.75$, $(\hat{\tau}_A' - \hat{\tau}_C') = 0.12$, $(\hat{\tau}_B' - \hat{\tau}_C') = -0.62$].

8.1.5. Reversal Designs

In some experiments period effects are likely to be large and to vary

Table 8.10. Analysis of Variance for Crossover Designs with Residual Effects of t Treatments in m Squares

Source of Variation	df	ss	ms	$E[MS]$
Squares	$m-1$	ss_S	ms_S	$\sigma^2+t\sigma_D^2+t^2\sum_{i=1}^{m}\theta_{i}^2/(m-1)$
Subjects/squares	$m(t-1)$	$ss_{D/S}$	$ms_{D/S}$	$\sigma^2+t\sigma_D^2$
Periods/squares	$m(t-1)$	$ss_{P/S}$	$ms_{P/S}$	$\sigma^2+t\Sigma\Sigma(\rho_{(i)k}^2)/m(t-1)$
Treatments		$ss_{T(\text{unadj})}+ss_{T'(\text{adj})}$		
Direct (unadj.)		$ss_{T(\text{unadj})}$		
Residual (adj.)	$t-1$	$ss_{T'(\text{adj})}$	$ms_{T'}$	$[\sigma^2+mt\Sigma(\tau_\ell')^2/(t-1)]^*$
Direct (adj.)	$t-1$	$ss_{T(\text{adj})}$	ms_T	$[\sigma^2+mt\Sigma\tau_\ell^2/(t-1)]^*$
Error	$(t-1)(mt-m-2)$	ss_E	ms_E	σ^2

*Not orthogonal (approximate).

Table 8.11. Analysis of Direct and Residual Effects of 3 Treatments in 2 Latin Squares (see Table 8.9)

Source of Variation	df	*ss*	*ms*	*f* ratio
Squares	1	14.222	14.222	
Subjects/squares	4	0.889	0.222	
Periods/squares	4	14.222	3.556	
Treatments		17.305		
Direct (unadj.)		16.444		
Residual (adj.)	2	0.861	0.430	1.509
Direct (adj.)	2	16.217	8.108	28.449*
Error	4	1.140	0.285	
Total	17	47.778		

*$P<0.01$.

markedly from one subject to another (e.g., lactation curves of cows differing greatly in genetic producing ability). Reversal (switchback) designs can be used to cope with the problem by applying 1 or more of the treatments twice to each subject in alternating order (e.g., *ABA* or *ABAB*). For 2 treatments, single reversal sequences (*ABA* and *BAB*) should be adequate if the nuisance trends are linear, as for milk production after the first 4-6 weeks of lactation; and double reversal sequences (*ABAB* and *BABA*) should be used if the nuisance trends are likely to be approximately quadratic in form (see Fig. 2.1). For example, in dairy cattle feeding trials some cows may be in the earliest part of lactation, where milk production tends to follow a convex curve with time. Then, 4 periods of 7-10 days duration each may suffice to compare 2 treatments, essentially free of nuisance trend.

For the single reversal experiment with 2 treatments, half the animals or subjects receive the sequence *ABA* and the other half *BAB* in 3 periods. The model is the same as for the simple crossover [(8.2)],

$$Y_{ijk} = \mu + D_i + \rho_j + \tau_k + E_{(ijk)} \tag{8.30}$$

except that $j = 1, 2, 3$; $k = 1, 2$; and $n = 3r$, where r is the number of subjects (blocks). Brandt (1938) showed that linear nuisance trends (period effects) may be accounted for by computing $y_{i1} - 2y_{i2} + y_{i3}$ for each subject, where the redundant subscript k is ignored. Estimated parameters included in such results are

$$(\hat{\mu} + \hat{D}_i + \hat{\rho}_1 + \hat{\tau}_A) - 2(\hat{\mu} + \hat{D}_i + \hat{\rho}_2 + \hat{\tau}_B) + (\hat{\mu} + \hat{D}_i + \hat{\rho}_3 + \hat{\tau}_A) = (\hat{\rho}_1 - 2\hat{\rho}_2 + \hat{\rho}_3) + 2(\hat{\tau}_A - \hat{\tau}_B)$$

for subjects given treatment sequence *ABA*. Similarly, for subjects given sequence *BAB*, the estimated parameters included are

$$(\hat{\rho}_1 - 2\hat{\rho}_2 + \hat{\rho}_3) - 2(\hat{\tau}_A - \hat{\tau}_B)$$

It is clear that the difference between effects of the 2 treatments may be estimated by

$$\hat{\tau}_A - \hat{\tau}_B = [\sum_{i=1}^{r/2} (y_{i1} - 2y_{i2} + y_{i3}) - \sum_{i=(r/2)+1}^{r} (y_{i1} - 2y_{i2} + y_{i3})]/[4(r/2)] \quad (8.31)$$

where it is understood that the first $r/2$ subjects are those that random allocation placed in the group receiving sequence *ABA* and the remaining subjects received *BAB*. The error sum of squares is computed from

$$ss_E = \sum_{i=1}^{r/2} (y_{i1} - 2y_{i2} + y_{i3})^2 - [\sum_{i=1}^{r/2} (y_{i1} - 2y_{i2} + y_{i3})]^2/(r/2) \quad (8.32)$$

$$+ \sum_{i=(r/2)+1}^{r} (y_{i1} - 2y_{i2} + y_{i3})^2 - [\sum_{i=(r/2)+1}^{r} (y_{i1} - 2y_{i2} + y_{i3})]^2/(r/2)$$

$$= \sum_{i=1}^{r} (y_{i1} - 2y_{i2} + y_{i3})^2 - \{[\sum_{i=1}^{r/2} (y_{i1} - 2y_{i2} + y_{i3})]^2 + [\sum_{i=(r/2)+1}^{r} (y_{i1} - 2y_{i2} + y_{i3})]^2\}/(r/2)$$

There are $r - 2$ df for error, i.e., $ms_E = ss_E/(r - 2)$, so the standard error of the difference between the two treatment effects is

$$s_{\hat{\tau}_A - \hat{\tau}_B} = \sqrt{2ms_E/(r/2)} = 2\sqrt{ms_E/r} \quad (8.33)$$

The hypothesis $H: \tau_A = \tau_B$ may be tested by comparing test statistic

$$t = (\hat{\tau}_A - \hat{\tau}_B)/s_{(\hat{\tau}_A - \hat{\tau}_B)} \quad (8.34)$$

with critical values $\pm t_{\alpha/2, r-2}$. For one-sided alternatives $\overline{H}: \tau_A > \tau_B$ or $\overline{H}: \tau_A < \tau_B$, use critical values $+t_{\alpha, r-2}$ or $-t_{\alpha, r-2}$, respectively.

As an example of a single reversal experiment, consider a study of the effects of injected folic acid (*A*) or physiological saline (*B*) on milk protein produced by Holstein cows. Ten cows were assigned at random, 5 to sequence *ABA* and 5 to *BAB*, where the 3 periods each were 10 days in length. Data (kg protein) and analysis are shown in Table 8.12.

For a double reversal trial with 2 treatments, subjects are randomly allocated, half to sequence *ABAB* and half to *BABA*. The model is the same as for single reversal data [(8.30)] except that there are 4 periods (j = 1, 2, 3, 4) and $n = 4r$, where r is the number of subjects. Both linear and quadratic

Table 8.12. Milk Protein Production (kg/10 days) by Holstein Cows in Single Reversal Trial (A=folic acid, B=saline control)

	Period			
	1	2	3	
Cow	A	B	A	$(y_{i1}-2y_{i2}+y_{i3})$
1	6.46	4.96	5.87	+2.41
2	8.62	7.85	8.38	+1.30
3	10.31	9.20	9.98	+1.89
4	8.89	7.30	8.18	+2.47
5	8.22	7.19	8.19	+2.03
	B	A	B	+10.10
6	8.38	9.34	8.75	-1.55
7	10.02	10.90	9.72	-2.06
8	7.13	7.64	6.10	-2.05
9	6.12	7.02	5.67	-2.25
10	7.85	8.60	7.63	-1.66
				-9.57

$$\hat{\tau}_A-\hat{\tau}_B=[10.10-(-9.57)]/[4(5)]=0.984$$

$$ss_E=[2.41^2+1.30^2+...+(-1.66)^2]-[10.10^2+(-9.57)^2]/5=1.2397$$

$$ms_E=1.2397/8=0.155;\ s_{\hat{\tau}_A-\hat{\tau}_B}=2\sqrt{0.155/10}=0.249$$

$$t=0.984/0.249=3.952>t_{0.01/2,8}=3.355;\ \text{reject } H:\tau_A=\tau_B$$

trends (period effects) are accounted for by computing $y_{i1} - 3y_{i2} + 3y_{i3} - y_{i4}$ for each subject (redundant subscript k ignored). For subjects given treatment sequence $ABAB$, estimated parameters included in the result are

$$(\hat{\rho}_1 - 3\hat{\rho}_2 + 3\hat{\rho}_3 - \hat{\rho}_4) + 4(\hat{\tau}_A - \hat{\tau}_B)$$

Similarly, for subjects given sequence $BABA$, the estimated parameters included are

$$(\hat{\rho}_1 - 3\hat{\rho}_2 + 3\hat{\rho}_3 - \hat{\rho}_4) - 4(\hat{\tau}_A - \hat{\tau}_B)$$

Clearly, the difference between effects of the two treatments may be estimated by

$$\hat{\tau}_A - \hat{\tau}_B = [\sum_{i=1}^{r/2} (y_{i1} - 3y_{i2} + 3y_{i3} - y_{i4}) - \sum_{i=(r/2)+1}^{r} (y_{i1} - 3y_{i2} + 3y_{i3} - y_{i4})]/[8(r/2)] \tag{8.35}$$

where it is understood that the first $r/2$ subjects received sequence *ABAB* and subjects numbered from $(r/2) + 1$ to r received *BABA*. The error sum of squares is

$$ss_E = \sum_{i=1}^{r} (y_{i1} - 3y_{i2} + 3y_{i3} - y_{i4})^2$$

$$- \{[\sum_{i=1}^{r/2} (y_{i1} - 3y_{i2} + 3y_{i3} - y_{i4})]^2$$

$$+ [\sum_{i=(r/2)+1}^{r} (y_{i1} - 3y_{i2} + 3y_{i3} - y_{i4})]^2\}/(r/2) \qquad (8.36)$$

with $r - 2$ df. The standard error of the difference between two treatment effects is

$$s_{\hat{\tau}_A - \hat{\tau}_B} = \sqrt{2ms_E/(r/2)} = 2\sqrt{ms_E/r} \qquad (8.37)$$

as for the single reversal, except that ms_E is derived from (8.36). The test of $H: \tau_A = \tau_B$ is performed as indicated by (8.34). A simple example of a double reversal experiment is given in the exercises at the end of this chapter. A more complex example involving the effects of NPH-insulin mixtures on blood sugar in rabbits, with initial level of blood sugar as a covariate, has been discussed by Ciminera and Wolfe (1953).

Extensions of reversal designs for experiments with 3 or more treatments have been given by Taylor and Armstrong (1953) and Lucas (1956).

The efficiency of reversal designs relative to ordinary crossover designs depends on the diversity of nuisance trends among the subjects involved. If the nuisance trends are relatively uniform within small groups of subjects, the Latin square crossover design is likely to be more efficient than the reversal designs. Having performed a reversal trial with 2 treatments, one could compare the error variance with error when the first 2 periods are analyzed as a simple crossover.

Some alternatives to reversal designs for the removal of individual nuisance trends have been proposed by C. P. Cox (1958a, b), Ogilvie (1963), and Taylor (1967). All these methods use orthogonal polynomials for removal of period effects.

8.1.6. Miscellaneous Designs for Removal of Trends

Balaam (1968) developed a design for 2-period experiments that enables the investigator to study interaction of treatments and periods, i.e., to discover if there is a time trend in the *differences* among treatments. For t treatments, t^2 subjects are required. For $t = 2$, 4 subjects (or multiples thereof) are randomly assigned to treatment sequences *AA*, *AB*, *BA*, and *BB*; for $t = 3$, 9 subjects are assigned to sequences *AA*, *AB*, *AC*, *BA*, *BB*, *BC*, *CA*, *CB*, and *CC*; etc. The model for such experiments is

$$Y_{ijk} = \mu + D_i + \rho_j + \tau_k + (\rho\tau)_{jk} + E_{(ijk)} \qquad (i = 1, 2, \ldots, t^2;\ j = 1, 2;\ k = 1, 2, \ldots, t) \qquad (8.38)$$

where D_i, ρ_j, τ_k, and $E_{(ijk)}$ represent effects of subject, period, treatment, and error, as in other crossover designs. The effect of interaction between periods and treatments is represented by $(\rho\tau)_{jk}$. Let a subject total be $y_{i..} = y_{i1k} + y_{i2k}$, a period total be $y_{.j.} = \sum_{i=1}^{t^2} y_{ijk}$, and the grand total be $y_{...} = \sum_{i=1}^{t^2}\sum_{j=1}^{2} y_{ijk}$ (redundant subscript k ignored). Then the sums of squares for total, subjects, and periods are

$$ss_y = \sum_{i=1}^{t^2}\sum_{j=1}^{2} y_{ijk}^2 - (y_{...}^2/2t^2) \quad (8.39)$$

$$ss_D = \sum_{i=1}^{t} (y_{i..}^2/2) - (y_{...}^2/2t^2) \quad (8.40)$$

$$ss_P = \sum_{j=1}^{2} (y_{.j.}^2/t^2) - (y_{...}^2/2t^2) \quad (8.41)$$

Let $y_{..k}$ be the usual treatment total; let $y^c_{.jk}$ be the total of t observations in the "complementary" (alternative) period, $j' \neq j$, for the same subjects that received treatment k in period j. For example, if subjects given sequences AA, AB, BA, and BB provide pairs of observations {13, 11}, {14, 8}, {10, 12}, and {6, 4}, then

$$y_{..1} = 13 + 11 + 14 + 12 = 50 \qquad y_{..2} = 8 + 10 + 6 + 4 = 28$$

$$y^c_{.11} = 11 + 8 = 19 \qquad y^c_{.21} = 13 + 10 = 23$$

$$y^c_{.12} = 12 + 4 = 16 \qquad y^c_{.22} = 14 + 6 = 20$$

Sums of squares for treatments and period-treatment interaction are

$$ss_T = \sum_{k=1}^{t} [y_{..k} - (y^c_{.1k} + y^c_{.2k})]^2/(4t) \quad (8.42)$$

$$ss_{PT} = [\sum_{k=1}^{t} (y_{.1k} - y_{.2k} - y^c_{.1k} + y^c_{.2k})^2/4t] - [(y_{.1.} - y_{.2.})^2/t^2] \quad (8.43)$$

where $y_{.11} = 13 + 14 = 27$, $y_{.21} = 11 + 12 = 23$, $y_{.12} = 10 + 6 = 16$, $y_{.22} = 8 + 4 = 12$, $y_{.1.} = 13 + 14 + 10 + 6 = 43$, and $y_{.2.} = 11 + 8 + 12 + 4 = 35$. The error sum of squares is obtained by difference as usual. Analysis of variance

Table 8.13. Analysis of Treatment-Period Interaction

Source of Variation	df	*ss*	*ms*
Subjects	t^2-1=3	61.5	20.5
Periods	1	8.0	8.0
Treatments	t-1=1	8.0	8.0
PT	t-1=1	0.0	0.0
Error	$(t-1)^2$=1	8.0	8.0
Total	$2t^2$-1=7	85.5	

for the artificial data above is shown in Table 8.13. Interaction sum of squares was zero for these data, which were artificially constructed without attempt to introduce interaction. Obviously, considerable replication of the basic design with t^2 subjects is needed to obtain adequate degrees of freedom for error. For additional details, see Balaam.

In some crossover experiments the number of periods is severely limited, the number of treatments or treatment combinations is large, or the number of subjects available is fewer than the number of treatments. For factorial experiments, the designs with double confounding (Sec. 7.8.4) may be of use in some cases. For nonfactorial experiments, the Youden squares, partially balanced incomplete Latin squares, and lattice squares (Secs. 7.8.1-7.8.3) may be used, or balanced sets of 2 × 2 Latin squares may be used if only 2 periods are to be considered (Gill and Magee 1976).

Patterson (1952) and Patterson and Lucas (1962) have investigated the incomplete block designs for crossover experiments. As an example of a situation requiring a Youden crossover design, consider a study of volatile fatty acid production in the rumen by steers fed t = 7 different silages (A, B, C, ..., G). In this case the number of animals rather than the number of periods is restrictive because the animals must be fitted with capped fistulas. Suppose only 3 fistulated animals are available, but each animal may be given the 7 treatments over a span of 7 experimental periods. Then the Youden design shown in Table 7.12 would be appropriate, where the factors listed as blocks and locations are considered periods and animals, respectively, in this study.

Two-period experiments with $t > 2$ treatments may be performed with complete balance in $t(t - 1)/2$ Latin squares of size 2 × 2, in which $t(t - 1)$ subjects (or multiples thereof) represent rows of the squares and the 2 periods are the columns. As well as being useful for multitreatment studies restricted to 2 periods, the designs also lend themselves well to 2-stage investigations. Suppose a set of $t(t - 1)$ subjects have been randomly assigned to t treatment groups in the first stage, and a repetition of the basic experiment is now desirable. The same subjects may be reassigned in a balanced fashion, thus conserving resources. Designs for t = 3 or 4 treatments are shown in Table 8.14. The design for t = 5 consists of the 6 squares shown for t = 4 plus 4 additional squares in which A, B, C, and D are compared with E in

Table 8.14. Two-Period Changeover Designs for 3 and 4 Treatments

t=3	Period							
	1	2		1	2		1	2
Subject 1:	A	B	3:	A	C	5:	B	C
2:	B	A	4:	C	A	6:	C	B

t=4	Period							
	1	2		1	2		1	2
Subject 1:	A	B	3:	A	C	5:	B	C
2:	B	A	4:	C	A	6:	C	B
7:	A	D	9:	B	D	11:	C	D
8:	D	A	10:	D	B	12:	D	C

one square each. Treatment means must be adjusted for subject differences, but treatment effects are orthogonal to period effects. The designs are completely balanced in the sense of incomplete blocks (Sec. 6.2.1.1) because each treatment is compared exactly $\lambda = 2$ times in the same block (subject) or period with each of the other treatments. The block size is $k = 2$, and replication in the basic plan (without repetition) is $r = 2(t - 1)$. Therefore the minimum efficiency of the designs relative to standard crossover designs is

$$E = t\lambda/kr = t(2)/[(2)(2)(t - 1)] = (1/2)t/(t - 1)$$

For $t = 3$, 4, 5, or 6 treatments, $E = 0.75$, 0.67, 0.62, or 0.6, with the limit of E being 0.5 for large t. The actual efficiencies achieved may be considerably higher if results in the 2 periods are relatively more homogeneous than could be achieved with the standard multiperiod designs.

A model for the 2-period design is

$$Y_{ijk} = \mu + D_i + \rho_j + \tau_k + E_{(ijk)} \qquad (i = 1, 2, \ldots, t(t-1);\ j = 1,2;\ k = 1, 2, \ldots, t) \tag{8.44}$$

where D_i, ρ_j, τ_k, and $E_{(ijk)}$ represent the effects of subject, period, treatment, and error as in other crossover designs. As in (6.8), estimates of treatment effects are obtained from

$$\hat{\tau}_k = (k/t\lambda)(r\bar{y}_{..k} - \sum_{i(k)}^{r} \bar{y}_{i..}) = [2(t - 1)\bar{y}_{..k} - \sum_{i(k)}^{2(t-1)} \bar{y}_{i..}]/t \tag{8.45}$$

where $\bar{y}_{..k}$ is an unadjusted treatment mean with $2(t - 1)$ observations and the block means ($\bar{y}_{i..}$) of 2 observations each are for those subjects to which the particular treatment was given (in either of the 2 periods). The $\hat{\tau}_k$ are unbiased if there are no residual effects of treatments. The adjusted treatment means are $\bar{y} + \hat{\tau}_k$ and have standard error

$$s_{\hat{\mu}+\hat{\tau}_k} = \sqrt{(ms_E/n)[1 + 2(t - 1)^2/t]} \tag{8.46}$$

The difference between two adjusted treatment means has standard error

$$s_{\hat{\tau}_1-\hat{\tau}_2} = \sqrt{2ms_E/t} \tag{8.47}$$

Let $y_{i..}$, $y_{.j.}$, and $y_{...}$ represent totals for subject, period, and all data (redundant subscript k ignored). Then, the sums of squares for total (y), subjects (D), periods (P), adjusted treatments (T), and error (E) are:

$$ss_y = \sum_{i=1}^{t(t-1)} \sum_{j=1}^{2} y_{ijk}^2 - (y_{...}^2/n) \quad [n = 2t(t - 1)] \tag{8.48}$$

$$ss_D = \sum_{i=1}^{t(t-1)} (y_{i..}^2/2) - (y_{...}^2/n) \tag{8.49}$$

$$ss_P = \sum_{j=1}^{2} y_{.j.}^2/[t(t - 1)] - (y_{...}^2/n) \tag{8.50}$$

$$ss_T = t \sum_{k=1}^{t} \hat{\tau}_k^2 \tag{8.51}$$

$$ss_E = ss_y - ss_D - ss_P - ss_T \tag{8.52}$$

The analysis of variance is summarized in Table 8.15.

The test of treatments, $f = ms_T/ms_E$ versus $f_{\alpha,t-1,t(t-2)}$, obviously is insensitive for $t = 3$ (only 3 df for error). Therefore repetition of the basic plan is advisable. Consider data on milk production (kg/day) by 12 Holstein cows, each fed 2 of the rations A, B, C, D in 2 successive periods of 30 days each. Rations B, C, and D involved various substitutions of nonprotein nitrogen for the protein of A. Data, along with adjusted and unadjusted means, are shown in Table 8.16. The analysis of variance is summarized in Table 8.17. Components of random variation are $\hat{\sigma}^2 = ms_E = 2.10$ within subjects and $\hat{\sigma}_D^2 = (ms_D - ms_E)/2 = (47.29 - 2.10)/2 = 22.60$ among subjects. The efficiency of the design relative to a completely randomized assignment of subjects to treatments for one period depends on $\hat{\sigma}_{CROSS}^2 = 2.10$, with $\nu_{CROSS} = \nu_T + \nu_E = 11$, and $\hat{\sigma}_{CRD}^2 = [520 + 12(2.10)]/23 = 23.70$, with $\nu_{CRD} = 23$. The effi-

Table 8.15. Analysis of Variance for 2-Period Changeover Designs for $t>2$ Treatments

Source of Variation	df	*ss*	*ms*	$E[MS]$
Subjects	t^2-t-1	ss_D	ms_D	$\sigma^2+2\sigma_D^2$
Periods	1	ss_P	ms_P	$\sigma^2+t(t-1)\sum_{j=1}^{2}\rho_j^2$
Treatments	$t-1$	ss_T	ms_T	$\sigma^2+2\sum_{k=1}^{t}\tau_k^2$
Error	$t(t-2)$	ss_E	ms_E	σ^2
Total	$2t^2-2t-1$	ss_y		

ciency is

$$E_{\mathrm{CROSS:CRD}} = (12/14)(26/24)(23.70/2.10)100\% = 1048\%$$

i.e., approximately 10.48(12) = 126 cows would be needed for a completely randomized design to provide the sensitivity permitted by this crossover design with 12 cows.

An unusual crossover design is a set of Latin squares in which all observations in each square are from the same subject. For example, Babcock (1889) described a study on the variations in yield and quality of milk. The treatments were the 4 quarters of the udder, the rows were 4 times of milking, and the columns were order of milking at each time.

A number of modifications of the basic designs for measuring residual effects of treatments (Sec. 8.1.4) have been devised. Lucas added an extra period to Latin squares balanced for residual effects. During the extra period $(t + 1)$, the treatment given in period t to a particular subject is continued. Thus, each treatment is preceded by itself as well as by each of the other treatments. Such designs permit orthogonal estimation of direct and residual effects, but the direct effects are not estimated as precisely as in the standard crossover design. Precision of residual effects is somewhat better than in the standard design and essentially equal to that of direct effects. These designs cannot be considered superior to the standard designs unless there is major interest in the residual effects themselves (τ_l') or in the permanent effects of treatments $(\tau_l + \tau_l')$. See Lucas (1957) or Patterson and Lucas (1959) for analysis and examples.

Berenblut (1964, 1967) created designs that permit orthogonal estimation of direct and residual effects but provide much better precision of direct effects than is obtainable in the extra-period designs. The designs generally are impractical for $t > 3$ treatments for they require $2t$ periods.

Federer and Atkinson (1964) constructed "tied-double-changeover" designs from orthogonal Latin squares. The designs are intended for use when direct and residual effects are of equal interest, as the precision achieved is about

Table 8.16. Milk Production (kg/day) in 2-Period Changeover Design for 4 Treatments (A,B,C,D)

	Period				Period				Period		
	1	2	$\bar{y}_{i..}$		1	2	$\bar{y}_{i..}$		1	2	$\bar{y}_{i..}$
Subject 1:	43.55(A)	32.17(B)	37.860	3:	35.33(A)	22.72(C)	29.025	5:	34.87(B)	23.98(C)	29.425
2:	38.03(B)	32.07(A)	35.050	4:	26.17(C)	28.78(A)	27.475	6:	24.77(C)	25.04(B)	24.905
7:	36.04(A)	29.50(D)	32.770	9:	25.14(B)	18.49(D)	21.815	11:	26.44(C)	22.88(D)	24.660
8:	29.62(D)	26.13(A)	27.875	10:	36.80(D)	32.70(B)	34.750	12:	36.86(D)	29.14(C)	33.000

Totals: Periods, $y_{.1.}=393.62$, $y_{.2.}=323.60$

Treatments, $y_{..1}=201.90$, $y_{..2}=187.95$, $y_{..3}=153.22$, $y_{..4}=174.15$

Means: Period, $\bar{y}_{.1.}=32.802$, $\bar{y}_{.2.}=26.967$

Treatments, $\bar{y}_{..1}=33.650$, $\bar{y}_{..2}=31.325$, $\bar{y}_{..3}=25.537$, $\bar{y}_{..4}=29.025$

$\hat{\tau}_1=[6(33.650)-(37.860+35.050+29.025+27.475+32.770+27.875)]/4=+2.961$ $\hat{\mu}+\hat{\tau}_1=32.84$

$\hat{\tau}_2=[6(31.325)-(37.860+35.050+29.425+24.905+21.815+34.750)]/4=+1.036$ $\hat{\mu}+\hat{\tau}_2=30.92$

$\hat{\tau}_3=[6(25.537)-(29.025+27.475+29.425+24.905+24.660+33.000)]/4=-3.817$ $\hat{\mu}+\hat{\tau}_3=26.07$

$\hat{\tau}_4=[6(29.025)-(32.770+27.875+21.815+34.750+24.660+33.000)]/4=-0.180$ $\hat{\mu}+\hat{\tau}_4=29.70$

Table 8.17. Analysis of Variance for Milk Production in 2-Period Changeover Design for 4 Treatments (see Table 8.16)

Source of Variation	df	*ss*	*ms*	*f* Ratio	$f_{0.01,3,8}$
Subjects	11	520.23	47.29		
Periods	1	204.28	204.28		
Treatments	3	97.78	32.59	15.52	7.59
Error	8	16.83	2.10		
Total	23				

Standard error of adjusted means: $\sqrt{(2.10/24)[1+2(3)^2/4]}=0.935$

Standard error of difference of 2 means: $\sqrt{2(2.10)/4}=1.025$

One-sided Dunnett's tests: $1.025t_{D,0.05,3,8}=1.025(2.42)=2.48$ (App. A.9.2)

$1.025t_{D,0.01,3,8}=1.025(3.51)=3.60$

$\hat{\tau}_1-\hat{\tau}_2=1.92$, $\hat{\tau}_1-\hat{\tau}_3=6.77^{**}$, $\hat{\tau}_1-\hat{\tau}_4=3.14^{*}$

$^{*}P<0.05$, $^{**}P<0.01$.

equal for the two kinds of effects.

Williams (1949, 1950) discussed construction of designs that are balanced for 2 periods of residual effects, i.e., the effects of a treatment linger for 2 periods beyond the period in which the treatment was applied. The designs require a completely orthogonal set of $t - 1$ Latin squares. Atkinson (1966) devised a method to study the residual effects of several different entire sequences of treatments.

Berenblut (1968, 1970), Patterson (1970), and Mason and Hinklemann (1971) have reported work on designs for equally spaced quantitative factors balanced for linear residual effects in such a way that one can test for additivity of direct and residual effects of treatments.

8.2. DESIGNS FOR STUDY OF TRENDS (Nonrandom Sampling)

Here subjects make up blocks of observations as in the changeover designs of the previous section, but the repeated factor of interest is not treatment applications over time but time itself (or space). The period effects that constituted a nuisance trend for crossover experiments are, for the present designs, the effects of interest. In complete block designs, subjects are treated or handled alike and monitored for response at various sampling intervals (Sec. 8.2.1). More commonly, subjects are assigned randomly to various treatment groups (factor A) as in a completely randomized design, but each subject is monitored over time (factor B) for response to treatment (Sec. 8.2.2). Then the variation among subjects (incomplete blocks) is relevant for comparing treatments, but the variation within subjects pertains to trends in response over time or to *AB* interaction. Such designs have been termed "split plots" from the agronomic tradition of experiments in which units that received a particular level of *A* were larger plots of ground than the subplots that received a particular level of *B*. In the context of designs for experiments with animals or humans, one might prefer the term "split subject," or more generally, "split unit." Discussion in this text will defer to the traditional terminology. The important thing to remember is that 2 (or more) distinct strata of random variation are involved. In some split-plot experiments the subject is not a block at all, since only one response is recorded per subject for a particular variable of interest (Sec. 8.3). The split-plot nature arises in those cases because a block of subjects that are homogeneous for one or more intrinsic nuisance variables or handled alike (e.g., measured the same day) differs from another block according to the level of one of the factors being investigated as well as by class or level of nuisance or blocking variable. That is, the main effect of one factor is confounded with blocks, as interactions were in the incomplete block designs of Secs. 6.3, 6.4, and 6.5.

8.2.1. Complete Block Designs (subject as block)

Complete block designs are appropriate for monitoring trends in time (or space) for subjects under some set of standardized conditions or when all subjects have been treated alike. For example, effects of stage of lactation or stage of estrus on various enzyme or hormone levels in serum can be estimated by sampling blood from each subject or animal at prescribed intervals over several weeks or days. Or histologic response of the endometrium to infusion of an antibiotic may be studied by taking biopsies at intervals of 2 or 3 days on each subject. An example involving space instead of time is the determination of radioactive iron in various subcellular fractions from liver cells of each subject.

The model indicated for random order changeover designs [(8.1)] is an appropriate model for such experiments:

$$Y_{ij} = \mu + \tau_i + D_j + (\tau D)_{ij} + E_{(ij)} \qquad (i = 1, 2, \ldots, t;\ j = 1, 2, \ldots, r) \qquad (8.53)$$

Here the τ_i represent "treatment" effects that involve fixed-order responses in time or space. As usual, the D_j are random effects of subjects and the treatment-subject interaction is inseparable from error; if trends in response over time or space are not likely to be consistent for all subjects, this design is not appropriate. It may be possible to group the subjects so that trends are likely to be relatively uniform within each group and perform the experiment as a split plot.

Intrablock correlation of errors is possible in any design but is most likely to be heterogeneous when the block is a subject, and the term "treatments" is merely a convenient name for fixed-order repeated measurement. A valid analysis of data from such experiments by standard univariate methods requires certain assumptions about the pattern of correlation. Sufficient conditions for valid f tests in the analysis of variance of normal data are that the variance among subjects be the same at each sampling point in time or space and the covariances for any two sampling points be homogeneous. In addition to the common tendency in biological data for larger variances to be associated with larger means is the tendency for responses in two adjacent periods or spaces to have higher correlation (or covariance) than those more distantly separated in time or space. Therefore more caution than usual is in order when analyzing data from this class of experiments. Huynh and Feldt (1970) have shown that the *sufficient* condition is not a *necessary* condition for valid f tests. The more flexible, necessary condition is only that all possible differences between responses in two different periods or sampling points be equally variable. Procedures for testing the validity of the sufficient conditions and alternative (multivariate) methods of analyzing data that do not fulfill the conditions are discussed in Sec. 8.4.

The sums of squares for total, treatments, subjects, and error are those of the standard analysis for complete block designs:

$$ss_y = \sum_{i=1}^{t} \sum_{j=1}^{r} y_{ij}^2 - (\sum_{i=1}^{t} \sum_{j=1}^{r} y_{ij})^2/n \tag{8.54}$$

$$ss_T = \sum_{i=1}^{t} y_{i.}^2/r - (\sum_{i=1}^{t} \sum_{j=1}^{r} y_{ij})^2/n \tag{8.55}$$

$$ss_D = \sum_{j=1}^{r} y_{.j}^2/t - (\sum_{i=1}^{t} \sum_{j=1}^{r} y_{ij})^2/n \tag{8.56}$$

$$ss_E = ss_y - ss_T - ss_D \tag{8.57}$$

The analysis of variance is summarized in Table 8.18 with expected mean

Table 8.18. Analysis of Repeated Measurements in Complete Block Designs

Source of Variation	df	ss	ms	$E[MS]$*
Treatments (time or space)	$t-1$	ss_T	ms_T	$\sigma^2(1-\rho)+r\sum_{i=1}^{t}\tau_i^2/(t-1)$
Blocks (subjects)	$r-1$	ss_D	ms_D	$\sigma^2[1+(t-1)\rho]$
Error	$(t-1)(r-1)$	ss_E	ms_E	$\sigma^2(1-\rho)$

*ρ=constant correlation between any two "treatments."

squares reflecting absence of block-treatment interaction, given sufficient conditions of homogeneous variance among subjects (σ^2) and constant correlation (ρ) between responses at any two sampling points in time or space. As usual, the standard error of a treatment mean is $\sqrt{ms_E/r}$. The hypothesis of zero treatment effects, $H{:}\tau_i = 0$, is a statement that no trend of increasing or decreasing response exists over time (or space). The test statistic, $f = ms_T/ms_E$, is compared with the usual critical value $f_{\alpha,t-1,\nu_E}$. As in other cases, the f test is robust against mildly heterogeneous variance, but it is not robust against heterogeneous covariance except for certain patterns of correlation. For cases in which $H{:}\tau_i = 0$ is rejected, Geisser and Greenhouse

(1958) suggested using a conservative critical value $f_{\alpha,1,r-1}$, as if only two sampling points were used, to see if rejection of the hypothesis should be reinforced. If $f > f_{\alpha,1,r-1}$, one may conclude that the evidence is sufficiently strong to reject H with the confidence specified despite possible heterogeneity of variances or covariances. However, if $f < f_{\alpha,1,r-1}$, one should be cautious in claiming high confidence that treatment differences exist. To obtain a more exact evaluation of confidence, it is necessary to examine the validity of assumptions and decide whether the original f test was accurate or proceed to a multivariate appraisal of the data (Sec. 8.4). For the special case where heterogeneous covariance can be described adequately by first order serial correlation, one could avoid the multivariate method by approximating the proper degrees of freedom for the f test (Box 1954b). Experience suggests that restriction of serial correlation to first order statistics often is not realistic for animal data.

Experimenters normally desire more specific inferences than those provided by the f test of overall treatment differences or trend. When the assumption of homogeneous variance-covariance structure is valid, the usual assortment of techniques for multiple comparisons and contrasts (Chap. 2) should be valid. Special attention to orthogonal polynomial contrasts may be in order for analysis of the shape of a regular (smooth) trend, especially if the sampling intervals are equal. When it is apparent that the variance-covariance structure is heterogeneous, paired tests of the mean differences between any two sampling points (treatments) should be used. If the tests to be performed are independent in the sense that data at a given sampling point are used in no more than one test, the ordinary paired t statistic may be used (Chap. 1). However, if multiple comparisons involving the same sampling points are desired, a modified Scheffé statistic should be used. The modification consists of substituting $f_{\alpha,1,r-1}$ for $f_{\alpha,t-1,n-t}$ and s_D^2/r for $V[\bar{q}_k]$, where $\bar{q}_k$ is the difference between the means at two sampling points, say $\bar{y}_{i.} - \bar{y}_{i'.}$ $(i \neq i')$, and s_D^2 is the variance of the sample *differences* on r subjects for the two sampling points. That is, if the approximate confidence interval,

$$\bar{y}_{i.} - \bar{y}_{i'.} \pm \sqrt{(t - 1)f_{\alpha,1,r-1}(s_D^2/r)} \qquad (i \neq i') \tag{8.58}$$

does not include zero, one may conclude that the mean difference is significant. The process may be repeated for each of the $t(t - 1)/2$ possible comparisons, or a Tukey-type procedure that allows heterogeneous variance may be used (Chap. 2).

Another procedure that may prove useful if the number of subjects r is not too small is based on contrast totals calculated for each subject, i.e., on $q_j = \sum_{i=1}^{t} c_i y_{ij}$ $(\Sigma c_i = 0)$. The test statistic, $t = \bar{q}/\sqrt{s_q^2/r}$, may be compared with critical values of Student's t, $\pm t_{\alpha/2,r-1}$ (App. A.4), for orthogonal contrasts (possibly polynomial) or with Bonferroni t, $\pm t_{B,\alpha/2,m,r-1}$ (App. A.10), for each of m nonorthogonal contrasts. This procedure largely avoids problems of correlated errors and questions about homogeneity of the variance-covariance structure.

As an example of a complete block design with repeated measurement, consider a study on the effects of different stages of estrus on serum ascorbic acid levels in rats. Data (μg/10 mL) are shown in Table 8.19. Although the

Table 8.19. Serum Ascorbic Acid (μg/10 ml) in Rats at Different Stages of Estrus

Block (rat)	1=Estrus	2=Metestrus	3=Diestrus	4=Proestrus	Total
1	54	50	55	56	215
2	56	47	50	52	205
3	53	58	51	60	222
4	52	60	58	43	213
5	50	52	58	53	213
6	50	46	44	56	196
7	50	54	49	50	203
8	50	42	50	50	192
9	45	44	45	44	178
10	48	50	45	43	186
11	55	45	52	50	202
12	55	52	47	55	209
13	52	43	49	49	193
14	53	49	54	55	211
15	52	45	49	49	195
16	49	52	52	50	203
Totals:	824	789	808	815	3236
Means:	51.50	49.31	50.50	50.94	50.56
Variances:	8.40	27.30	18.13	23.80	

Covariances: s_{12}=1.90, s_{13}=4.20, s_{14}=7.30, s_{23}=10.37, s_{24}=3.29, s_{34}=2.43

Correlations: r_{12}=0.12, r_{13}=0.34, r_{14}=0.52, r_{23}=0.47, r_{24}=0.13, r_{34}=0.12

variances at the 4 different stages of estrus are not independent, an f_{max} test (Chap. 2) may have some value for indicating validity of the assumption of homogeneous variance. The test statistic is f_{max} = 27.30/8.40 = 3.25. Pertinent critical values (App. A.6.1) are $f_{max,0.25,4,15}$ = 2.67 and $f_{max,0.10,4,15}$ = 3.41. One may have (approximately) 75% (but not 90%) confidence that the population variances are not equal. No simple test exists for equality of the 6 correlations because they are not independent and tend to be similar in magnitude. However the largest and smallest correlations in a set of sample correlations tend to differ considerably by chance alone if the samples are small. A rather crude approximate test of the largest difference requires the z transformation (Chap. 1). The largest correlation, $r_{14} = 0.52$, becomes $z_{14} = 0.576$ and the smallest, $r_{12} = 0.12$ (or r_{34}), becomes $z_{12} = 0.121$. Then the test statistic is $z = (0.576 - 0.121)/\sqrt{2/(16 - 3)} = 1.16$, which is smaller than the standard normal value $z_{1-\alpha/2} = z_{0.9} = 1.28$ but exceeds $z_{0.875} = 1.15$. Therefore, one could have approximately 75% confidence that the two population correlations differ. As often happens in such experiments, it is not clear whether the assumption of homogeneous variance-covariance structure is valid or not. (See Sec. 8.4 for a more rigorous evaluation procedure.) The analysis of variance is summarized in Table 8.20. The test

Table 8.20. Analysis of Variance of Serum Ascorbic Acid at Stages of Estrus in Rats (see Table 8.19)

Source of Variation	df	*ss*	*ms*	$E[MS]$
Treatments (stages of estrus)	3	41.38	13.79	$\sigma^2(1-\rho)+16\Sigma\tau_i^2/3$
Blocks (rats)	15	512.25	34.15	$\sigma^2(1+3\rho)$
Error	45	652.12	14.49	$\sigma^2(1-\rho)$

of $H{:}\tau_i = 0$ requires $f = 13.79/14.49 = 0.95$, which is much smaller than $f_{0.05,3,45} = 2.82$. Very little evidence exists that serum ascorbic acid of rats is affected by stage of estrus, and the question of validity of assumptions holds little interest. Note that the conservative critical value, $f_{0.05,1,15} = 4.54$, would require considerably larger sample differences in treatment means for significance to be claimed in the face of heterogeneous variance-covariance structure. Had the stages of estrus been significantly different in effect, one might have selected Dunnett's procedure (Chap. 2) to compare results at the first stage of estrus (control) with results at each of the other stages.

If one equates the mean squares for blocks and error to their respective expectations, a set of two equations may be solved to estimate σ^2 and ρ:

$$\hat{\sigma}^2(1 + 3\hat{\rho}) = 34.15 \qquad \hat{\sigma}^2(1 - \hat{\rho}) = 14.49$$

By subtracting the second equation from the first, one obtains $\hat{\rho}\hat{\sigma}^2 = (34.15 - 14.49)/4 = 4.92$. Substitution of that result into the second equation provides $\hat{\sigma}^2 = 14.49 + 4.92 = 19.41$; then $\hat{\rho} = 1 - (14.49/19.41) = 0.25$. Note that the average of the 4 variances in Table 8.19 provides precisely the same estimate of σ^2, and the average of the 6 covariances divided by the average of the 4 variances provides the same estimate of ρ. Of course those estimates are realistic only if the variance-covariance structure of the population is truly homogeneous. Little can be said about efficiency, except that the ratio, $(\hat{\sigma}^2/ms_E)100\% = (19.41/14.49)100\% = 134\%$, indicates 34% better efficiency from measuring the same rats instead of different rats at each stage of estrus. A more complete evaluation, with adjustment for differences in degrees of freedom (Chap. 5) would reduce the apparent efficiency slightly.

8.2.2. Split-Plot Designs (subject as incomplete block)

Experiments in which animals or subjects are assigned randomly to the levels of a treatment factor (or to treatment combinations of 2 or more factors) and then are measured for trend at several sampling times or spaces (a repeated factor) have become increasingly common. In some disciplines, such as reproductive physiology, they now are the most frequently used designs. The structure of the underlying causes of variation often is oversimplified in analysis of such data. Unjustified simplification is of two kinds: (1) analysis of the factorial data as if the design were completely randomized instead of split plot and (2) ignorance of the correlation of errors induced by repeated measurement. When the split-plot structure is not recognized, the experimenter performs the analysis as if each observation were measured on a different subject and fails to separate the random error into variation among and variation within subjects. Consequently, the apparent significance of differences among the means of treatments to which the subjects were assigned often will be grossly exaggerated, whereas the sensitivity of the tests for trend or interaction (different trends over time or space for different treatments) will be seriously reduced. When the correlation of errors is ignored, inferences may or may not be distorted (in a probability sense), depending on the heterogeneity or homogeneity of the pattern of variances and covariances of data taken at the several sampling points on the same subjects.

A model for the simplest experiment is

$$Y_{ijk} = \mu + \alpha_i + D_{(i)j} + \beta_k + (\alpha\beta)_{ik} + (D\beta)_{(i)jk} + E_{(ijk)} \tag{8.59}$$

where $i = 1, 2, \ldots, a$; $j = 1, 2, \ldots, r$ per i; $k = 1, 2, \ldots, b$; and $n = arb$. The α_i are effects of the various levels or classes of factor A to which subjects are randomly assigned. The $D_{(i)j}$ represent random effects of subjects nested within A and correspond to the $E_{(i)j}$ of the model for a completely randomized design without the repeated measurement. The β_k are effects of time or space at the various sampling points in the process of repeated measurement of the subjects. They are cross classified with the α_i because the same set of sampling points is used for all levels of A. Therefore interaction of A and B, $(\alpha\beta)_{ik}$, is possible and indeed may be the most important aspect of the experiment, for significant interaction indicates nonparallel trends in response over time or space for the several levels of A. The interaction of subjects with time or space, $(D\beta)_{(i)k}$, ordinarily is not separable from the residual error within subjects, $E_{(ijk)}$, unless 2 or more random observations are taken from each subject at each sampling point specified by factor B. In many experiments an assumption of uniform trend in response (for animals treated alike with respect to factor A) is reasonable, and random samples or duplicates at each point are considered unnecessary. If subjects are not likely to show uniform response, they should be blocked with random assignments of subjects to levels of A within each block (Sec. 8.2.3). Alternatively, the random sampling process could be used at each point specified by B to provide "pure" error to test the $(D\beta)_{(i)jk}$, or some post facto methods of Kosswig (1970, 1971) may be used to partition the $DB + E$ sum of squares into interaction and residual error without having duplicate observations on each animal at each sampling point.

Suppose the assumptions of uniform trend among animals treated alike, uniform pattern of variances and covariances among the sampling points within each treatment (A), and uniform variance-covariance structure from level to level of A are valid. Then the appropriate partition of the total sum of squares may be determined from the general rules of Chap. 2. For total (y), factor (A), subjects within that factor (D/A), time or space factor (B), interaction (AB), and residual error (E), the respective sums of squares are

$$ss_y = \sum_{i=1}^{a} \sum_{j=1}^{r} \sum_{k=1}^{b} y_{ijk}^2 - (y_{...}^2/n) \tag{8.60}$$

$$ss_A = (\sum_{i=1}^{a} y_{i..}^2/br) - (y_{...}^2/n) \tag{8.61}$$

$$ss_{D/A} = (\sum_{i=1}^{a} \sum_{j=1}^{r} y_{ij.}^2/b) - (\sum_{i=1}^{a} y_{i..}^2/br) \tag{8.62}$$

$$ss_B = (\sum_{k=1}^{b} y_{..k}^2/ar) - (y_{...}^2/n) \tag{8.63}$$

$$ss_{AB} = (\sum_{i=1}^{a} \sum_{k=1}^{b} y_{i.k}^2/r) - (\sum_{i=1}^{a} y_{i..}^2/br) - (\sum_{k=1}^{b} y_{..k}^2/ar) + (y_{...}^2/n) \tag{8.64}$$

$$ss_E = ss_y - ss_A - ss_{D/A} - ss_B - ss_{AB} \tag{8.65}$$

If the numbers of subjects differ from treatment to treatment (classes of A), one may use the rules for analysis of unweighted means (Chap. 2) to compute approximate sums of squares for A and AB. Also,

$$ss_{D/A} = (\sum_{i=1}^{a} \sum_{j=1}^{r_i} y_{ij.}^2/b) - (y_{...}^2/n) - ss_A$$

in that case, but ss_B and ss_E may be computed as usual. Also, see McNee and Crump (1962) and Snee (1972). Note that some computer programs for analysis of variance do not have sufficient flexibility to handle such problems.

The analysis of variance is summarized in Table 8.21, with expected mean squares reflecting constant correlation (ρ) between responses at any two sampling points in time or space for the subjects within any treatment group. In most cases the correlation is positive, so error_b is considerably smaller than error_a. Therefore, inferences about B and AB (the trends) may be made with greater precision than those about the average effects of A (the treatments).

Table 8.21. Analysis of Variance of Split-Plot Data with Repeated Measurement (A=treatments, B=time or space)

Source of Variation	df	ss	ms	$E[MS]$
Among subjects	$ar-1$			
Treatments (A)	$a-1$	ss_A	ms_A	$\sigma^2[1+(b-1)\rho]+br\sum_{i=1}^{a}\alpha_i^2/\nu_A$
Subjects/A ($error_a$)	$a(r-1)$	$ss_{D/A}$	$ms_{D/A}$	$\sigma^2[1+(b-1)\rho]$
Within subjects	$ar(b-1)$			
Time or space (B)	$b-1$	ss_B	ms_B	$\sigma^2(1-\rho)+\sigma_{DB}^2+ar\sum_{k=1}^{b}\beta_k^2/\nu_B$
AB	$(a-1)(b-1)$	ss_{AB}	ms_{AB}	$\sigma^2(1-\rho)+\sigma_{DB}^2+r\sum_{i=1}^{a}\sum_{k=1}^{b}(\alpha\beta)_{ik}^2/\nu_{AB}$
(Subjects/A) x B ($error_b$)	$a(r-1)(b-1)$	ss_E	ms_E	$\sigma^2(1-\rho)+\sigma_{DB}^2$

Let the average sample variance among the observations at one sampling point for r subjects in the same treatment group be designated $\overline{s_k^2}$, and let the average sample covariance for two sampling points within a treatment group be designated $\overline{s}_{kk'}$ $(k \neq k')$. Then, $ms_{D/A} = \overline{s_k^2} + (b - 1)\overline{s}_{kk'}$ and $ms_E = \overline{s_k^2} - \overline{s}_{kk'}$.

The standard errors of a treatment A mean, a B mean, and a combination AB mean are

$$s_{\bar{y}_{i..}} = \sqrt{ms_{D/A}/(br)} \tag{8.66}$$

$$s_{\bar{y}_{..k}} = \sqrt{ms_E/(ar)} \tag{8.67}$$

$$s_{\bar{y}_{i.k}} = \sqrt{ms_E/r} \tag{8.68}$$

Note that the standard error of the difference between two combination means sharing the same level or class of A is $\sqrt{2ms_E/r}$, whereas the standard error of the difference between two combination means at the same sampling point B is $\sqrt{2ms_{D/A}/r}$. The standard error of treatment means may be much larger than the standard error of means at the various sampling points B, even when the same number of observations is involved in each type of mean. For example, if there are 4 treatments and 4 sampling points ($a = b = 4$), the ratio of (8.66) to (8.67) is 2, 3, or 4 if the correlation between responses at any two sampling points is $\rho = 0.43$, 0.67, or 0.79, respectively.

The hypothesis of zero interaction, $H{:}(\alpha\beta)_{ik} = 0$, should be tested before testing main effects. The test statistic

$$f = ms_{AB}/ms_E \tag{8.69}$$

should be compared with $f_{\alpha,\nu_{AB},\nu_E}$. If the validity of assuming equality of variance-covariance structure from level to level of A is doubtful, the Geisser-Greenhouse (1958) conservative critical value $f_{\alpha,a-1,a(r-1)}$ may be used. See Sec. 8.4 for a test of equality of variance-covariance structures. If the interaction is judged to be nonsignificant, the trends in response over time or space may be relatively parallel for the various levels of factor A. Furthermore, the usual tests of significance for A and B may be carried out independently without exaggerating the generality of inferences thus achieved. For $H{:}\alpha_i = 0$, compare

$$f = ms_A/ms_{D/A} \tag{8.70}$$

with critical value $f_{\alpha,a-1,a(r-1)}$ and extend the inferences to more specific comparisons by using the various techniques for multiple comparisons of means. For $H{:}\beta_k = 0$, compare

$$f = ms_B/ms_E \tag{8.71}$$

ith critical value $f_{\alpha,b-1,\nu_E}$ and likewise make specific comparisons. If the ample variance-covariance structure is relatively uniform from one level of A o the next but is heterogeneous among the sampling points in time or space, he validity of the test of $H{:}\alpha_i = 0$ or specific comparisons among treatment eans is not affected but the test statistic of (8.71) for $H{:}\beta_k = 0$ should be ompared with critical value $f_{\alpha,1,a(r-1)}$ for a conservative test or one should urn to a more accurate multivariate test (Sec. 8.4). In the heterogeneous ase, specific differences between the means of all pairs of sampling points ay be tested approximately by a modified Scheffé interval,

$$\bar{y}_{..k} - \bar{y}_{..k'} \pm \sqrt{(b - 1)f_{\alpha,1,a(r-1)}(\overline{s_D^2}/ar)} \quad (k \neq k') \tag{8.72}$$

here $\overline{s_D^2}$ is the average of sample variances of differences from the several evels of factor A, each variance being computed from the r sample differences or the two sampling points in question.

A special question, often of interest whether or not interaction is significant, concerns the difference in mean trend (over time or space) between 2 reatments. Here the question involves not just 2 means but 2 profiles or ectors of means. Such questions may be resolved accurately by multivariate rocedures (Sec. 8.4) if the variance-covariance structure is the same from treatment to treatment. The structure need not be homogeneous across sampling oints.

Consider procedures for making inferences about treatments (A) and the time or space factor (B) when interaction has been judged significant. If interaction is important, one should make comparisons of the treatment means separately at various sampling points. Also, one may wish to compare sampling oints for trend separately for each treatment. Selection of procedures best fitted for those comparisons depends on the variance-covariance structure. First, consider testing the treatments (A) at each sampling point in time or space. There will be b separate tests of $H{:}\mu_{1.k} = \mu_{2.k} = \cdots = \mu_{a.k}$ ($k = 1,$ 2, ..., b), where $\mu_{i.k} = \alpha_i + \beta_k + (\alpha\beta)_{ik}$. If the treatment group variances are equal at a given sampling point ($\sigma_{1k}^2 = \sigma_{2k}^2 = \cdots = \sigma_{ak}^2$), one may analyze the ar observations at any point as a one-way analysis of variance followed by selected specific comparisons (see Chap. 2). The b different analyses of variance are not independent, so the overall tendency will be toward claiming more "significant" results erroneously than would be indicated by the nominal Type I error unless one uses Scheffé's method or Bonferroni t statistics evaluated for the total number of comparisons. If the variances σ_{1k}^2, σ_{2k}^2, etc., are not equal, Box's (1954a) approximations for degrees of freedom for the critical value may be used [see (2.43)]. Remember that at a single sampling point, $n = ar$ observations and $a = t$ for the one-way analysis. The Box procedure may be applied even when interaction is not significant if the variance-covariance structure differs from treatment to treatment. For specific comparisons one may use the Box degrees of freedom to modify Scheffé's procedure or use the Dunnett-type or Tukey-type procedures for heterogeneous variance (Chap. 2).

Second, consider comparing the sampling points (B) for trend within each treatment group. There will be a separate tests of $H: \mu_{i.1} = \mu_{i.2} = \cdots = \mu_{i.b}$ ($i = 1, 2, \ldots, a$), where $\mu_{i.k} = \alpha_i + \beta_k + (\alpha\beta)_{ik}$ as before. Obviously, these tests cannot be independent of the tests of treatments at each sampling point so caution is in order before strongly claiming significance of effects. If the variance-covariance structure is homogeneous for any given treatment (the correlation between responses at any two sampling points is relatively constant for that treatment), the hypothesis may be tested separately for each treatment group, as in the complete block analysis of Sec. 8.2.1, followed by the usual specific comparisons. One may use the Scheffé procedure or Bonferroni t statistics evaluated for the entire set of comparisons. When constant correlation cannot be claimed, the Geisser-Greenhouse conservative critical value $f_{\alpha,1,r-1}$ should be used, or one should turn to a more accurate multivariate test (Sec. 8.4). Lacking constant correlation, one may make pair-wise comparisons of means at any two sampling points (within one treatment group) by using the modified Scheffé interval of (8.58). Note that the mean differences here are $\bar{y}_{i.k} - \bar{y}_{i.k'}$ ($k \neq k'$), and $b = t$. These procedures may be applied even when interaction is not significant if the variance-covariance structure differs from treatment to treatment.

A number of questions about trends may be addressed by computing contrast totals across time (possible polynomial) for individual animals or subjects:

$$q_{ij} = \sum_{k=1}^{b} c_k y_{ijk} \qquad (\Sigma c_k = 0)$$

This procedure largely eliminates difficulties with heterogeneous variance-covariance structure across time. However, if variances differ from treatment to treatment, additional modifications are required to compare treatments for trend. One may test differences among treatments for any contrast in question (e.g., linear trend or curvature via orthogonal polynomials) by performing the usual procedures for analysis of variance and multiple comparisons (Chap. 2) for the model

$$Q_{ij} = \mu + \alpha_i + D_{(i)j}$$

where $i = 1, \ldots, a$ treatments, $j = 1, \ldots, r$ subjects per group, using the q_{ij} as data and the $\bar{q}_{i.}$ as means. Box's adjustments for degrees of freedom may be used if variances ($s^2_{q_i}$) indicate heterogeneity from treatment to treatment. Furthermore, in the presence of significant interaction between treatments and time, one may wish to test the significance of a specific trend contrast within each treatment group by comparing $t = \bar{q}_{i.}/\sqrt{s^2_{q_i}/r}$ with $\pm t_{\alpha/2,r-1}$. For a set of m such contrasts that are not orthogonal, use the Bonferroni critical values $\pm t_{B,\alpha/2,m,r-1}$ (App. A.10).

As an example of a split-plot experiment with repeated measurement, consider a study of the influence of pregnancy on concurrent lactational performance of rats, measured by litter weight gains (Paape and Tucker 1969). Litter size was adjusted to maintain 1 suckling pup per mammary gland and litters were replaced every 4 days to maintain intense suckling stimulus. Gill and Hafs (1971) examined the data in considerable detail. The two treatments (*A*) were pregnant and nonpregnant. The 4 sampling points (*B*) were 8-12, 12-16, 16-20, and 20-24 days of lactation. Data are presented in Table 8.22. To il-

Table 8.22. Litter Weight Gains (*g*) for Pregnant and Nonpregnant Lactating Rats (split-plot design with repeated measurement)

		B=Period of Lactation (days)					
A=Treatment Group	Rat Number	8-12	12-16	16-20	20-24	Total	Mean
Pregnant	1	7.5	8.6	6.9	0.8	23.8	5.95
	2	10.6	11.7	8.8	1.6	32.7	8.18
	3	12.4	13.0	11.0	5.6	42.0	10.50
	4	11.5	12.6	11.1	7.5	42.7	10.68
	5	8.3	8.9	6.8	0.5	24.5	6.12
	6	9.2	10.1	8.6	3.8	31.7	7.92
	Total	59.5	64.9	53.2	19.8	197.4	
	Mean	9.92	10.82	8.87	3.30		8.23
	Variance	3.62	3.57	3.55	8.03		
Nonpregnant	7	13.3	13.3	12.9	11.1	50.6	12.65
	8	10.7	10.8	10.7	9.3	41.5	10.38
	9	12.5	12.7	12.0	10.1	47.3	11.82
	10	8.4	8.7	8.1	5.7	30.9	7.72
	11	9.4	9.6	8.0	3.8	30.8	7.70
	12	11.3	11.7	10.0	8.5	41.5	10.38
	Total	65.6	66.8	61.7	48.5	242.6	
	Mean	10.93	11.13	10.28	8.08		10.11
	Variance	3.40	3.17	4.01	7.70		
Total		125.1	131.7	114.9	68.3	440.0	
Mean		10.42	10.98	9.58	5.69		9.17

lustrate the folly of ignoring the split-plot structure, compare the analysis of variance shown in Table 8.23, in which it has been assumed (falsely) that the data came from a completely randomized design (CRD) with 6 different rats for each of the 8 factorial combinations (see Chap. 2), to the split-plot analysis shown in Table 8.24. The erroneous analysis of Table 8.23 indicates highly significant differences between pregnant and nonpregnant rats and among periods but little evidence of interaction, i.e., the trends over time for the two groups appear to be relatively parallel. However, when the analysis is performed properly (Table 8.24), one obtains quite different conclusions. The average difference between pregnant and nonpregnant rats is not significant but the interaction is highly significant, indicating nonparallel trends over time for the two groups. The striking difference in inferences from the two

Table 8.23. Analysis of Litter Weight Gains as if Experiment Were Completely Randomized (see Table 8.22)

Source of Variation	df	*ss*	*ms*	*f* Ratio	*E*[*MS*]
Treatments (*A*)	1	42.56	42.56	9.17**	$\sigma^2+24\Sigma\alpha^2$
Periods (*B*)	3	205.15	68.38	14.74**	$\sigma^2+12\Sigma\beta^2/3$
AB	4	35.50	11.87	2.56	$\sigma^2+6\Sigma\Sigma(\alpha\beta)^2/3$
Error	40	185.60	4.64		σ^2

**$P<0.01$ claimed falsely.

Table 8.24. Analysis of Litter Weight Gains as a Split-Plot Experiment (see Table 8.22)

Source of Variation	df	*ss*	*ms*	*f* Ratio	*E*[*MS*]
Treatments (*A*)	1	42.56	42.56	2.54	$\sigma^2(1+3\rho)+24\Sigma\alpha^2$
Rats/trt. (*D*/*A*)	10	167.54	16.75		$\sigma^2(1+3\rho)$
Periods (*B*)	3	205.15	68.38	113.59**	$\sigma^2(1-\rho)+\sigma_{DB}^2+12\Sigma\beta^2/3$
AB	3	35.50	11.87	19.72**	$\sigma^2(1-\rho)+\sigma_{DB}^2+6\Sigma\Sigma(\alpha\beta)^2/3$
Error_b	30	18.06	0.60		$\sigma^2(1-\rho)+\sigma_{DB}^2$

**$P<0.01$.

analyses arises from failure to split the random error among and within rats in the first analysis. Note that the sums of squares, $ss_{D/A}$ and ss_E, of the split-plot analysis together equal the error sum of squares in the first analysis. A comparison of standard errors for the various means of interest is shown in Table 8.25. See (2.138)-(2.140) for the CRD and (8.66)-(8.68) for the split-plot design. For the example at hand, standard errors from the CRD analysis are too small by a factor of 2 for treatment means. They are too

Table 8.25. Comparison of Standard Errors of Means for CRD and Split-Plot Design (see Tables 8.23, 8.24)

Design	Treatment Means (*A*)	Period Means (*B*)	Combination Means (*AB*)
CRD (erroneous)	±0.44	±0.62	±0.88
Split-plot	±0.84	±0.22	±0.32

arge by a factor of almost 3 for period means or combination means. For the ata of Table 8.22, the average of the 8 variances, each computed within an AB ombination, is $\overline{s_k^2} = 4.640$, and the average of the covariances (for each two eriods, within treatment) is $\overline{s}_{kk'} = 4.038$. Note that

$$ms_{D/A} = 4.640 + 3(4.038) = 16.75 \qquad ms_E = 4.640 - 4.038 = 0.60$$

s in Table 8.24. If σ_{DB}^2 may be assumed to be negligible, one may equate $s_{D/A}$ and ms_E to their respective expectations to form a set of two equations o estimate σ^2 and ρ:

$$\hat{\sigma}^2[1 + 3\hat{\rho}] = 16.75 \qquad \hat{\sigma}^2(1 - \hat{\rho}) = 0.60$$

y subtracting the second equation from the first, one obtains $\hat{\rho}\hat{\sigma}^2 = (16.75 -$ $.60)/4 = 4.038$, which is precisely the same as $\overline{s}_{kk'}$. Substitution of that esult into the second equation provides $\hat{\sigma}^2 = 0.60 + 4.038 = 4.64$, which is he same as $\overline{s_k^2}$. Finally $\hat{\rho} = 1 - (0.60/4.64) = 0.87$, which is the same as $\overline{s}_{kk'}/\overline{s_k^2}$. Note that

$$(\hat{\sigma}^2/ms_E)100\% = (4.64/0.60)100\% = 773\%$$

fficiency for testing period effects and interaction, relative to a CRD with ifferent rats in each period. Also,

$$(\hat{\sigma}^2/ms_{D/A})100\% = (4.63/16.75)100\% = 28\%$$

he apparent efficiency for testing treatments. However, the number of rats equired per treatment was 6 instead of 24, only 25% as many as in a CRD. herefore, split plots and CRDs with an equal number of rats per treatment hould provide essentially the same sensitivity for testing treatments.

The question of validity of the analysis in Table 8.24 still remains. he f ratio for interaction ($f = 19.72$) is so large that it exceeds the conservative critical value $f_{0.01,1,10} = 10.0$, leaving little doubt that the time rends differ for pregnant and nonpregnant rats. It can be shown via multivariate techniques that the two vectors of means over time are significantly ifferent. The apparent significance of interaction suggests that comparisons f treatments within periods and of periods within treatments are in order. he variances at a given period (Table 8.22) are not significantly different or the pregnant and nonpregnant rats (see Sec. 8.4 for a more comprehensive est of equality of variance-covariance structure for the two groups). Therefore, ordinary t tests (using $ms_{D/A}$) should provide reasonably valid evidence f treatment differences at each period. For the 4 periods, t statistics are .427, 0.131, 0.597, and 2.023; and the 10% critical values are $\pm t_{0.05,10} =$

±1.812 in each case. Because of lack of independence of the four tests, one may feel uneasy in claiming that pregnancy restricted gains in period 4.

Section 8.4 shows that whereas the variance-covariance structure is not significantly different for pregnant and nonpregnant rats, the structure is heterogeneous with respect to time (variances and covariances involving responses in period 4 are larger than others). Therefore, the period means should be compared separately for each treatment group (as in Table 8.18), and the conservative critical value $f_{\alpha,1,5}$ should be applied. Results are shown in Table 8.26. The f ratios for period effects are 98.55 and 23.08 for pregnant and nonpregnant rats, respectively. Since $f_{0.05,1,5} = 6.61$ and $f_{0.01,1,5} = 16.3$, time trends are clearly highly significant in both cases, although the sample differences among periods were much larger for pregnant rats (see the low mean for pregnant rats in period 4). More specifically, the modified Scheffé procedure [(8.58)] may be used to compare any two periods within a treatment group. The 95% confidence interval,

Table 8.26. Analysis of Period Effects on Litter Weight Gains for Pregnant and Nonpregnant Lactating Rats (see Table 8.22)

Source of Variation	df	*ms* (pregnant)	*ms* (nonpregnant)
Periods	3	68.492	11.725
Rats	5	16.685	16.824
Error	15	0.695	0.508

$$\bar{y}_{i.k} - \bar{y}_{i.k'} \pm \sqrt{(b - 1)f_{0.05,1,r-1}(s_D^2/r)}$$

requires two means to differ by $\sqrt{3(6.61)(s_D^2/6)} = \sqrt{(3.3)s_D^2}$ for significance, where s_D^2 is the variance of 6 differences for the two periods in a given comparison. A 99% confidence interval requires a difference of $\sqrt{8.15s_D^2}$. The means and variances of differences are shown in Table 8.27. For pregnant rats, all period differences are highly significant except 1 versus 3, which was nearly significant at 95% confidence. However, for nonpregnant rats only period 4 was significantly different from the others.

The split-plot analysis may be extended for experiments in which subjects are assigned to treatment combinations instead of treatments by using the rules of Chap. 2 to obtain proper formulas for sums of squares. For example, consider a study of the effects of 4 types of diets (α) in combination with 3 types of sugars (β) on blood glucose (Y) in rats (D/AB), with sequential measurements on each rat at short intervals (γ) after feeding. A model for such an experiment may be written

Table 8.27. Means and Variances of Differences of Litter Weight Gains between Periods, within Treatment Groups (see Table 8.22)

	Pregnant		
	Period 2	3	4
Period	$\bar{y}_D(s_D^2)$	$\bar{y}_D(s_D^2)$	$\bar{y}_D(s_D^2)$
1	0.90** (0.060)	1.05 (0.343)	6.62** (3.090)
2		1.95** (0.279)	7.52** (2.974)
3			5.57** (1.595)
	Nonpregnant		
	2	3	4
Period	$\bar{y}_D(s_D^2)$	$\bar{y}_D(s_D^2)$	$\bar{y}_D(s_D^2)$
1	0.20 (0.020)	0.65 (0.323)	2.85* (2.063)
2		0.85 (0.427)	3.05* (2.183)
3			2.20* (1.084)

*$P<0.05$, **$P<0.01$.

$$Y = \mu + \alpha_i + \beta_j + (\alpha\beta)_{ij} + D_{(ij)k} + \gamma_l + (\alpha\gamma)_{il} + (\beta\gamma)_{jl}$$

$$+ (\alpha\beta\gamma)_{ijl} + (D\gamma)_{(ij)kl} + E_{(ijkl)} \tag{8.73}$$

ee the exercises at the end of this chapter for another example.

Kosswig (1970, 1971) developed a method of partitioning the error sum of quares into portions for subject-period interaction and residual error. ithin each treatment or treatment combination one may reduce the data on r ubjects over b periods to $r - 1$ sets of b observations each by subtracting ata for each of subjects 2, 3, etc., from data of the first subject. For ach of the $r - 1$ sets, one may compute orthogonal polynomial contrasts up to egree $p \leq (b - 2)$ if desired. For a set of p contrasts then, there are $a(r - $ $)p$ df across $r - 1$ sets in a treatment groups. The sums of squares for conrasts,

$$ss_R = \sum_{i=1}^{a} \sum_{j=1}^{r-1} \sum_{m=1}^{p} [(q_m)_{ij}^2 / b \sum_{k=1}^{b} (\xi'_{km})^2] \tag{8.74}$$

ay be subtracted from ss_E (Table 8.21) to provide residual error ss'_E. Then he hypothesis of no subject-period interaction may be tested by comparing the test statistic,

$$f = [ss_R/a(r-1)p]/[ss_E'/a(r-1)(b-p-1)] \tag{8.75}$$

with critical value $f_{\alpha,a(r-1)p,a(r-1)(b-p-1)}$. Ideally, if significant subject-period interaction is considered to be a mere nuisance, one should design subsequent experiments by blocking the subjects into homogeneous groups with separate random assignment to treatments in each block (Sec. 8.2.3).

Most split-plot experiments with repeated measurement involve time trends (period effects); a few involve repetition in space. For example, an experiment was designed to study testosterone production in slices of bull testes incubated in vitro with various doses of luteinizing hormone (LH). Eight bull calves were castrated at birth, 8 each at 1 month, 3 months, and 5 months of age. Six slices of testis were taken from each bull, one for each of 6 doses of LH (0, 0.005, 0.02, 0.25, 0.5, 1.0 μg). Model (8.59) adequately describes the situation with α_i = age effects and β_j = dose effects instead of period effects. The preceding discussion applies; an advantage here is the ability to assign doses to slices randomly, which may minimize difficulties with variance-covariance structure so that the standard analysis (Table 8.21) is more likely to be adequate than in cases where time is involved. A slightly more complex case involved the inoculation of pork loins with different organisms and measurement of response at different temperatures and lengths of incubation. Five organisms (A) were selected for study and 20 animal carcasses (D/A) were randomly assigned, 4 to each organism. Fourteen samples of loin were taken from each carcass. Within each set of 14, samples were randomly assigned to 14 treatment combinations of factor B (high or low temperature) and factor C (7 different lengths of incubation). An appropriate model is

$$Y_{ijkl} = \mu + \alpha_i + D_{(i)j} + \beta_k + \gamma_l + (\alpha\beta)_{ik} + (\alpha\gamma)_{il} + (\beta\gamma)_{kl} + (\alpha\beta\gamma)_{ikl}$$

$$+ (D\beta)_{(i)jk} + (D\gamma)_{(i)jl} + (D\beta\gamma)_{(i)jkl} + E_{(ijkl)} \tag{8.76}$$

See the rules of Chap. 2 to determine sums of squares and find which components belong in the expected mean squares. The coefficients of components specified by the rules also will be correct for all components except the random components involving subjects (D).

8.2.3. Other Designs

Many possibilities exist for variations in design structure for experiments with repeated measurement. Most involve the split-plot or split-error idea in one way or another. Some involve double or triple splitting in time, space, or both. Others involve split error in the cells of changeover designs (Sec. 8.1.3) or changeovers inside the common split-plot designs. The structures, so many and varied as to almost defy classification, are illustrated here by case studies. The following experimental situations with brief descriptions and models are real biological problems, selected to represent the broad scope of possibilities with repeated measurement.

Case 1: Split Plot with Blocking of Subjects (split block)

Sixty cows were paired (D) by prior milk production records. Dietary treatments (α) were thyroprotein and control, assigned randomly within each pair of similar cows. Milk production was recorded by month for 10 months (β). The model was

$$Y_{ijk} = \mu + \alpha_i + D_j + (\alpha D)_{ij} + \beta_k + (\alpha\beta)_{ik} + (D\beta)_{jk} + (\alpha D\beta)_{ijk} + E_{(ijk)} \quad (8.77)$$

where pair-treatment interaction (αD) served as error$_a$ to test treatments, as in an RCBD, and $D\beta + \alpha D\beta + E$ served as error$_b$ to test months and interaction of treatments with months.

Case 2: Split Plot with Subjects Treated in Groups

In a poultry experiment with 4 treatments (α), 48 birds were grouped homogeneously, 4 to a cage (D/A). The 12 cages were assigned randomly to treatments. Birds [$B/(D/A)$] were measured individually at 4 sampling times (γ). The model was

$$Y_{ijkl} = \mu + \alpha_i + D_{(i)j} + B_{(ij)k} + \gamma_l + (\alpha\gamma)_{il} + (D\gamma)_{(i)jl} + (B\gamma)_{(ij)kl} + E_{(ijkl)} \quad (8.78)$$

where $D_{(i)j}$ was error$_a$ for testing treatments and the interaction of cage and time served as error$_c$ for comparisons of sampling points and interaction of treatments with time. Of course, any physical differences or differences in handling from cage to cage were confounded with treatments (see Sec. 5.9).

Case 3: Split Plot with Replication and Subsampling at Each Sampling Point

A microbiological study involved 4 chickens (D/A) in each of 3 sensitization groups (α). Two plates (E) were cultured for each of 4 antigens (β) for each bird. Subsampling consisted of 8 spots (U) on each plate. The model was

$$Y_{ijklm} = \mu + \alpha_i + D_{(i)j} + \beta_k + (\alpha\beta)_{ik} + (D\beta)_{(i)jk} + E_{(ijk)l} + U_{(ijkl)m} \quad (8.79)$$

where $D_{(i)j}$ was error$_a$ for comparing sensitization groups and $D\beta + E$ was error$_b$ for inferences concerning antigens.

Case 4: Split Plot within Subjects

Four cows (D) were treated with 15 mg prostaglandin-$F_{2\alpha}$. Three samples of blood were drawn from each cow. In one sample prostaglandin was measured in plasma at 4°C, the second in serum at 4°C, and the third in serum at 22°C (classes of factor A). For each sample, readings were made at 5, 10, 15, 20, 25, 30, 40, 50, 60, and 90 minutes (classes of factor B). The model was

$$Y_{ijk} = \mu + \alpha_i + D_j + (\alpha D)_{ij} + \beta_k + (\alpha\beta)_{ij} + (D\beta)_{jk} + (\alpha D\beta)_{ijk} + E_{(ijk)} \quad (8.80)$$

which is the same as (8.77) except that here D_j is a single cow instead of a pair or block of cows.

Case 5: Double Split Plot in Time

Serum glucose was measured in 16 rats (D/AB) randomly assigned to 8 treatment combinations of A (oral contraceptive or control) and B (4 different dietary sugars). Measurements were taken in each of 2 weeks (γ). Once each week, blood samples were taken from each rat at 3 times (τ) after feeding (0, 20, 40 min). The model was

$$\begin{aligned} Y_{ijklm} &= \mu + \alpha_i + \beta_j + (\alpha\beta)_{ij} + D_{(ij)k} + \gamma_l + (\alpha\gamma)_{il} + (\beta\gamma)_{jl} \\ &+ (\alpha\beta\gamma)_{ijl} + (D\gamma)_{(ij)kl} + \tau_m + (\alpha\tau)_{im} + (\beta\tau)_{jm} + (\alpha\beta\tau)_{ijm} \\ &+ (\gamma\tau)_{lm} + (\alpha\gamma\tau)_{ilm} + (\beta\gamma\tau)_{jlm} + (\alpha\beta\gamma\tau)_{ijlm} + (D\tau)_{(ij)km} \\ &+ (D\gamma\tau)_{(ij)klm} + E_{(ijklm)} \end{aligned} \tag{8.81}$$

where $D_{(ij)k}$ is error_{ab} for testing AB treatment combinations, $(D\gamma)_{(ij)kl}$ is error_c for testing effects concerning weeks, and the last three terms jointly serve as error_t to test effects concerning time after feeding.

Case 6: Double Split Plot in Space

Prolactin released by bovine pituitaries was studied in vitro. Pituitaries were removed from intact and castrated males (λ); 4 subjects were of each type (S/L). Each pituitary was divided into central and peripheral regions (ρ); then each region was divided into 8 slices. The experimenter wished to examine 16 treatment combinations (A = 1 or 5 particles per slice, B = 2 or 4 hours preculture time, C = 199 or KRBB culture medium, and D = 1 or 2 hours culture time). Since the pituitaries were too small for each region to provide 16 slices of sufficient size, the experimenter used only one-half replicate of the 2^4 treatment combinations to estimate the main effects (see Sec. 6.3.4). Main effects had 3-factor aliases and the 2-factor interactions were confounded among themselves ($AB = CD$, $AC = BD$, $AD = BC$). The model was

$$\begin{aligned} Y_{ijklmno} &= \mu + \lambda_i + S_{(i)j} + \rho_k + (\lambda\rho)_{ik} + (S\rho)_{(i)jk} + \alpha_l + \beta_m \\ &+ \gamma_n + \delta_o + (\lambda\alpha)_{il} + (\lambda\beta)_{im} + (\lambda\gamma)_{in} + (\lambda\delta)_{io} + (\rho\alpha)_{kl} \\ &+ (\rho\beta)_{km} + (\rho\gamma)_{kn} + (\rho\delta)_{ko} + (\lambda\rho\alpha)_{ikl} + (\lambda\rho\beta)_{ikm} + (\lambda\rho\gamma)_{ikn} \\ &+ (\lambda\rho\delta)_{iko} + (\alpha\beta + \alpha\gamma + \alpha\delta) + E_{(ijklmno)} \end{aligned} \tag{8.82}$$

where $S_{(i)j}$ is error_l for testing intact versus castrate; $(S\rho)_{(i)jk}$ is error_r for testing effects concerning region of the pituitary; and the residual error $E_{(ijklmno)}$, which includes interactions of $S_{(i)j}$ and $(S\rho)_{(i)jk}$ with estimable effects of the 2^4 factorial, is used to test the main effects and their interactions with type of male and region of pituitary.

Case 7: Double Split Plot in Space and Time

Pituitaries of 16 thyroidectomized rats (D/A) were used in an in vitro study of prolactin. Whole pituitaries were randomly assigned to 4 doses of T_3 (factor A). For each pituitary, half was subjected to one dose of TRH, the other half to a second dose (factor B). Response for each half was measured at 2, 4, 8, and 12 hours (factor C). The model was

$$Y_{ijkl} = \mu + \alpha_i + D_{(i)j} + \beta_k + (\alpha\beta)_{ik} + (D\beta)_{(i)jk} + \gamma_l + (\alpha\gamma)_{il}$$
$$+ (\beta\gamma)_{kl} + (\alpha\beta\gamma)_{ikl} + (D\gamma)_{(i)jl} + (D\beta)_{(i)jkl} + E_{(ijkl)} \quad (8.83)$$

where $D_{(i)j}$ is error$_a$ for testing dose of T_3, $(D\beta)_{(i)jk}$ is error$_b$ for testing dose of TRH and interaction of T_3 with TRH, and the last three terms combined are error$_c$ for testing effects involving time.

Case 8: Triple Split Plot in Time

Serum testosterone was measured in bull calves that were 2, 4, or 6 months of age at the start of the experiment. Six calves (D/AB) of each age (A) were injected twice daily for 28 days, 3 with 40 μg gonadotropin-releasing hormone, 3 with saline (classes of factor B). Blood samples were drawn one day a week for 5 weeks (γ), A.M. and P.M. each day (ρ), at 0, 0.5, and 3 hours (τ) after injection. The model was

$$Y_{ijklmn} = \mu + \alpha_i + \beta_j + (\alpha\beta)_{ij} + D_{(ij)k} + \gamma_l + (\alpha\gamma)_{il} + (\beta\gamma)_{jl}$$
$$+ (\alpha\beta\gamma)_{ijl} + (D\gamma)_{(ij)kl} + \rho_m + (\alpha\rho)_{im} + (\beta\rho)_{jm} + (\alpha\beta\rho)_{ijm}$$
$$+ (\gamma\rho)_{lm} + (\alpha\gamma\rho)_{ilm} + (\beta\gamma\rho)_{jlm} + (\alpha\beta\gamma\rho)_{ijlm} + (D\rho)_{(ij)km}$$
$$+ (D\gamma\rho)_{(ij)klm} + \tau_n + (\alpha\tau)_{in} + (\beta\tau)_{jn} + (\alpha\beta\tau)_{ijn} + (\gamma\tau)_{ln}$$
$$+ (\alpha\gamma\tau)_{iln} + (\beta\gamma\tau)_{jln} + (\alpha\beta\gamma\tau)_{ijln} + (\rho\tau)_{mn} + (\alpha\rho\tau)_{imn}$$
$$+ (\beta\rho\tau)_{jmn} + (\alpha\beta\rho\tau)_{ijmn} + (\gamma\rho\tau)_{lmn} + (\alpha\gamma\rho\tau)_{ilmn} + (\beta\gamma\rho\tau)_{jlmn}$$
$$+ (\alpha\beta\gamma\rho\tau)_{ijlmn} + (D\tau)_{(ij)kn} + (D\gamma\tau)_{(ij)kln} + (D\rho\tau)_{(ij)kmn}$$
$$+ (D\gamma\rho\tau)_{(ij)klmn} + E_{(ijklmn)} \quad (8.84)$$

where $D_{(ij)k}$ is error$_{ab}$, $(D\gamma)_{(ij)kl}$ is error$_c$, $(D\rho)_{(ij)km} + (D\gamma\rho)_{(ij)klm}$ is error$_r$, and the last five terms make up error$_t$.

In each of the preceding cases the rules of Chap. 2 may be employed to find appropriate sums of squares and components that belong in the expected mean squares. The coefficients of components specified by the rules also will be correct for all components except the random components involving subjects (D, or S in Case 6). However, the remaining cases in this section all involve changeover structure with Latin squares and the procedures of Sec. 8.1.3 must be incorporated into the framework of the more complex structure. See Winer (1962, pp. 538-77) for 13 variants of the basic structure.

Case 9: Latin Square Crossover with Repeated Measurement in Cells

Serum luteinizing hormone (LH) was measured in 9 bull calves, where 3 calves of similar age made up the rows ($D_{(i)j}$) of a Latin square (θ_i). The treatments (τ_l) were 200, 400, or 800 μg gonadotropin-releasing hormone injected intramuscularly on 3 successive days, the columns (γ_k) of each square. On each day, blood samples were taken from each animal at 11 periods (ρ_m) following injection (0.25 to 12 hours). The model was

$$Y_{ijklm} = \mu + \theta_i + D_{(i)j} + \gamma_k + \tau_l + E_{(ijkl)} + \rho_m + (\theta\rho)_{im} + (\gamma\rho)_{km} + (\tau\rho)_{km} + (D\rho)_{(i)jm} + U_{(ijklm)} \tag{8.85}$$

where $E_{(ijkl)}$ is error$_t$ appropriate for testing treatments, $(D\rho)_{(i)jm}$ is error$_r$ for tests of effects of sampling periods after injection, and $U_{(ijklm)}$ is error$_{tr}$ for testing treatment-period interaction. See Westlake (1974) for balanced incomplete block crossover designs with repeated measurement in the cells.

Case 10: Latin Square Crossover with Split-Plot Structure in Cells

An investigation of lipoprotein lipase activity involved 8 pregnant cows as rows (D_i) of an 8 × 8 Latin square. Columns (ρ_j) were periods, prepartum and postpartum, in which biopsy samples of tissue were collected. Treatments were 2^3 combinations of A = prolactin, B = progesterone, and C = corticosterone applied to in vitro tissue cultures. On each day of biopsy, both mammary and mesenteric adipose tissues (τ_n) were collected from each animal. Length of culture time (λ_o) was the final variable involved. The model was

$$\begin{aligned} Y_{ijklmno} = {} & \mu + D_i + \rho_j + \alpha_k + \beta_l + \gamma_m + (\alpha\beta)_{kl} + (\alpha\gamma)_{km} + (\beta\gamma)_{lm} \\ & + (\alpha\beta\gamma)_{klm} + E_{(ijklm)} + \tau_n + (\rho\tau)_{jn} + (\alpha\tau)_{kn} + (\beta\tau)_{ln} \\ & + (\gamma\tau)_{mn} + (\alpha\beta\tau)_{klm} + (\alpha\gamma\tau)_{kmn} + (\beta\gamma\tau)_{lmn} + (\alpha\beta\gamma\tau)_{klmn} \\ & + (D\tau)_{in} + (E\tau)_{(ijklm)n} + \lambda_o + (\rho\lambda)_{jo} + (\alpha\lambda)_{ko} + (\beta\lambda)_{lo} \\ & + (\gamma\lambda)_{mo} + (\alpha\beta\lambda)_{klo} + (\alpha\gamma\lambda)_{kmo} + (\beta\gamma\lambda)_{lmo} + (\alpha\beta\gamma\lambda)_{klmo} \\ & + (\tau\gamma)_{no} + (\rho\tau\gamma)_{jno} + (\alpha\tau\lambda)_{kno} + (\beta\tau\lambda)_{lno} + (\gamma\tau\lambda)_{mno} \\ & + (\alpha\beta\tau\lambda)_{klno} + (\alpha\gamma\tau\lambda)_{kmno} + (\beta\gamma\tau\lambda)_{lmno} + (\alpha\beta\gamma\tau\lambda)_{klmno} \\ & + (D\lambda)_{io} + (E\lambda)_{(ijklm)o} + (D\tau\lambda)_{ino} + (E\tau\lambda)_{(ijklm)no} \\ & + E'_{(ijklmno)} \end{aligned} \tag{8.86}$$

where $E_{(ijklm)}$ is error$_{abc}$ for testing the 2^3 hormonal effects, $(D\tau)_{ij}$ + $(E\tau)_{(ijklm)n}$ is error$_t$ for testing effects of tissues and interactions of tissues with hormones, and the last five terms are error$_l$ for testing effects involving length of culture time.

Case 11: Latin Square Crossovers within One Factor of Split-Plot Design

Twelve beef steers (D/A) were randomly assigned to 3 dietary levels of protein (A = 1, 2, or 4 times NRC requirement). Within each level of protein, 4 steers made up the rows of a Latin square and 4 feeding periods were the columns (ρ_k). Treatments (B) were types of protein (soybean meal, urea, ammonium lactate, ammonium acetate). Average daily gains were recorded for each period. The model was

$$Y_{ijkl} = \mu + \alpha_i + D_{(i)j} + \rho_k + \beta_l + (\alpha\rho)_{ik} + (\alpha\beta)_{il} + E_{(ijkl)} \qquad (8.87)$$

where $D_{(i)j}$ is error$_a$ for testing levels of protein and $E_{(ijkl)}$ is error$_b$ for testing types of protein.

Case 12: Latin Square Crossover, Repeated within One Factor with Repeated Measurement in Cells

Eight cows (D) were measured for serum prolactin in euthyroid condition, hyperthyroid condition (following dietary thyroprotein), and hypothyroid condition (following removal of dietary thyroprotein). Let the thyroid conditions be factor A. Within each condition, the same 8 cows made up the rows of 4 Latin squares (2 × 2), with 2 days (periods) as columns (ρ). The treatments (factor B) were infusion of 100 μg TRH or saline. Blood samples were taken at 8 sampling times (T = hours and fractions thereof) for each cow on each day in infusion. The model was

$$\begin{aligned} Y_{ijklm} &= \mu + \alpha_i + D_j + (\alpha D)_{ij} + \rho_{(i)k} + \beta_l + E_{(ijkl)} + \tau_m \\ &+ (\alpha\tau)_{im} + (\rho\tau)_{(i)km} + (\beta\tau)_{lm} + (D\tau)_{jm} + (\alpha D\tau)_{ijm} \\ &+ (E\tau)_{(ijkl)m} + U_{(ijklm)} \end{aligned} \qquad (8.88)$$

where $(\alpha D)_{ij}$ is error$_a$ for testing thyroid conditions, $E_{(ijkl)}$ is error$_b$ for testing TRH versus saline, and the last four terms combined are error$_t$ for testing effects involving time of sampling.

Case 13: Latin Square Crossover Repeated within Two Factors of Split Plot

Forty dairy calves (D/A) were measured for growth hormone, 10 in each of 4 assay groups (A). Repeatedly, at 1, 3, 5, and 7 weeks of age (B), 5 Latin squares (2 × 2) were formed in each group with calves as rows and days as columns ($\rho_{(ik)l}$). Treatments (C) were absolute dose of TRH or dose of TRH based on body weight. Data were recorded as differences between posttreatment and pretreatment measurements. The model was

$$\begin{aligned} Y_{ijklm} &= \mu + \alpha_i + D_{(i)j} + \beta_k + (\alpha\beta)_{ik} + (D\beta)_{(i)jk} + \rho_{(ik)l} + \gamma_m \\ &+ (\alpha\gamma)_{im} + (\beta\gamma)_{km} + (\alpha\beta\gamma)_{ikm} + E_{(ijklm)} \end{aligned} \qquad (8.89)$$

where $D_{(i)j}$ was error$_a$, $(D\beta)_{(i)jk}$ was error$_b$ for testing age effects, and $E_{(ijklm)}$ was error$_c$ for testing effects involving the treatments and their interaction with assay groups and age.

The assumptions involved in the foregoing cases are obviously many and varied. Before embarking on a particular course of experimentation, one should consider the possibilities with respect to resources available, practical restrictions, time trends (both nuisance and primary), residual effects of treatments, etc. Before and during analysis of data from such experiments, questions about variance-covariance structure should be raised and answered as far as possible to validate the standard analysis or to point to alternative procedures more likely to provide accurate inferences. The assistance of professional statisticians and competent computer specialists undoubtedly will be needed in many if not most cases.

8.3. SPLIT-PLOT INCOMPLETE BLOCK DESIGNS (Subjects within Blocks)

Repeated measurement is not necessarily involved in the designs of this section, but the split-plot or split-error principle (usually without a serious problem of correlated errors) is involved. Also, these designs are incomplete block designs for factorial experiments, thus sharing the confounding principles of Secs. 6.3-6.5. The incomplete blocks in designs of this section are groups of homogeneous subjects rather than the subjects themselves; the members of a block all receive the same level of one factor, whereas that does not happen when one or more interactions are confounded with the incomplete blocks of Chap. 6. Thus the split-plot designs involve confounding of a main effect with blocks. Two basic motivations for using split-plot designs of this type are: (1) practical restrictions dictate that one factor must be applied to a larger amount of experimental material or set of units, whereas the other factor(s) may be applied to smaller amounts or to individual subjects, and (2) a "known" main effect of a factor is less interesting than interaction and may be confounded deliberately with blocks to achieve greater homogeneity of subjects in small blocks, thus reducing experimental error.

An example of practical restriction is a 2^3 factorial calf-feeding experiment in which the factors were A = level of dry matter (10%, 15%), B = weaning age (4 or 6 weeks), and C = feeding level (10% of body weight or ad libitum). The feeding machinery restricted the experiment to use of only one level of dry matter at one time. Therefore, a replicate consisted of 8 calves in 2 incomplete blocks of 4 homogeneous animals. The first block received 10% dry matter, the second 15%. Within each block, calves were assigned randomly to the BC combinations. To obtain adequate sensitivity, 6 replicates were used or 48 calves in all. The first blocks of the replicates were performed at one time and the second blocks followed. Thus, the dry matter effect (A) was confounded with block effects, which included time effects (if any) as well as differences between groups of 4 homogeneous calves. However, interactions of dry matter with the other 2 factors were clearly estimable.

An example of deliberate confounding of a main effect by choice was a 2 × 4 factorial experiment on the influence of methionine on milk production. Factor A was level of protein (high or low) and factor B was daily equivalent dose (0 or 40 g) of methionine (control, DL-methionine, MHA, or processed methionine). The main effect of A was already well established and was much less interesting than the interaction with methionine. Also, considerable heterogeneity existed among the cows available. Therefore, a replicate consisted of 8 cows in 2 incomplete blocks of 4 homogeneous animals. The first block received one level of protein, the second the other. Within each block, cows were assigned randomly to the 4 methionine treatments. Five replicates were performed sequentially over time. Thus, the main effect of A was confounded with block effects but not with time.

An appropriate model for the split-plot experiments is

$$Y_{ijk} = \mu + \alpha_i + R_j + (\alpha R)_{ij} + \beta_k + (\alpha\beta)_{ik} + (R\beta)_{jk} + (\alpha R\beta)_{ijk} + E_{(ijk)} \quad (8.90)$$

where $i = 1, 2, \ldots, a$; $j = 1, 2, \ldots, r$; $k = 1, 2, \ldots, b$; and $n = abr$. The format of the model is the same as (8.77) except that R_j is the random effect of a complete replicate of ab cows, instead of a block of a cows measured at b sampling times. Note that the incomplete block effects (if nonzero) are completely confounded with the main effects of $A(\alpha_i)$. Therefore, although it is traditional to list the effects as treatments rather than blocks, remember that the effects of A and groups of r blocks are inseparable. The rationale for listing the effects as α_i is that truly random replicates, in which levels of A are randomly associated with blocks, may lead to counterbalancing block effects for a given level of A measured over r blocks in r replicates. Also, note that the structure of the model is the same as for a 3-factor experiment in a completely randomized design with one factor (replicates here) having random effects and only one observation per combination. Because of the equivalent structure, the rules of Chap. 2 may be used to derive sums of squares and expected mean squares. Note that residual error and ARB interaction cannot be separated.

Sums of squares for total, factor A (or blocks), replicates (R), and the interaction AR (error_a) are

$$ss_y = \sum_{i=1}^{a} \sum_{j=1}^{r} \sum_{k=1}^{b} y_{ijk}^2 - (y_{\cdots}^2/n) \quad (8.91)$$

$$ss_A = \left(\sum_{i=1}^{a} y_{i\cdot\cdot}^2/br\right) - (y_{\cdots}^2/n) \quad (8.92)$$

$$ss_R = \left(\sum_{j=1}^{r} y_{\cdot j\cdot}^2/ab\right) - (y_{\cdots}^2/n) \quad (8.93)$$

$$ss_{AR} = \left(\sum_{i=1}^{a} \sum_{j=1}^{r} y_{ij\cdot}^2/b\right) - \left(\sum_{i=1}^{a} y_{i\cdot\cdot}^2/br\right) - \left(\sum_{j=1}^{r} y_{\cdot j\cdot}^2/ab\right) + (y_{\cdots}^2/n) \quad (8.94)$$

Computations involving factor A are said to be among "whole plots" (incomplete blocks of subjects) while the effects of factor B, its interactions with A and R, and residual error (error_b) are said to be computed among "subplots" (individual subjects) within blocks. Sums of squares for the subplot effects are

$$ss_B = (\sum_{k=1}^{b} y^2_{..k}/ar) - (y^2_{...}/n) \quad (8.95)$$

$$ss_{AB} = (\sum_{i=1}^{a}\sum_{k=1}^{b} y^2_{i.k}/r) - (\sum_{i=1}^{a} y^2_{i..}/br) - (\sum_{k=1}^{b} y^2_{..k}/ar) + (y^2_{...}/n) \quad (8.96)$$

$$ss_{RB} = (\sum_{j=1}^{r}\sum_{k=1}^{b} y^2_{.jk}/a) - (\sum_{j=1}^{r} y^2_{.j.}/ab) - (\sum_{k=1}^{b} y^2_{..k}/ar) + (y^2_{...}/n) \quad (8.97)$$

$$ss_E = ss_y - ss_A - ss_R - ss_{AR} - ss_B - ss_{AB} - ss_{RB} \quad (8.98)$$

The analysis of variance is summarized in Table 8.28, with expected mean squares reflecting the assumption that the average of the effects of the r incomplete blocks associated with any level of A is sufficiently near zero (measured as deviations from μ) as to be negligible. If the nuisance variation (block effect) is known to be large, one could improve the accuracy of information about A (especially desirable when the split plot is involuntary) by using exactly a replicates as rows of a Latin square (Chap. 7), a complete blocks as columns, and a levels of factor A as treatments. One complete block (column) would be made up of a incomplete blocks sharing similar nuisance values in each replicate. For example, if the blocking factor were body weight, the animals of lower weight in each replicate would constitute the first block of each replicate and the several blocks of low weight animals would jointly make up the first column or complete block. That would change the "whole-plot" structure of Table 8.28 from $A + R + AR$ to $A + R + D + E_a$, where D stands for blocks (columns) and E_a is the residual "cell" error with $(a - 1)(a - 2)$ df as in an ordinary Latin square. Of course, in this case each cell of the Latin square would be subdivided into b subplots (subjects) for random assignment of the levels of factor B. Such strategy is not necessary (or even desirable) when the effects of A are of no interest per se or are well known a priori.

For those cases in which A means are of interest, the standard error is

$$s_{\bar{y}_{i..}} = \sqrt{(ms_{AR}/br)} \quad (8.99)$$

Standard errors for B means are

$$s_{\bar{y}_{..k}} = \sqrt{(ms_{E_b}/ar)} \quad (8.100)$$

Differences between two treatment combination means having the same levels of A or same levels of B, respectively, have standard errors

$$s_{\bar{y}_{i.k}-\bar{y}_{i.k'}} = \sqrt{2ms_{E_b}/r} \qquad s_{\bar{y}_{i.k}-\bar{y}_{i'.k}} = \sqrt{2ms_{AR}/r} \quad (8.101)$$

As in the split-plot designs of Sec. 8.2, ms_{E_b} normally is smaller than ms_{AR} so that superior precision is achieved for B and AB at the expense of A.

Table 8.28. Analysis of Variance for Split-Plot Design (subject=subplot)

Source of Variation	df	ss	ms	E[MS]
Blocks of subjects (whole plots)	[ar-1]			
A	$a-1$	ss_A	ms_A	$\sigma^2+b\sigma^2_{AR}+br\Sigma\alpha^2_i/\nu_A$
Replicates (R)	$r-1$	ss_R	ms_R	$\sigma^2+ab\sigma^2_R$
AR (error_a)	$(a-1)(r-1)$	ss_{AR}	ms_{AR}	$\sigma^2+b\sigma^2_{AR}$
Subjects (subplots)/blocks	$[n-ar]$			
B	$b-1$	ss_B	ms_B	$\sigma^2+a\sigma^2_{RB}+ar\Sigma\beta^2_k/\nu_B$
AB	$(a-1)(b-1)$	ss_{AB}	ms_{AB}	$\sigma^2+\sigma^2_{ARB}+r\Sigma\Sigma(\alpha\beta)^2_{ik}/\nu_{AB}$
RB	$(r-1)(b-1)$	ss_{RB}	ms_{RB}	$\sigma^2+a\sigma^2_{RB}$
Error_b	$(a-1)(r-1)(b-1)$	ss_E	ms_E	$\sigma^2+\sigma^2_{ARB}$

Testing of hypotheses should begin with a test of AB interaction. If sufficient degrees of freedom are available for ms_{E_b}, then $H{:}(\alpha\beta)_{ik} = 0$ should be tested by comparing

$$f = ms_{AB}/ms_{E_b} \tag{8.102}$$

with critical value $f_{\alpha,\nu_{AB},\nu_{E_b}}$. However, in small experiments ν_{E_b} may be very small. For example, in a 2^2 factorial with 4 replicates, $\nu_{E_b} = 3$. In such cases, Harter (1961) suggested a preliminary two-tailed test of $H{:}\sigma^2_{ARB} = a\sigma^2_{RB}$, using test statistic

$$f = ms_{RB}/ms_{E_b} \quad \text{or} \quad f = ms_{E_b}/ms_{RB} \tag{8.103}$$

whichever is larger, against critical value $f_{\alpha/2,\nu_{RB},\nu_E}$ or $f_{\alpha/2,\nu_{E_b},\nu_{RB}}$ with $a \geq 0.2$. Acceptance of H suggests either $\sigma^2_{RB} = \sigma^2_{ARB} = 0$ or $E[MS_{RB}] = E[MS_{E_b}]$. No matter which conclusion is essentially correct, a reasonably practical procedure is to pool the two sums of squares and their degrees of freedom to form a composite error,

$$ms_{E_b'} = (ss_{RB} + ss_{E_b})/(\nu_{RB} + \nu_{E_b}) \tag{8.104}$$

with $\nu_{E_b'} = a(r - 1)(b - 1)$ df. Then, tests of $H{:}(\alpha\beta)_{ik} = 0$ and $H{:}\beta_k = 0$ may be made with greater sensitivity.

If interaction is judged to be nonsignificant, a test of $H{:}\beta_k = 0$ would be meaningful. Compare test statistic

$$f = ms_B/ms_{RB} \tag{8.105}$$

with $f_{\alpha,\nu_B,\nu_{RB}}$, or use $ms_{E_b'}$ in the denominator if pooling seems advisable. Also, if factor A is of interest, $H{:}\alpha_i = 0$ may be tested approximately (acknowledging possible block confounding) by comparing

$$f = ms_A/ms_{AR} \tag{8.106}$$

with $f_{\alpha,\nu_A,\nu_{AR}}$.

If interaction is judged to be significant, overall tests of the main effects are not sufficiently specific to be very meaningful. Treatment combination means become the items of chief interest. Levels of B may be compared separately for each level of A, as in a one-way analysis of variance with contrasts (Chap. 2), or specific contrasts among the entire set of treatment combinations may be in order (Kirk 1968, p. 269).

Computation of the efficiency of a split-plot design (SPD) for comparing B means relative to a complete block design (replicates = blocks) requires estimation of the random variation that would have occurred in a randomization trial with replicates and error only, i.e.,

$$\hat{\sigma}^2_{\text{RCBD}} = [(\nu_A + \nu_{AR})ms_{AR} + (\nu_B + \nu_{AB} + \nu_{E'_b})ms_{E'_b}]/(n - r) \tag{8.107}$$

with $\nu_{\text{RCBD}} = n - r$. Then, using $\hat{\sigma}^2_{\text{SPD}} = ms_{E'_b}$ (pooled error) with $\nu_{\text{SPD}} = n - ar$, one obtains

$$E_{\text{SPD:RCBD}} = [(n - ar + 1)/(n - ar + 3)][(n - r + 3)/(n - r + 1)](\hat{\sigma}^2_{\text{RCBD}}/\hat{\sigma}^2_{\text{SPD}})100\% \tag{8.108}$$

The number of replicates (r) that should be used in split-plot experiments may be estimated by using App. Figs. A.15.17-A.15.21, with specified error rates (α and β) for Type I and Type II errors for the overall test of B means or for orthogonal f tests of contrasts among the B means. For the overall test, the abscissa scale of the power charts is

$$\phi = \sqrt{(ra/b)\Sigma(\beta_k/\hat{\sigma})^2} \tag{8.109}$$

if one assumes that $\sigma^2_{RB} = 0$. Degrees of freedom are $\nu_1 = b - 1$ and $\nu_2 = (r - 1)(b - 1)$. One must specify a set of b effects $\{\beta_k\}$ desirable to detect, such that $\Sigma\beta_k = 0$, and provide an estimate of error $\hat{\sigma}$. The estimate of σ may be $\sqrt{ms_{RB}}$ from a prior split-plot experiment of similar nature (in which case the assumption that $\sigma^2_{RB} = 0$ is irrelevant) or may be obtained by estimating the potential range of response of homogeneous subjects treated alike. The potential range divided by a constant in the range from 3 to 6 (see App. A.14) provides a crude estimate of σ. For a contrast, $\Sigma c_i y_{i..k}$ ($\Sigma c_i = 0$) among B means with Δ_d = the mean difference desirable to detect, the abscissa scale is

$$\phi = (\Delta_d/\hat{\sigma})\sqrt{(ra/8)[(\Sigma|c_i|)^2/\Sigma c_i^2]} \tag{8.110}$$

and degrees of freedom are $\nu_1 = 1$ and $\nu_2 = (r - 1)(b - 1)$. The latter process usually is easier because specifying Δ_d is less difficult than listing the array $\{\beta_k\}$ of interest.

Suppose an animal dies or the result for one cell of the split plot is missing for other reasons. Then, one may reduce ν_{RB} and ν_{E_b} by 1 df each and perform the usual analysis after estimating the missing value from

$$\hat{y}_{ijk} = (ry'_{ij.} + by'_{i.k} - y'_{i..})/[(r - 1)(b - 1)] \tag{8.111}$$

where $y'_{ij.}$ and $y'_{i.k}$ are the totals of actual data in the block and treatment combination concerned and $y'_{i..}$ is the total of actual data for the level of factor A involved. All sums of squares except ss_E are biased upward slightly, but the bias usually is ignored. Anderson (1946) has indicated minor adjustments in standard errors of treatment means. When two or more cells are missing, one may employ the covariance technique (Chap. 3), iteration (Kirk, Chap. 6), or minimum error variance (Sec. 5.4 or Haseman and Gaylor 1973) to estimate the missing values.

Model (8.90) may be extended to permit extra factors to be randomized either over the whole plot (incomplete blocks) or over the subplots (subjects), usually the latter. As an example of a split-plot experiment with combinations of 2 factors randomized over the subjects of each block, consider an experiment described by D. R. Cox (1958). A biochemical experiment was performed to assess the effect of injection of either or both of 2 experimental preparations (A, B) on response of a serum component. A third factor was sex of the mice (C), making the experiment a 2^3 factorial. Day-to-day variation in experimental technique was a known nuisance variable, suggesting a complete block design with block = day = 8 mice. However, the complexity of measurement prevented evaluating more than 4 or 5 mice per day, suggesting the use of incomplete blocks. The main effect of sex was well known, but interactions of sex with the other factors were of interest. Therefore, a split-plot design in which sex was confounded with days was more desirable than the ordinary incomplete block design in which ABC interaction would be confounded. Four replications (8 mice each) were used. Within each sex (M = male, F = female) each daily set of 4 mice was selected to have mice as uniform as possible. The 4 mice in each block were assigned randomly to combinations of A and B. Data are shown in Table 8.29. A model for the experiment is

$$\begin{aligned} Y_{ijkl} &= \mu + \gamma_i + R_j + (\gamma R)_{ij} + \alpha_k + \beta_l + (\alpha\beta)_k + (\alpha\gamma)_{ik} + (\beta\gamma)_{il} \\ &+ (\alpha\beta\gamma)_{ikl} + (R\alpha)_{jk} + (R\beta)_{jl} + (R\alpha\beta)_{ikl} + (R\alpha\gamma)_{ijk} \\ &+ (R\beta\gamma)_{ijl} + (R\alpha\beta\gamma)_{ijkl} + E_{(ijkl)} \end{aligned} \tag{8.112}$$

where α, β, and γ are the effects of factors A, B, and C (2 levels each), and R_j is the random effect of replicate (j = 1, 2, 3, 4). The term $(\gamma R)_{ij}$ may be referred to as error$_c$, pertaining to C = sex = blocks; and the last two terms jointly make up error$_{ab}$ if one assumes σ^2_{RABC} to be trivial. Computation of sums of squares follows the rules for 4-factor experiments (replicates as the fourth factor) in Chap. 2. A summary of the analysis of variance is shown in Table 8.30, with expected mean square for sex reflecting the fact that sex effects (γ_i) are confounded with differences between effects of odd and even days (δ_i).

One difficulty with the analysis is that error$_{ab}$ (and each of the other random effects as well) has only 3 df. If one could establish that some or

Table 8.29. Split-Plot 2^3 Experiment on Mice with Factor C=Sex (M=male, F=female) Confounded with Blocks (days)

	Replicate 1		2		3		4		
A B	Day 1(M)	2(F)	3(M)	4(F)	5(M)	6(F)	7(M)	8(F)	Total
0 0	4.4	3.7	1.8	5.3	5.3	6.5	5.4	5.2	37.6
0 1	4.8	6.2	3.1	4.3	1.9	5.7	6.2	7.9	40.1
1 0	2.8	5.9	2.6	7.0	3.3	5.4	6.9	6.8	40.7
1 1	6.8	5.1	4.8	7.2	8.7	6.7	9.3	7.9	56.5
Total	18.8	20.9	12.3	23.8	19.2	24.3	27.8	27.8	174.9
Rep. Total	39.7		36.1		43.5		55.6		

Totals: A_0, 77.7, A_1, 97.2, B_0, 78.3, B_1, 96.6, $C_0(M)$, 78.1, $C_1(F)$, 96.8

all the interactions of replicates with A and B or combinations involving them are clearly nonsignificant, a pooled error could be calculated. Storm's (1962) pooling criterion may be used. Effects for which $f < 2f_{0.50}$ may be judged clearly nonsignificant. For the preliminary pooling tests in Table 8.30, $2f_{0.50,3,3} = 2.0$. The effects RA, RB, RAC, and RBC when tested with error$_{ab}$ each have $f < 2.0$ and may be pooled with error$_{ab}$. The sample effect of RAB is a bit too large for the criterion. The pooled error becomes

$$ms_{E'_{ab}} = (ss_{RA} + ss_{RB} + ss_{RAC} + ss_{RBC} + ss_{E_{ab}})/15 = 0.943$$

which is only 65% as large as $ms_{E_{ab}}$ and has 5 times as many degrees of freedom. Factorial effects involving A or B and having f ratios exceeding $f_{0.05,1,15} = 4.54$ may be judged significant. The effects of A, B, AB, and ABC appear to be substantial. However, $E[MS_{AB}]$ involves σ^2_{RAB}, which may not be trivial. Therefore, a "pure" test of AB interaction requires $f = ms_{AB}/ms_{RAB} = 1.68$, which fails to exceed $f_{0.05,1,3} = 10.1$. The significance of AB is doubtful. In any event, the significance of ABC interaction suggests that one should examine treatment combination means rather than means of broader classifications. The means for combinations A_0B_0, A_0B_1, A_1B_0, and A_1B_1 are (4.22, 4.00, 3.90, and 7.40) for males and (5.18, 6.02, 6.28, and 6.72) for females (4 observations per mean). Factors A and B appear to be synergistic in males but not in females. That is, in males the addition of B in the presence of A gives much stronger response than addition of B in the absence of A. Consider the following contrasts among means:

1. Males: $(A_0B_0 + A_1B_1)$ versus $(A_0B_1 + A_1B_0)$, interaction

Table 8.30. Analysis of Variance for 2^3 Factorial Experiment in Split-Plot Design (see Table 8.29)

Source of Variation	df	*ms*	$E[MS]$
Blocks of mice (7)			
Sex (C)=days	1	10.928	$\sigma^2+4\sigma^2_{CR}+16\Sigma\gamma^2+8\ (\Sigma \text{ odd } \delta_i - \Sigma \text{ even } \delta_i)^2$
Replicates (R)	3	8.975	$\sigma^2+8\sigma^2_R$
CR ($\text{error}_{\hat{c}}$)	3	3.135	$\sigma^2+4\sigma^2_{CR}$
Mice/blocks (24)			
A	1	11.883	$\sigma^2+4\sigma^2_{RA}+16\Sigma\alpha^2$
B	1	10.465	$\sigma^2+4\sigma^2_{RB}+16\Sigma\beta^2$
AB	1	5.528	$\sigma^2+2\sigma^2_{RAB}+8\Sigma\Sigma(\alpha\beta)^2$
AC	1	0.813	$\sigma^2+2\sigma^2_{RAC}+8\Sigma\Sigma(\alpha\gamma)^2$
BC	1	1.950	$\sigma^2+2\sigma^2_{RBC}+8\Sigma\Sigma(\beta\gamma)^2$
ABC	1	8.508	$\sigma^2+4\Sigma\Sigma\Sigma(\alpha\beta\gamma)^2$
RA	3	0.755	$\sigma^2+4\sigma^2_{RA}$
RB	3	0.668	$\sigma^2+4\sigma^2_{RB}$
RAB	3	3.295	$\sigma^2+2\sigma^2_{RAB}$
RAC	3	1.309	$\sigma^2+2\sigma^2_{RBC}$
RBC	3	0.533	$\sigma^2+2\sigma_{ABC}$
Error_{ab}	3	1.449	σ^2

2. Females: $(A_0B_0 + A_0B_1)$ versus $(A_1B_0 + A_1B_1)$, A effect
3. Females: $(A_0B_0 + A_1B_0)$ versus $(A_0B_1 + A_1B_1)$, B effect
4. Females: $(A_0B_0 + A_1B_1)$ versus $(A_0B_1 + A_1B_0)$, interaction

The contrasts are orthogonal but were selected after examining the data considerably. A conservative approach is to use Scheffé's interval (Chap. 2) with $ms_{E'_{ab}}$, since comparisons are within sex. Contrast coefficients are $\{c_i\}$ = $\{+1, +1, -1, -1\}$ and replication is $r = 4$ in each case, so $\hat{V}[\bar{q}_k] = (\Sigma c^2/4)ms_{E'_{ab}} = ms_{E'_{ab}} = 0.943$. Then the half-width of the confidence interval for each mean difference is

$$\pm\sqrt{(t-1)f_{0.05,7,15}(0.943)} = \pm\sqrt{7(2.71)(0.943)} = \pm 1.34$$

where $t = 8$ treatment combinations of A, B, and C. The four contrast means are 3.72, 1.80, 1.28, and 0.40. Therefore, interaction is significant among males but not among females (as suspected). The main effect of A is significant for females and B is nearly so. Standard errors of AB combination means for males are $\pm\sqrt{0.943/4} = 0.49$. Standard errors for either A or B means for females are $\pm\sqrt{0.943/8} = 0.34$. Computation of the efficiency of the split-plot design (with respect to factors A and B) relative to a complete block design (RCBD) requires

$$\hat{\sigma}^2_{\text{RCBD}} = [r(c-1)ms_{CR} + (n-cr)ms_{E'_{ab}}]/(n-r) = [4(3.135) + 24(0.943)]/28 = 1.256$$

with $\nu_{\text{RCBD}} = n - r = 28$. For the split plot, $\hat{\sigma}^2_{\text{SPD}} = ms_{E'_{ab}} = 0.943$, with $\nu_{\text{SPD}} = n - cr = 24$. Then,

$$E_{\text{SPD:RCBD}} = (25/27)(31/29)(1.256/0.943)100\% = 132\%$$

which indicates that (1.32)(32) = 42 mice would be required for the RCBD to have equal sensitivity to the split-plot design with 32 mice.

Several variations of the split-plot design are possible. The whole-plot factor (factor confounded with blocks) may not be randomized in each replicate for practical reasons, or the subplot factor(s) may not be randomized within an incomplete block (whole plot). In either case more precision is lent to interaction(s) at the expense of main effects. Robinson (1967) has discussed designs having fewer subjects per block than the number of subplot treatments or treatment combinations. Also, double splitting is possible for experiments with 3 or more factors; i.e., a hierarchy of 3 (or more) levels of randomization and hence error terms may be devised. A special case involves groups of subjects in incomplete blocks in the first split, and repeated measurement of the subjects (as in Sec. 8.2.2) in the second split. As an example, consider a study of pesticide residues in milk. Factors were A (phenobarbital or none), B (dieldren extended or not), and C (repeated information at 10 sampling times). Two replicates consisted of 4 dieldren-contaminated cows each (number of AB combinations) with 10 observations per cow. In each replicate, 2 blocks of 2 cows were established. One block received phenobarbital; the other did not. Within each block, one cow received extended dieldren contamination; the other did not. Each cow was sampled at 10 points in time. The model was

$$\begin{aligned} Y_{ijkl} = {} & \mu + \alpha_i + R_j + (\alpha R)_{ij} + \beta_k + (\alpha\beta)_{ik} + (R\beta)_{jk} + (\alpha R\beta)_{ijk} \\ & + \gamma_l + (\alpha\gamma)_{il} + (\beta\gamma)_{kl} + (\alpha\beta\gamma)_{ikl} + (R\gamma)_{jl} + (\alpha R\gamma)_{ijl} \\ & + (R\beta\gamma)_{jkl} + (\alpha R\beta\gamma)_{ijkl} + E_{(ijkl)} \end{aligned} \tag{8.113}$$

where $(\alpha R)_{ij}$ may be taken as error$_a$, $(R\beta)_{jk} + (\alpha R\beta)_{ijk}$ as error$_b$, and the last five terms as error$_c$. Obviously, the sub-subplots involve correlated errors because of repeated measurement. Therefore, the discussion of variance-covariance structure in Sec. 8.2.2 is pertinent to inferences about the time factor.

8.4. CORRELATED ERRORS AND MULTIVARIATE ANALYSIS

The designs for repeated measurements discussed in this chapter and in particular the designs for the study of trends (Sec. 8.2) have in common correlated error structure. Fixed-order trend experiments in which time is the repeated factor are especially likely to result in heterogeneity of the variance-covariance structure. For example, periods with larger means commonly have larger variances also, and adjacent periods commonly have correlations (or covariances) larger than those associated with periods more separated in time. Also, in some cases the variance-covariance structure differs significantly from one treatment group to another (Sec. 8.2.2). The objective of this section is to discuss procedures for examining the assumptions of homogeneous variance-covariance and to indicate multivariate techniques for analyzing data in which the assumptions are not valid.

8.4.1. One Sample

Consider the complete block designs of Sec. 8.2.1, in which all animals are untreated (or treated alike) and each animal is measured p times to monitor changes in variable Y. For example, one may examine concentration of serum growth hormone at several periods during a particular stage of growth or measure concentration of prolactin during different stages of estrus. Among the r animals measured, one may compute p different variances ($s_i^2 = s_{ii}$, $i = 1, 2, \ldots, p$), one for each of the p periods of measurement. Also, one may compute a covariance ($s_{ii'}$, $i \neq i'$) for any two periods. Because $s_{12} = s_{21}$, $s_{13} = s_{31}$, etc., there are only $p(p - 1)/2$ different covariances. For univariate analysis to provide appropriate probability statements about period effects, the p variances each must be estimates of a common population variance (σ^2) and the $p(p - 1)/2$ covariances each must be estimates of a common population covariance. If the population variances are equal, any population covariance is equal to the common correlation in the population (ρ) multiplied by the common variance, i.e., $\rho\sigma^2$. From the sample of data, one may estimate σ^2 from the average sample variance, $\overline{s_i^2} = \sum_{i=1}^{p} s_i^2/p$, and $\rho\sigma^2$ from the average sample covariance, $\overline{s_{ii'}} = \sum\sum_{i<i'} s_{ii'}/[p(p - 1)/2]$. The $p \times p$ matrix of common variances and covariances of the population is an "ideal" matrix Σ_0 to which the real population variance-covariance matrix Σ may or may not conform; i.e., one hypothesizes $H{:}\Sigma = \Sigma_0$. The sample variance-covariance matrix S is an estimate of Σ, and the matrix of averaged sample variances and covariances S_0 is an estimate of Σ_0 if H is true:

$$\Sigma = \begin{bmatrix} \sigma_1^2 & \sigma_{12} & \cdots & \sigma_{1p} \\ \sigma_{21} & \sigma_2^2 & \cdots & \sigma_{2p} \\ \cdot & \cdot & \cdots & \cdot \\ \sigma_1 & \sigma_2 & \cdots & \sigma_p^2 \end{bmatrix} \qquad \Sigma_0 = \begin{bmatrix} \sigma^2 & \rho\sigma^2 & \cdots & \rho\sigma^2 \\ \rho\sigma^2 & \sigma^2 & \cdots & \rho\sigma^2 \\ \cdot & \cdot & \cdots & \cdot \\ \rho\sigma^2 & \rho\sigma^2 & \cdots & \sigma^2 \end{bmatrix}$$

$$S = \begin{bmatrix} s_1^2 & s_{12} & \cdots & s_{1p} \\ s_{21} & s_2^2 & \cdots & s_{2p} \\ \cdot & \cdot & \cdots & \cdot \\ s_{p1} & s_{p2} & \cdots & s_p^2 \end{bmatrix} \qquad S_0 = \begin{bmatrix} \overline{s_{\cdot i}^2} & \bar{s}_{\cdot ii'} & \cdots & \bar{s}_{\cdot ii'} \\ \bar{s}_{\cdot ii'} & \overline{s_{\cdot i}^2} & \cdots & \bar{s}_{\cdot ii'} \\ \cdot & \cdot & \cdots & \cdot \\ \bar{s}_{\cdot ii'} & \bar{s}_{\cdot ii'} & \cdots & \overline{s_{\cdot i}^2} \end{bmatrix}$$

Box (1949, 1950) described two methods of testing $H{:}\Sigma = \Sigma_0$. For an experiment with r subjects, each measured for Y at p times, let

$$h_1 = [p(p + 1)^2(2p - 3)]/[6(r - 1)(p - 1)(p^2 + p - 4)] \tag{8.114}$$

$$h_2 = [(p - 1)(p)(p + 1)(p + 2)]/[6(r - 1)^2(p^2 + p - 4)] \tag{8.115}$$

$$h_3 = [-(r - 1)\log_e\{|S|/|S_0|\}] \tag{8.116}$$

where $|S|$ and $|S_0|$ are determinants of the matrices S and S_0. Two tests are available: a chi-square test with ν_1 df and an f test with ν_1 and ν_2 df, where

$$\nu_1 = (p^2 + p - 4)/2 \tag{8.117}$$

$$\nu_2 = (\nu_1 + 2)/(h_2 - h_1^2) \tag{8.118}$$

For $p \leq 5$, use test statistic

$$q = (1 - h_1)h_3 \tag{8.119}$$

versus χ^2_{α,ν_1} (App. A.3). The test is not as accurate as is desirable if $r <$ 20. For $p > 5$, use approximate test statistic

$$f = h_3[1 - h_1 - (\nu_1/\nu_2)]/\nu_1 \tag{8.120}$$

versus f_{α,ν_1,ν_2} (App. A.5). In either case, $H{:}\Sigma = \Sigma_0$ is rejected if the appropriate test statistic exceeds the corresponding critical value for moderately small α. If H is not rejected, a univariate analysis (Sec. 8.2.1) should be adequate for comparing the means of the sampling times.

Suppose an experiment is performed with $r = 30$ animals, each measured at $p = 3$ times following injection of a releasing hormone; and the sample variances and covariances are $s_1^2 = 76.8$, $s_2^2 = 42.8$, $s_3^2 = 14.8$, $s_{12} = 53.2$, $s_{13} = 29.2$, and $s_{23} = 15.8$. The average variance and covariance are $\overline{s_{\cdot i}^2} = 44.8$ and

$\bar{s}_{ii'} = 32.7$. Then

$$\mathbf{S} = \begin{bmatrix} 76.8 & 53.2 & 29.2 \\ 53.2 & 42.8 & 15.8 \\ 29.2 & 15.8 & 14.8 \end{bmatrix} \qquad \mathbf{S}_0 = \begin{bmatrix} 44.8 & 32.7 & 32.7 \\ 32.7 & 44.8 & 32.7 \\ 32.7 & 32.7 & 44.8 \end{bmatrix}$$

It appears that variances and covariances decrease with time following injection. Determinants of the two matrices are $|\mathbf{S}| = 183.9$ and $|\hat{\Sigma}_0| = 16{,}134.4$. Then

$$h_1 = [3(4^2)(3)]/[6(29)(2)(8)] = 0.0517$$

$$h_3 = [-(29)\log_e(183.9/16{,}134.4)] = (-29)(-4.477) = 129.833$$

$$\nu_1 = (3^2 + 3 - 4)/2 = 4$$

$$q = (1 - 0.0517)(129.833) = 123.12$$

which far exceeds $\chi^2_{0.01,4} = 13.277$ (App. A.3). Therefore one should conclude that the variance-covariance matrix is heterogeneous and a univariate test of means for the 3 sampling times is not appropriate.

For p sampling points, the corresponding means (involving the same r subjects in each mean) constitute a set of estimates $\{\hat{\mu} + \hat{\tau}_i\}$ for an overall mean plus time parameters of model (8.53). In matrix notation, a row vector of p estimates may be denoted $\bar{\mathbf{y}}'$. One may test the hypothesis $H{:}\boldsymbol{\tau} = 0$ by using Hotelling's T^2 statistic in the form

$$T^2 = r(\mathbf{C}'\bar{\mathbf{y}})'(\mathbf{C}'\mathbf{S}\mathbf{C})^{-1}(\mathbf{C}'\bar{\mathbf{y}}) \qquad (8.121)$$

In (8.121), $\mathbf{S}$ is the $p \times p$ matrix of variances and covariances (each based on the same r subjects) and $\mathbf{C}'$ is a contrast matrix with $p - 1$ rows and p columns, such that the elements in any row sum to zero (Williams 1970). If T^2 exceeds the critical value $T^2_{\alpha,p-1,r-p}$ (App. A.18), one should reject $H{:}\boldsymbol{\tau} = 0$ and conclude that the response variable is not stable over the times sampled; i.e., some kind of trend exists. It is necessary that $r > p$ for the multivariate test. For the experiment above ($p = 3$, $r = 30$), suppose the estimated time means are $\bar{\mathbf{y}}' = [33.87, 33.07, 26.06]$. Any one of the contrast matrices,

$$\mathbf{C}_1' = \begin{bmatrix} +1 & -1 & 0 \\ +1 & 0 & -1 \end{bmatrix} \qquad \mathbf{C}_2' = \begin{bmatrix} +1 & -1 & 0 \\ 0 & +1 & -1 \end{bmatrix} \qquad \mathbf{C}_3' = \begin{bmatrix} +1 & +1 & -2 \\ +1 & -1 & 0 \end{bmatrix}$$

etc., may be used to compute T^2, which is invariant with the choice. Using the variance-covariance matrix $\mathbf{S}$ shown above and $\mathbf{C}' = \mathbf{C}_1'$, one obtains

$$\mathbf{C'SC} = \begin{bmatrix} +1 & -1 & 0 \\ +1 & 0 & -1 \end{bmatrix} \begin{bmatrix} 76.8 & 53.2 & 29.2 \\ 53.2 & 42.8 & 15.8 \\ 29.2 & 15.8 & 14.8 \end{bmatrix} \begin{bmatrix} +1 & +1 \\ -1 & 0 \\ 0 & -1 \end{bmatrix} = \begin{bmatrix} 13.2 & 10.2 \\ 10.2 & 33.2 \end{bmatrix}$$

The inverse of that 2×2 matrix $[(\mathbf{C'SC})^{-1}]$ and

$$\mathbf{C'\bar{y}} = \begin{bmatrix} +1 & -1 & 0 \\ +1 & 0 & -1 \end{bmatrix} \begin{bmatrix} 33.87 \\ 33.07 \\ 26.06 \end{bmatrix} = \begin{bmatrix} 0.80 \\ 7.81 \end{bmatrix}$$

are utilized in computing

$$T^2 = (30)[0.80 \quad 7.81] \begin{bmatrix} 0.099342 & -0.030521 \\ -0.030521 & 0.039497 \end{bmatrix} \begin{bmatrix} 0.80 \\ 7.81 \end{bmatrix} = 62.74$$

(See Chap. 4 for multiplication and inversion of matrices.) The test statistic exceeds (by far) $T^2_{0.01,2,27} = 11.478$. Therefore one may claim high confidence that the mean response of the hormone is not constant for the 3 times at which sampling occurred.

Because the multivariate procedure is relatively easy to use (especially for small t), one may ask why the univariate analysis should be used at all. Morrison (1967) has indicated that the power of the univariate test is better than that of the multivariate test when both are valid and the absolute magnitude of the constant correlation (ρ) is less than 0.5. The reverse is true for $|\rho| > 0.5$. Therefore, lowly (but uniformly) correlated data should be analyzed by the univariate method. Gupta and Perlman (1974) have noted that the power of the T^2 test may be drastically diminished if a large number of sampling points, some of which differ only slightly from each other in mean response, are included. Suppose one intends to monitor a variable (such as hormone response after treatment) and expects to find a relatively rapid rise and fall in response (a "spike") followed by a relatively slow decline to pretreatment level. In such situations, the inclusion of too many sampling points in the long tail of the decline may dilute the apparent significance of the T^2 test, leading to only low confidence at best that the response shows any trend at all. Of course, one usually prefers to have more specific answers if some overall trend is indicated. Simultaneous confidence intervals (with $1 - \alpha$ confidence for the *set* of intervals) may be used for any linear function $\mathbf{c}_k'\tau$, where τ contains the true time effects, $\tau_1, \tau_2, \ldots, \tau_t$, and $\mathbf{c}_k'$ contains t arbitrary constants, $c_1, c_2, \ldots, c_t$, such that $\Sigma c_i = 0$. If a desired contrast is included in the specification of matrix $\mathbf{C'}$, one of the diagonal elements of ($\mathbf{C'SC}$) may be utilized directly in the confidence interval. For example, if $\mathbf{c}_k'$ is [+1, 0, -1] for comparing time 1 with time 3 and $\mathbf{c}_k'$ is the second row of matrix $\mathbf{C'}$, one may use the diagonal element in the second row of ($\mathbf{C'SC}$), i.e., 33.2. The 99% confidence interval is

$$\bar{y}_{1.} - \bar{y}_{3.} \pm \sqrt{(1/r)(33.2)T^2_{0.01,2,27}}$$

$$= (33.87 - 26.06) \pm \sqrt{(1/30)(33.2)(11.478)} = 7.81 \pm 3.56$$

Because the interval does not include zero, one may conclude that mean response is different for times 1 and 3. Similarly, a confidence interval for comparison of time 1 and time 2 would utilize the diagonal element in the first row of (C'SC). However, a confidence interval for comparison of time 2 with time 3 would require recomputation of (C'SC) using $C' = C_2'$, or an interval for times 1 and 2 jointly versus time 3 would require (C'SC) using $C' = C_3'$, where C_2 and C_3 are the alternative matrices mentioned above. The selected contrasts need not be orthogonal nor are they limited in number.

8.4.2. Two or More Samples (treatment groups)

The split-plot experiments of Sec. 8.2.2, where r subjects are assigned randomly to each of t treatments (or treatment combinations of 2 or more factors) and then measured in p successive periods following treatment, often do not conform to the assumptions required for valid univariate tests. Because the errors are correlated from period to period and the correlation between data taken in two periods often is not constant, the assumption of uniformity of variance-covariance among periods often is not justified. Further, because of the natural tendency of variance to be related to magnitude of the mean for many biological variables, the variance-covariance matrices associated with t different treatment groups frequently are not equal ($\Sigma_1 \neq \Sigma_2 \neq \Sigma_3$, etc.).

To test $H{:}\Sigma_1 = \Sigma_2 = \cdots = \Sigma_t$, one may use the chi-square test (for t and $p \leq 5$) or f test (t or $p > 5$) described by Box (1949, 1950). These tests are explained in Chap. 4 in conjunction with the T^2 test for comparing vectors of p response variables measured on n subjects. The details are relevant for the present discussion if one remembers that p stands for the number of periods in which the same variable (Y) is measured on each animal and n stands for the total number of subjects instead of the total number of observations. If one rejects $H{:}\Sigma_1 = \Sigma_2 = \cdots = \Sigma_t$, neither the univariate tests nor the multivariate tests are valid. One of three alternatives for analysis may provide reasonably good inference: (1) transform data to logs and retest $H{:}\Sigma_1 = \Sigma_2 = \cdots = \Sigma_t$; (2) use a nonparametric analysis described by Koch (1970); or (3) test treatment differences separately for each period with one of the procedures described in Sec. 2.1.5.5, and test period differences separately for each treatment with the procedure described in Sec. 8.4.1. One could test period differences averaged over all treatments by using approximate degrees of freedom for a multivariate test as described by Yao (1965). If one accepts $H{:}\Sigma_1 = \Sigma_2 = \cdots = \Sigma_t$, the next step is to pool the variances and covariances from the t separate matrices ($S_1, S_2, \ldots, S_t$) to form a variance-covariance matrix (S) that estimates the variance-covariance structure common to all treatment groups (Σ). For balanced replication, $S = \sum_{i=1}^{t} S_i/t$. For unbalanced replication with r_i subjects in the ith treatment group, $S = \sum_{i=1}^{t} (r_i - 1)S_i / \sum_{i=1}^{t} (r_i - 1)$.

To test $H:\Sigma = \Sigma_0$, where Σ is the population variance-covariance matrix ommon to all treatments and Σ_0 is the "ideal" uniform matrix with all vari-nces equal to σ^2 and all covariances equal to $\rho\sigma^2$ (ρ = correlation), one may se the chi-square test (for $p \leq 5$ periods) or f test ($p > 5$) described by ox. These tests are explained in Sec. 8.4.1; the details are relevant for he present discussion if one substitutes $\sum_{i=1}^{t} (r_i - 1)$ for $r - 1$ in (8.114)-8.116) to account for degrees of freedom for the total number of subjects ithin treatments. If one accepts $H:\Sigma = \Sigma_0$, the univariate tests of Sec. .2.2 should be valid. If one rejects $H:\Sigma = \Sigma_0$, the univariate test of over-ll treatment differences is valid; but multivariate tests are required for ccurate inferences about overall period differences or about treatment-period nteraction, often the most important effect.

To illustrate assessment of the nature of the variance-covariance struc-ure, consider the example of Table 8.22 where the $t = 2$ treatments were preg-ant or nonpregnant lactating rats measured for litter weight gain in $p = 4$ eriods. There were $r = 6$ rats per treatment or $n = 12$ in all. Variance-co-ariance matrices S_1, S_2, S, and S_0 are shown in Table 8.31 with their respec-ive determinants. Equations (4.121), (4.123), (4.124), and (4.126) are re-uired for testing $H:\Sigma_1 = \Sigma_2$. For $t = 2$ and $p = 4$, one obtains

$$g_1 = [(t + 1)/(n - t)][(2p^2 + 3p - 1)/6(p + 1)]$$

$$= [(2 + 1)/(12 - 2)]\{[2(4)^2 + 3(4) - 1]/6(4 + 1)\} = 0.43$$

$$g_3 = [(n - t)\log_e|S| - \sum_{i=1}^{t} (r_i - 1)\log_e|S_i|]$$

$$= [(12 - 2)\log_e(0.0188) - (6 - 1)\{\log_e(0.0086) + \log_e(0.0022)\}]$$

$$= 10(-3.974) - 5[(-4.758) + (-6.134)] = 14.72$$

$$\nu_1 = [p + p(p - 1)/2](t - 1) = [4 + 4(4 - 1)/2](2 - 1) = 10$$

$$q = g_3(1 - g_1) = 14.72(1 - 0.43) = 8.39$$

Because $q < \chi^2_{0.30,10} = 11.78$ (App. A.3), one may accept $H:\Sigma_1 = \Sigma_2$. Equations (8.114), (8.116), (8.117), and (8.119) are required for testing $H:\Sigma = \Sigma_0$. Re-member that $r - 1$ is to be replaced by $\Sigma(r_i - 1) = 10$ df for animals within treatments in (8.114) and (8.116):

$$h_1 = 4(4 + 1)^2[2(4) - 3]/[6(12 - 2)(4 - 1)(4^2 + 4 - 4)] = 0.1736$$

$$h_3 = [-10 \log_e(0.0188/3.6551)] = -10(-5.273) = 52.73$$

Table 8.31. Variance-Covariance Matrices (among periods) and Determinants for Pregnant (S_1) and Non-pregnant (S_2) Lactating Rats, Averaged across Treatments (S) and Periods (S_0)

	P_1	P_2	P_3	P_4		
S_1 =	3.622	3.564	3.415	4.282	P_1	
	3.564	3.566	3.419	4.312	P_2	$\lvert S_1\rvert$=0.0086
	3.415	3.419	3.551	4.994	P_3	
	4.282	4.312	4.994	8.032	P_4	
S_2 =	3.403	3.277	3.543	4.555	P_1	
	3.277	3.171	3.375	4.379	P_2	$\lvert S_2\rvert$=0.0022
	3.543	3.375	4.006	5.346	P_3	
	4.555	4.379	5.346	7.770	P_4	
S =	3.512	3.420	3.479	4.418	P_1	
	3.420	3.368	3.397	4.346	P_2	$\lvert S\rvert$=0.0188
	3.479	3.397	3.778	5.170	P_3	
	4.418	4.346	5.170	7.901	P_4	
S_0 =	4.640	4.038	4.038	4.038	P_1	
	4.038	4.640	4.038	4.038	P_2	$\lvert S_0\rvert$=3.6551
	4.038	4.038	4.640	4.038	P_3	
	4.038	4.038	4.038	4.640	P_4	

$$\nu_1 = (4^2 + 4 - 4)/2 = 8 \qquad q = (1 - 0.1736)(52.73) = 43.58$$

Because $q > \chi^2_{0.001,8} = 26.12$, one should reject $H{:}\Sigma = \Sigma_0$; i.e., evidence is strong that the variance-covariance structure is not uniform across periods. Therefore, the univariate analysis of Table 8.24 is valid (in a probability sense) for treatments but not for periods or for treatment-period interaction.

The multivariate technique of Cole and Grizzle (1966) provides the most unified approach to the analysis of treatments, periods, and interaction when assumptions for univariate analysis are not fulfilled. In matrix notation, let $\mathbf{y}'_j$ be a row vector of p observations on subject j ($j = 1, 2, \ldots, n = \sum_{i=1}^{t} r_i$). Then, the $n \times p$ matrix $\mathbf{Y}$ contains all the data for the response variable Y. Model (8.59) is appropriate for the simplest case, in which n animals are randomly assigned among t treatments and measured in p periods. The effects of treatment (α_i), period (β_k), and interaction $(\alpha\beta)_{ik}$ may be added to the general mean (μ), so the effect of each treatment combination may be denoted by

$$\mu_{ik} = \mu + \alpha_i + \beta_k + (\alpha\beta)_{ik} \qquad (i = 1, 2, \ldots, t;\ k = 1, 2, \ldots, p) \quad (8.122)$$

Without loss of generality, one may use the μ_{ik} for experiments in which the treatments are factorial in nature and t is the number of treatment combinations. The $t \times p$ array of mean parameters (μ_{ik}) may be symbolized in matrix notation by the capital Greek letter M (mu). A *design matrix* (or incidence matrix) X with dimensions $n \times t$ may be used to denote which subjects are in which treatment group. Each row of the matrix represents one subject and has t elements, of which one is unity and the remainder are zero. Then, given random effects of subjects and residual error, the expected value of matrix Y is XM. For the example of Table 8.22 ($t = 2$, $n = 12$, $p = 4$), the expected value is shown explicitly in Table 8.32.

Table 8.32. Expected Value of Matrix of Data [$E(\mathsf{Y})=\mathsf{XM}$] for Example of Table 8.22

$$E\begin{bmatrix} 7.5 & 8.6 & 6.9 & 0.8 \\ 10.6 & 11.7 & 8.8 & 1.6 \\ 12.4 & 13.0 & 11.0 & 5.6 \\ 11.5 & 12.6 & 11.1 & 7.5 \\ 8.3 & 8.9 & 6.8 & 0.5 \\ 9.2 & 10.1 & 8.6 & 3.8 \\ 13.3 & 13.3 & 12.9 & 11.1 \\ 10.7 & 10.8 & 10.7 & 9.3 \\ 12.5 & 12.7 & 12.0 & 10.1 \\ 8.4 & 8.7 & 8.1 & 5.7 \\ 9.4 & 9.6 & 8.0 & 3.8 \\ 11.3 & 11.7 & 10.0 & 8.5 \end{bmatrix} = \begin{bmatrix} 1 & 0 \\ 1 & 0 \\ 1 & 0 \\ 1 & 0 \\ 1 & 0 \\ 1 & 0 \\ 0 & 1 \\ 0 & 1 \\ 0 & 1 \\ 0 & 1 \\ 0 & 1 \\ 0 & 1 \end{bmatrix} \begin{bmatrix} \mu_{11} & \mu_{12} & \mu_{13} & \mu_{14} \\ \mu_{21} & \mu_{22} & \mu_{23} & \mu_{24} \end{bmatrix}$$

Hypotheses about treatments, periods, and interaction all may be stated in the form $H\colon \mathsf{T'MP} = \mathsf{0}$, where $\mathsf{0}$ is a matrix of 0s and $\mathsf{T'}$ and $\mathsf{P'}$ are matrices of 1s and 0s whose rows are independent and limited in number to no more than $t - 1$ and $p - 1$, respectively. To test any such hypothesis, the total number of subjects (n) must be no less than t plus the number of rows in P. Usually P will have $p - 1$ rows when one tests effects of periods and interactions. Therefore one should ensure that $n \geq t + p - 1$. Two forms of T and two forms of P are required to test the general hypotheses about treatments, periods, and interaction. Matrix T_1' is a $1 \times t$ row vector of 1s and matrix P_1' is a $1 \times p$ row vector of 1s. Matrix T_2' has $t - 1$ independent rows of t elements each such that the elements in each row sum to zero. Matrix P_2' has $p - 1$ independent rows of p elements each such that the elements in each row sum to zero.

Then, the hypothesis of equal treatment effects may be stated H_T:$T_2'MP_1 = 0$, the hypothesis of equal period effects may be stated H_P:$T_1'MP_2 = 0$, and the hypothesis of no interaction may be stated H_{TP}:$T_2'MP_2 = 0$. Specific matrices suitable for the example of Table 8.22 ($t = 2$, $p = 4$) are shown in Table 8.33. The matrix T_2' is quite simple, [1 -1], when there are only 2 treatment groups as in the example. In general, T_2' usually has $t - 1$ rows and t columns. For example, for 5 treatments, let

$$T_2' = \begin{bmatrix} 1 & 0 & 0 & 0 & -1 \\ 0 & 1 & 0 & 0 & -1 \\ 0 & 0 & 1 & 0 & -1 \\ 0 & 0 & 0 & 1 & -1 \end{bmatrix}$$

The test statistics for the three hypotheses each involve a matrix for the hypothesis S_H and a matrix for error S_E. The general format for the hypothesis matrix is

$$S_H = P'Y'X(X'X)^{-1}T[T'(X'X)^{-1}T]^{-1}T'(X'X)^{-1}X'YP \quad (8.123)$$

and the format for the error matrix is

$$S_E = P'Y'[I - X(X'X)^{-1}X']YP \quad (8.124)$$

where I is $n \times n$ for n subjects. In (8.123) and (8.124), the matrices T, T', P, and P' each may take two forms, depending on the hypothesis being tested (Table 8.33). For the test of treatments, one uses T_2, T_2', P_1, and P_1'. Then two matrices are to be inverted: (X'X), which is $t \times t$ in any case (also diagonal unless treatment combinations are partitioned into main effects and interactions), and $[T_2'(X'X)^{-1}T_2]$, which is $(t - 1) \times (t - 1)$ in this case. Because P_1' is $1 \times p$, both S_H and S_E are scalars (1×1) in the test of treatments for which one rejects H if the test statistic

$$U = S_E/(S_H + S_E) \quad (8.125)$$

is smaller than critical value U_{α,p,ν_1,ν_2} (App. A.19), where $\nu_1 = t - 1$ and $\nu_2 = n - t - p + 1$. For the test of periods, use T_1, T_1', P_2, P_2' in (8.123) and (8.124). Then, because T_1' is only $1 \times t$, $[T_1'(X'X)^{-1}T_1]$ is scalar (1×1) and may be inverted by taking the reciprocal. However, P_2' is $(p - 1) \times p$, so S_H and S_E have dimensions $(p - 1) \times (p - 1)$. Then to test periods, one rejects H if the test statistic

Table 8.33. Hypotheses about Treatments, Periods, and Interaction for Example of Table 8.22

H_T: $T_2'MP_1=0$

$$[1 \quad -1]\begin{bmatrix}\mu_{11} & \mu_{12} & \mu_{13} & \mu_{14}\\ \mu_{21} & \mu_{22} & \mu_{23} & \mu_{24}\end{bmatrix}\begin{bmatrix}1\\1\\1\\1\end{bmatrix}=\{(\mu_{11}+\mu_{12}+\mu_{13}+\mu_{14})-(\mu_{21}+\mu_{22}+\mu_{23}+\mu_{24})\}=0$$

H_P: $T_1'MP_2=0$

$$[1 \quad 1]\begin{bmatrix}\mu_{11} & \mu_{12} & \mu_{13} & \mu_{14}\\ \mu_{21} & \mu_{22} & \mu_{23} & \mu_{24}\end{bmatrix}\begin{bmatrix}1 & 0 & 0\\0 & 1 & 0\\0 & 0 & 1\\-1 & -1 & -1\end{bmatrix}=\left[\begin{Bmatrix}(\mu_{11}+\mu_{21})\\-(\mu_{14}+\mu_{24})\end{Bmatrix}\begin{Bmatrix}(\mu_{12}+\mu_{22})\\-(\mu_{14}+\mu_{24})\end{Bmatrix}\begin{Bmatrix}(\mu_{13}+\mu_{23})\\-(\mu_{14}+\mu_{24})\end{Bmatrix}\right]=[0 \quad 0 \quad 0]$$

H_{TP}: $T_2'MP_2=0$

$$[1 \quad -1]\begin{bmatrix}\mu_{11} & \mu_{12} & \mu_{13} & \mu_{14}\\ \mu_{21} & \mu_{22} & \mu_{23} & \mu_{24}\end{bmatrix}\begin{bmatrix}1 & 0 & 0\\0 & 1 & 0\\0 & 0 & 1\\-1 & -1 & -1\end{bmatrix}=\left[\begin{Bmatrix}(\mu_{11}-\mu_{21})\\-(\mu_{14}-\mu_{24})\end{Bmatrix}\begin{Bmatrix}(\mu_{12}-\mu_{22})\\-(\mu_{14}-\mu_{24})\end{Bmatrix}\begin{Bmatrix}(\mu_{13}-\mu_{23})\\-(\mu_{14}+\mu_{24})\end{Bmatrix}\right]=[0 \quad 0 \quad 0]$$

$$U = |S_E|/|[S_H + S_E]| \quad (8.126)$$

is smaller than critical value U_{α,p,ν_1,ν_2}, where $\nu_1 = p - 1$, $\nu_2 = (n - t - 1)\cdot(p - 1)$. Note: (8.126) requires determinants of S_E and $[S_H + S_E]$, both of which are $(p - 1) \times (p - 1)$. High-speed computation will be needed if there are more than 4 or 5 periods (see Chap. 4). For the test of treatment-period interaction, use T_2, T_2', P_2, and P_2' in (8.123) and (8.124). Matrix $[T_2'(X'X)^{-1}T_2]$ is $(t - 1) \times (t - 1)$ and must be inverted. Matrices S_H and S_E are $(p - 1) \times (p - 1)$. To test interaction, compare test statistic

$$U = |S_E|/|[S_H + S_E]| \quad (8.127)$$

with critical value U_{α,p,ν_1,ν_2}, where $\nu_1 = (t - 1)(p - 1)$ and $\nu_2 = (n - t - 1)\cdot(p - 1)$. Reject H_{TP} only if the test statistic is smaller than the critical value. Note: test statistic U requires determinants of two $(p - 1) \times (p - 1)$ matrices. Table A.19 provides critical values for $p \leq 7$, $\nu_1 \leq 12$, and $\nu_2 \leq 1000$. For $12 < \nu_1 \leq 120$, consult Kramer (1972). For large p and ν_1, an alternative approximate test of any of the hypotheses may be made by comparing test statistic

$$q = [(p + 1 - 3\nu_1)/2](\log_e U) \quad (8.128)$$

with critical value $\chi^2_{\alpha,\nu}$ (Table A.3), where $\nu = p\nu_1$. Reject H for $q > \chi^2_{\alpha,\nu}$.

Consider the data of Table 8.22 with $t = 2$ treatments, $p = 4$ periods, and $n = 12$ animals. Using matrices Y and X from Table 8.32, one obtains

$$X'X = \begin{bmatrix} r_1 & 0 \\ 0 & r_2 \end{bmatrix} = \begin{bmatrix} 6 & 0 \\ 0 & 6 \end{bmatrix} \qquad (X'X)^{-1} = \begin{bmatrix} 1/6 & 0 \\ 0 & 1/6 \end{bmatrix}$$

$$(X'Y) = \begin{bmatrix} y_{1.1} & y_{1.2} & y_{1.3} & y_{1.4} \\ y_{2.1} & y_{2.2} & y_{2.3} & y_{2.4} \end{bmatrix} = \begin{bmatrix} 59.5 & 64.9 & 53.2 & 19.8 \\ 65.6 & 66.8 & 61.7 & 48.5 \end{bmatrix}$$

For the test of treatments

$$T_2'(X'X)^{-1}T_2 = [1 \;\; -1] \begin{bmatrix} 1/6 & 0 \\ 0 & 1/6 \end{bmatrix} \begin{bmatrix} 1 \\ -1 \end{bmatrix} = [1/3]$$

and $[T_2'(X'X)^{-1}T_2]^{-1} = 1/[1/3] = [3]$. Also, because $P_1' = [1, 1, 1, 1]$ and $(Y'X) = (X'Y)'$, then $(P_1'Y'X) = [y_{1..}, y_{2..}] = [197.4, 242.6]$, the two treatment totals. From (8.123) one obtains

$$S_H = [197.4 \quad 242.6] \begin{bmatrix} 1/6 & 0 \\ 0 & 1/6 \end{bmatrix} \begin{bmatrix} 1 \\ -1 \end{bmatrix} [3][1 \quad -1] \begin{bmatrix} 1/6 & 0 \\ 0 & 1/6 \end{bmatrix} \begin{bmatrix} 197.4 \\ 242.6 \end{bmatrix} = 170.2$$

In the process of computing S_E from (8.124), one obtains $X(X'X)^{-1}X'$, which is $n \times n$. This $n \times n$ matrix is made up of t^2 submatrices of size $r \times r$. All submatrices off the diagonal consist of 0s only (unless treatment combinations are partitioned into main effects and interactions in constructing X); the t submatrices on the diagonal each have r^2 elements equal to corresponding diagonal elements of $(X'X)^{-1}$, i.e., $1/r = 1/6$ for the example. On subtraction of $X(X'X)^{-1}X'$ from the $n \times n$ identity matrix I, the diagonal elements of diagonal submatrices become $1 - (1/r) = 5/6$ for the example; the off-diagonal elements of those submatrices become $-1/r = -1/6$, and all other elements are 0. For the test of treatments, $P_1'Y'$ is made up of n elements, each of which is the total of the observations on one subject over p periods. Then $P_1'Y'[I - X(X'X)^{-1}X']$ is 1×12, where each element is $1/r$ times the difference between $r - 1$ times the total of responses for one subject and the total of the observations on the remaining $r - 1$ subjects in the same treatment group. On multiplying that result by (YP_1), which is the same as $(P_1'Y')'$, one obtains S_E = 1006. For the test statistic of (8.125), $U = 1006/(170.2 + 1006) = 0.855$, which is not smaller than $U_{0.05,4,1,7} = 0.1354$. Therefore, H_T is accepted and one must conclude that the mean response is not significantly different for pregnant and nonpregnant rats.

For the test of periods (H_P), one obtains

$$T_1'(X'X)^{-1}T_1 = [1 \quad 1] \begin{bmatrix} 1/6 & 0 \\ 0 & 1/6 \end{bmatrix} \begin{bmatrix} 1 \\ 1 \end{bmatrix} = [1/3]$$

and its inverse [3]. Also,

$$P_2'Y'X = \begin{bmatrix} 39.7 & 17.1 \\ 45.1 & 18.3 \\ 33.4 & 13.2 \end{bmatrix}$$

where each column represents subtraction of the period 4 total from each of the other period totals for a given treatment. Then, from (8.123), one obtains

$$S_H = \begin{bmatrix} 39.7 & 17.1 \\ 45.1 & 18.3 \\ 33.4 & 13.2 \end{bmatrix} \begin{bmatrix} 1/6 & 0 \\ 0 & 1/6 \end{bmatrix} \begin{bmatrix} 1 \\ 1 \end{bmatrix} [3][1 \quad 1] \begin{bmatrix} 1/6 & 0 \\ 0 & 1/6 \end{bmatrix} \begin{bmatrix} 39.7 & 45.1 & 33.4 \\ 17.1 & 18.3 & 13.2 \end{bmatrix}$$

$$= \begin{bmatrix} 268.872 & 300.114 & 220.590 \\ 300.114 & 334.983 & 246.222 \\ 220.590 & 246.222 & 180.978 \end{bmatrix}$$

In computing S_E, use $[I - X(X'X)^{-1}X']$ exactly as in the test of treatments. Also, $P_2'Y'$ is 3×12, containing three elements for each rat (period 4 response subtracted from each of the others):

$$\begin{bmatrix} 6.7 & 9.0 & 6.8 & 4.0 & 7.8 & 5.4 & 2.2 & 1.4 & 2.4 & 2.7 & 5.6 & 2.8 \\ 7.8 & 10.1 & 7.4 & 5.1 & 8.4 & 6.3 & 2.2 & 1.5 & 2.6 & 3.0 & 5.8 & 3.2 \\ 6.1 & 7.2 & 5.4 & 3.6 & 6.3 & 4.8 & 1.8 & 1.4 & 1.9 & 2.4 & 4.2 & 1.5 \end{bmatrix}$$

Then, from (8.124), one obtains

$$S_E = \begin{bmatrix} 25.75 & 25.56 & 17.90 \\ 25.56 & 25.77 & 17.81 \\ 17.90 & 17.81 & 13.38 \end{bmatrix}$$

The determinant of S_E is $|S_E| = 9.57$. The matrix sum

$$[S_H + S_E] = \begin{bmatrix} 294.62 & 325.67 & 238.49 \\ 325.67 & 360.75 & 264.03 \\ 238.49 & 264.03 & 194.36 \end{bmatrix}$$

has determinant $|[S_H + S_E]| = 243.13$. For the test statistic of (8.126), $U = 9.57/243.13 = 0.0394$, which is smaller than $U_{0.01,4,3,27} = 0.3632$. Therefore H_P is rejected and one must conclude that mean response changes with time (averaged over pregnant and nonpregnant rats).

For the test of treatment-period interaction (H_{TP}), one obtains

$$T_2'(X'X)^{-1}T_2 = [1 \quad -1] \begin{bmatrix} 1/6 & 0 \\ 0 & 1/6 \end{bmatrix} \begin{bmatrix} 1 \\ -1 \end{bmatrix} = [1/3]$$

with inverse [3], as in the test of treatments. Also $P_2'Y'X$ is required, as in the test of periods (above). Then from (8.123) one obtains

$$S_H = \begin{bmatrix} 39.7 & 17.1 \\ 45.1 & 18.3 \\ 33.4 & 13.2 \end{bmatrix} \begin{bmatrix} 1/6 & 0 \\ 0 & 1/6 \end{bmatrix} \begin{bmatrix} 1 \\ -1 \end{bmatrix} [3][1 \quad -1] \begin{bmatrix} 1/6 & 0 \\ 0 & 1/6 \end{bmatrix} \begin{bmatrix} 39.7 & 45.1 & 33.4 \\ 17.1 & 18.3 & 13.2 \end{bmatrix}$$

$$= \begin{bmatrix} 42.5709 & 51.6117 & 38.0505 \\ 51.6117 & 62.5725 & 46.1313 \\ 38.0505 & 46.1313 & 34.0101 \end{bmatrix}$$

The error matrix $\mathbf{S}_E$ is exactly the same as for the test of periods (above). The determinant of $\mathbf{S}_E$ is 9.57 as before. The matrix sum

$$[\mathbf{S}_H + \mathbf{S}_E] = \begin{bmatrix} 68.32 & 77.17 & 55.95 \\ 77.17 & 88.34 & 63.94 \\ 55.95 & 63.94 & 47.39 \end{bmatrix}$$

has determinant $|[\mathbf{S}_H + \mathbf{S}_E]| = 88.25$. For the test statistic of (8.127), $U = 9.57/88.25 = 0.1084$, which is smaller than $U_{0.01,4,3,27} = 0.3632$. Therefore, one may conclude that the time trend for pregnant rats is not parallel to the time trend for nonpregnant rats.

Orthogonal contrasts ($\nu_1 = 1$ df) among t treatments (or treatment combinations) may be tested by placing the appropriate $\mathbf{T}$ vector for each contrast in turn and $\mathbf{P}_1$ into (8.123). For example, suppose treatments consist of 2 sources of protein (factor A) and 2 dietary levels (factor B), such that the 4 combinations are (1) U, 14%; (2) U, 16%; (3) N, 14%; and (4) N, 16%. Then vectors for testing source, dose, and interaction of source with dose are

$$\mathbf{T}'_S = [+1 \quad +1 \quad -1 \quad -1] \qquad \mathbf{T}'_D = [-1 \quad +1 \quad -1 \quad +1] \qquad \mathbf{T}'_{SD} = [-1 \quad +1 \quad +1 \quad -1]$$

Similarly, one may wish to use orthogonal polynomials (App. A.7) to characterize the shape of response over time if the response is regular. If response is erratic, some other set of contrasts should be used. One places the appropriate $\mathbf{P}$ vector for each contrast in turn into (8.123) and (8.124), using $\mathbf{T}_1$ in (8.123). For example, if there are measurements from 3 periods equally spaced in time, vectors for linear and quadratic trends are

$$\mathbf{P}'_L = [-1 \quad 0 \quad +1] \qquad \mathbf{P}'_Q = [+1 \quad -2 \quad +1]$$

Finally, one may examine specific aspects of treatment-period interaction by combining appropriate contrast vectors $\mathbf{T}$ and $\mathbf{P}$ in (8.123) and (8.124).

See Jensen (1972) for some alternative methods that utilize Hotelling's T^2 statistics and permit multivariate as well as univariate response over time. See Eaton (1969) for multiple comparisons of several response profiles over time.

EXERCISES

8.1. A complete block, random order, changeover experiment was performed.

Three vasopressor drugs were used on each of 5 dogs (blocks). Thirty minutes after injection of a drug, the drop in arterial pressure was recorded (mmHg). After the pressure returned to normal, another drug was tested (Huntsberger and Leaverton 1970).

Block (dog)	Treatment (drug) 1	2	3	Total
1	13	32	22	67
2	40	61	54	155
3	34	65	47	146
4	16	25	30	71
5	23	40	41	104
Total	126	223	194	543
Mean	25.2	44.6	38.8	36.2

(a) Test for treatment differences, including Tukey's comparisons of all pairs.
(b) Estimate the efficiency of this design relative to a completely randomized design with 15 dogs.

8.2. Samples of plasma from 8 subjects were divided into 4 parts per subject. The 4 parts were randomly assigned to 4 treatments for each subject (block). Clotting time (minutes) was recorded in each case (Armitage 1971).
(a) Test for treatment differences, including Dunnett's tests against the control (treatment 1).
(b) Estimate the efficiency of this design relative to a completely randomized design with 32 subjects.

Block (subject)	Treatment 1	2	3	4	Total
1	8.4	9.4	9.8	12.2	39.8
2	12.8	15.2	12.9	14.4	55.3
3	9.6	9.1	11.2	9.8	39.7
4	9.8	8.8	9.9	12.0	40.5
5	8.4	8.2	8.5	8.5	33.6
6	8.6	9.9	9.8	10.9	39.2
7	8.9	9.0	9.2	10.4	37.5
8	7.9	8.1	8.2	10.0	34.2
Total	74.4	77.7	79.5	88.2	319.8
Mean	9.30	9.71	9.94	11.02	9.99

8.3. The effects of dose of TRH (A = 10 or 25 μg) and method of injection (B = intravenous, intramuscular, or subcutaneous) of serum prolactin in cattle were studied in a random order design with animal = block. The data are differences between peak response (samples taken for 30 minutes, at 2 minute intervals) and pretreatment response.

	10 μg			25 μg			
Block	iv	im	sc	iv	im	sc	Total
1	74.7	33.5	38.0	243.9	118.3	59.5	567.9
2	156.7	2.8	2.4	236.3	34.2	33.0	465.4
3	310.9	38.7	75.2	313.9	237.1	314.9	1290.7
4	15.3	6.8	15.7	112.6	190.9	50.8	392.1
5	195.9	26.6	12.7	132.8	35.0	75.3	478.3
6	24.1	1.6	4.5	74.5	87.4	11.7	203.8
Total	777.6	110.0	148.5	1114.0	702.9	545.2	3398.2
Mean	129.6	18.3	24.8	185.7	117.1	90.9	94.4

(a) Test factorial effects and compute standard errors for the most useful means.
(b) Test for nonadditivity of block and treatment effects.
(c) Test for dependence of variance on level of response.

8.4. Consider a simple changeover experiment with cows to study serum prolactin response to multiple washing of the udder. The known nuisance trend is decline of prolactin with stage of lactation. Suppose 10 cows, at the same (early) stage of lactation are available; 5 are assigned to sequence *AB* and 5 to sequence *BA*, with washing for 30 seconds at *A* = 5 minute intervals and *B* = 15 minute intervals, for 1 hour. Data represent hypothetical differences in prolactin measured 5 minutes after last washing and prolactin measured pretreatment.

Cow	1(*A*)	2(*B*)	Total	Cow	1(*B*)	2(*A*)	Total
1	19.5	10.7	30.2	6	8.9	11.9	20.8
2	8.3	2.2	10.5	7	61.4	68.2	129.6
3	78.2	70.5	148.7	8	40.2	46.0	86.2
4	10.5	2.2	12.7	9	13.4	17.8	31.2
5	38.2	33.8	72.0	10	4.2	9.4	13.6
Total	154.7	119.4		Total	128.1	153.3	

Period 1: 282.8; Period 2: 272.7; *A*: 308.0; *B*: 247.5; $y_{...} = 555.5$

(a) Compare mean effectiveness of the two washing procedures.
(b) Estimate the number of animals required for equal sensitivity without repeated measurement.
(c) Estimate the efficiency of accounting for period effects (relative to a random order design).

8.5. A Latin square changeover design was used to assay insulin from the blood sugar responses of rabbits (Young and Romans 1948). Each of 4 rabbits was used on 4 different days for 4 factorial treatments (A = standard, low dose, B = standard, high dose, C = unknown, low dose, D = unknown, high dose). Data are mg% blood sugar/mL from an ear vein 50 minutes after injection.

Rabbit	Period 1	2	3	4	Total
1	27(B)	36(A)	54(C)	38(D)	155
2	54(C)	40(D)	61(A)	36(B)	191
3	48(D)	72(C)	60(B)	60(A)	240
4	63(A)	50(B)	50(D)	59(C)	222
Total	192	198	225	193	808

Treatments: (A) 220, (B) 173, (C) 239, (D) 176

(a) Test for parallel response for the two preparations and for mean difference by dose and between preparations.
(b) Estimate the number of rabbits required for equal sensitivity without repeated measurement.

8.6. A study of the possible effects of order of administration (treatments A-F), in a series of inoculations of testicular diffusing factor on the same animal, was performed as a Latin square changeover design in space, using 6 positions on the skin of the backs of 6 rabbits (Bacharach et al. 1940). Data are blister areas (cm^2) following inoculation.
(a) Should order of administration be considered a nuisance variable for future assays? Why?
(b) Estimate the number of rabbits required for equal assay sensitivity if each rabbit were given only one inoculation.

Rabbit	Position 1	2	3	4	5	6	Total
1	7.9(C)	6.1(D)	7.5(A)	6.9(F)	6.7(B)	7.3(E)	42.4
2	8.7(E)	8.2(B)	8.1(C)	8.5(A)	9.9(D)	8.3(F)	51.7
3	7.4(D)	7.7(F)	6.0(E)	6.8(C)	7.3(A)	7.3(B)	42.5
4	7.4(A)	7.1(E)	6.4(F)	7.7(B)	6.4(C)	5.8(D)	40.8
5	7.1(F)	8.1(C)	6.2(B)	8.5(D)	6.4(E)	6.4(A)	42.7
6	8.2(B)	5.9(A)	7.5(D)	8.5(E)	7.3(F)	7.7(C)	45.1
Total	46.7	43.1	41.7	46.9	44.0	42.8	265.2

Treatments: (A) 43.0, (B) 44.3, (C) 45.0, (D) 45.2, (E) 44.0, (F) 43.7

8.7. Bliss (1967) analyzed a dosage-response experiment for the mydriatic action of a complex amine in rabbits in 3 Latin squares (4×4). Four animals in each square were given doses A = 12.5 mg, B = 50 mg, C = 200 mg, and D = 800 mg on 4 different days. The days (periods) were common to all 3 squares. Data are increases in pupil diameter (mm).

	Square 1 Period						Square 2 Period						Square 3 Period				
	1	2	3	4	Total		1	2	3	4	Total		1	2	3	4	Total
1	7_D	5_C	2_B	1_A	15	5	4_C	5_D	1_A	2_B	12	9	3_B	0_A	5_D	3_C	11
2	4_C	6_D	1_A	3_B	14	6	6_D	4_C	2_B	0_A	12	10	0_A	4_D	3_C	2_B	9
3	3_B	1_A	6_C	7_D	17	7	1_A	3_B	4_C	5_D	13	11	7_D	3_C	2_B	0_A	12
4	1_A	3_B	6_D	3_C	13	8	2_B	2_A	7_D	4_C	15	12	4_C	2_B	0_A	6_D	12
Total	15	15	15	14		Total	13	14	14	11		Total	14	9	10	11	

Periods: (1) 42, (2) 38, (3) 39, (4) 36

Treatments: (A) 8, (B) 29, (C) 47, (D) 71; $y_{...} = 155$

(a) The doses are equally spaced on a fourfold ($\log_4$) scale. On that scale, use orthogonal polynomials to test linear and nonlinear effects. Pool periods across squares.

(b) Estimate the efficiency of the design relative to a complete block random order design with repeated measurement, ignoring period effects.

8.8. An experiment was performed to determine effects on physical performance of drinking A = none, B = 30 mL, or C = 60 mL of 200 proof alcohol. Six subjects were divided into groups of three by weight. The heavier subjects were assigned to one Latin square and the lighter ones to another; the two squares were selected for balanced residuals. Each subject was given the 3 treatments in 3 successive weeks. Response was number of seconds to negotiate a very short obstacle course. The week between treatments was sufficient to dissipate physiological residual effects but possibly not long enough to eliminate psychological residual effects (learning).

	Heavy					Light			
Subject	1	2	3	Total	Subject	1	2	3	Total
1	2.09(A)	2.10(B)	2.06(C)	6.25	4	2.49(A)	2.46(C)	2.45(B)	7.40
2	2.24(B)	2.26(C)	2.18(A)	6.68	5	2.64(C)	2.46(B)	2.42(A)	7.52
3	2.05(C)	2.03(A)	2.07(B)	6.15	6	2.36(B)	2.35(A)	2.42(C)	7.13
Total	6.38	6.39	6.31	19.08	Total	7.49	7.27	7.29	22.05

Treatments: (A) 13.56, (B) 13.68, (C) 13.89; $y_{...} = 41.13$

(a) Complete an analysis of variance, including residual effects. Remove period effects within squares.

(b) Estimate the mean differences between (1) A and B and (2) A and C (with standard errors) for direct and residual effects.

(c) Estimate the efficiency of allocating subjects to squares by weight (removing period effects within squares rather than across squares).
(d) Estimate the efficiency (for direct effects) relative to an RCBD without repeated measurement (blocks = weight).
(e) What type of design would you recommend for future experiments of this type? Why?

8.9. A dairy feeding experiment to compare A = whey supplemented ration with B = control ration was performed in a double reversal design with 4 periods of 30 days in early lactation. Twenty-four cows from a high producing Guernsey herd were randomly allocated: 12 to sequence *ABAB* and 12 to sequence *BABA*. Data are kg milk/day. Totals ($-q_3$) for subjects are $y_{i1} - 3y_{i2} + 3y_{i3} - y_{i4}$.

	ABAB						*BABA*				
Cow	1	2	3	4	$-q_3$	Cow	1	2	3	4	$-q_3$
1	30.8	31.0	29.6	21.4	+5.2	13	31.2	31.3	29.2	28.1	-3.2
2	35.4	33.3	29.2	30.2	-7.1	14	35.3	36.7	34.4	31.9	-3.5
3	35.5	35.5	37.0	29.9	+10.1	15	28.2	26.8	24.2	22.0	-1.6
4	31.3	28.0	29.0	26.3	+8.0	16	26.4	26.7	27.0	25.8	+1.5
5	30.2	27.3	26.2	24.9	+2.0	17	21.7	21.4	20.3	19.1	-0.7
6	30.2	32.2	30.4	29.5	-4.7	18	25.6	27.1	25.8	25.1	-3.4
7	22.7	22.6	20.4	19.1	-3.0	19	31.3	30.2	26.8	25.5	-4.4
8	31.7	30.3	27.3	24.5	-1.8	20	29.7	27.2	25.6	22.2	+2.7
9	34.1	29.5	27.4	25.7	+2.1	21	30.0	28.6	26.8	23.5	+1.1
10	32.4	29.5	28.9	25.8	+4.8	22	37.6	36.4	33.5	29.2	-0.3
11	36.3	33.8	30.0	26.7	-1.8	23	32.6	27.9	26.5	22.7	+5.7
12	25.6	23.9	20.1	19.2	-5.0	24	25.0	25.4	25.7	24.1	+1.8
Total	376.2	356.9	335.5	303.2	+8.8	Total	354.6	345.7	325.8	299.2	-4.3

(a) Compute an estimate of mean treatment difference with standard error. Is the difference significant?
(b) Analyze data in the first 2 periods as a simple crossover design. Compare estimates of mean difference and mean square error with the reversal results.

8.10. An experiment was proposed to study dry matter digestibility of rations containing A = 0, B = 20%, C = 60%, or D = 80% dehydrated chicken feces by feeding mature male sheep. Complexity of equipment and facilities limited the study to 6 sheep.
(a) Devise a Youden crossover design in which each sheep will be fed each of the rations in a sequence of 4 two-week periods.
(b) What is the minimum efficiency for the design?

8.11. Consider a 2-period changeover experiment to study methionine requirement in dairy calves. Suppose 5 doses are studied (A = 70%, B = 85%,

C = 100%, D = 115%, E = 130% of methionine content of milk). Then 20 calves in 10 Latin squares (2 × 2) will permit a balanced design to be used. Data represent hypothetical response in serum methionine (mg/ 100 mL).

(a) Compute adjusted treatment means and standard errors and test for treatment differences.

(b) Use first and second degree orthogonal polynomials to estimate the dose that maximizes serum methionine.

(c) Estimate the number of calves required in a completely randomized design to equal the precision provided by 20 calves in this design.

Calf	1	2	Total	Calf	1	2	Total	Calf	1	2	Total
1	0.49_A	0.80_B	1.29	3	0.45_A	0.92_C	1.37	5	0.20_A	0.67_D	0.87
2	0.42_B	0.45_A	0.87	4	0.73_C	0.48_A	1.21	6	0.66_D	0.40_A	1.06
7	0.54_A	0.96_E	1.50	9	0.71_B	0.80_C	1.51	11	0.80_B	0.97_D	1.77
8	0.81_E	0.69_A	1.50	10	0.76_C	0.69_B	1.45	12	0.65_D	0.50_B	1.15
13	0.48_B	0.74_E	1.22	15	0.86_C	1.02_D	1.88	17	0.87_C	0.93_E	1.80
14	0.70_E	0.77_B	1.47	16	1.18_D	1.22_C	2.40	18	0.63_E	0.67_C	1.30
19	1.20_D	1.26_E	2.46								
20	1.19_E	1.38_D	2.57								

Totals: Period 1: 14.33, Period 2: 16.32, $y_{...}$ = 30.65

(A) 3.70, (B) 5.17, (C) 6.83, (D) 7.73, (E) 7.22

Means: (A) 0.462, (B) 0.646, (C) 0.854, (D) 0.966, (E) 0.902

8.12. An experiment was conducted to study serum prolactin changes following TRH injection on days 0, 2, 4, 7, 15, and 18 of the estrous cycle of dairy heifers. Data represent differences between average response during 30 minutes injection and response prior to injection.

(a) Compute the analysis of variance and discuss validity of f tests.

(b) Use Scheffé's procedure to test contrasts: (1) days 2 and 4 vs. others and (2) day 0 vs. day 18.

	Days of Estrus						
	(1)	(2)	(3)	(4)	(5)	(6)	
Animal	0	2	4	7	15	18	Total
1	8.0	21.6	10.9	73.4	50.5	59.2	223.6
2	94.8	14.3	48.9	58.3	49.8	97.8	363.9
3	28.3	20.7	26.2	41.5	85.7	28.7	231.1
4	78.4	56.9	53.6	44.3	34.2	64.9	332.3
5	65.9	79.1	28.0	25.8	63.0	67.1	328.9
6	1.3	4.8	24.1	37.2	40.0	45.6	153.0

Animal	(1) 0	(2) 2	(3) 4	(4) 7	(5) 15	(6) 18	Total
	Days of Estrus						
Total	276.7	197.4	191.7	280.5	323.2	363.3	1632.8
Mean	45.12	32.90	31.95	46.75	53.87	60.55	
Variance	1517	825	262	281	341	538	

Covariances: $s_{12} = 538$, $s_{13} = 541$, $s_{14} = -85$, $s_{15} = -94$, $s_{16} = 674$,
$s_{23} = 110$, $s_{24} = -242$, $s_{25} = 13$, $s_{26} = 108$, $s_{34} = -46$,
$s_{35} = -107$, $s_{36} = 206$, $s_{45} = -63$, $s_{46} = 111$, $s_{56} = -196$

8.13. An experiment was designed to compare radioactive iron in (1) nuclei, (2) mitochondria, (3) microsomes, and (4) supernatant from liver cells of 16 animals (Koch and Sen 1968). The animals were randomly assigned to receive or not to receive ethionine in their diets and were pair-fed to equalize intake. Data are ratios of radioactive iron in treated and control animals of each pair. Therefore, the subject in this experiment is a pair of animals.
(a) Compute the analysis of variance and discuss validity of f tests.
(b) Test all possible pairs of mean ratios.

Pair	Fraction ("treatment") 1	2	3	4	Total
1	1.73	1.08	2.60	1.67	7.08
2	2.50	2.55	2.51	1.80	9.36
3	1.17	1.47	1.49	1.47	5.60
4	1.54	1.75	1.55	1.72	6.56
5	1.53	2.71	2.51	2.25	9.00
6	2.61	1.37	1.15	1.67	6.80
7	1.86	2.13	2.47	2.50	8.96
8	2.21	1.06	0.95	0.98	5.20
Total	15.15	14.12	15.23	14.06	58.56
Mean	1.894	1.765	1.904	1.758	
Variance	0.256	0.407	0.473	0.214	

Covariances: $s_{12} = -0.008$, $s_{13} = -0.058$, $s_{14} = -0.036$
$s_{23} = 0.268$, $s_{24} = 0.214$, $s_{34} = 0.232$

8.14. A split-plot experiment was designed to study trends in methemoglobin (M) in sheep following treatment with 3 equally spaced levels of NO_2 (factor A). Four sheep were assigned to each level and each animal was

measured at 6 sampling times (factor *B*), 5 of them following treatment. Data are log(M + 5).
(a) Compute the analysis of variance and discuss validity of *f* tests.
(b) Use Scheffé's procedure to test orthogonal polynomials for NO_2 at each sampling time.
(c) Use Scheffé's procedure to compare all possible pairs of means for the sampling points at each level of NO_2.

NO_2 (*A*)	Sheep	Sampling Time (*B*) 1	2	3	4	5	6	Total
1	1	2.197	2.442	2.542	2.241	1.960	1.988	13.370
	2	1.932	2.526	2.526	2.152	1.917	1.917	12.970
	3	1.946	2.251	2.501	1.988	1.686	1.841	12.213
	4	1.758	2.054	2.588	2.197	2.140	1.686	12.423
	Total	7.833	9.273	10.157	8.578	7.703	7.432	50.976
	Mean	1.958	2.318	2.539	2.144	1.926	1.858	2.124
	Variance	0.0326	0.0443	0.0013	0.0122	0.0349	0.0168	
2	5	2.230	3.086	3.357	3.219	2.827	2.534	17.253
	6	2.398	2.580	2.929	2.874	2.282	2.303	15.366
	7	2.054	3.243	3.653	3.811	3.816	3.227	19.804
	8	2.510	2.695	2.996	3.246	2.565	2.230	16.242
	Total	9.192	11.604	12.935	13.150	11.490	10.294	68.665
	Mean	2.298	2.901	3.234	3.288	2.872	2.574	2.861
	Variance	0.0397	0.0989	0.1135	0.1505	0.4452	0.2066	
3	9	2.140	3.896	4.246	4.461	4.418	4.331	23.492
	10	2.303	3.822	4.109	4.240	4.127	4.084	22.685
	11	2.175	2.907	3.086	2.827	2.493	2.230	15.718
	12	2.041	3.824	4.111	4.301	4.206	4.182	22.665
	Total	8.659	14.449	15.552	15.829	15.244	14.827	84.560
	Mean	2.165	3.612	3.888	3.957	3.811	3.707	3.523
	Variance	0.0117	0.2222	0.2900	0.5764	0.7872	0.9796	
Total		25.684	35.326	38.644	37.557	34.437	32.553	204.201
Mean		2.140	2.944	3.220	3.130	2.870	2.713	2.836

8.15. The length of estrous cycles of rats was studied in a split-plot design with rats assigned to different diets: (1) modified cereal, (2) high roughage, and (3) high alcohol. Each rat was measured for 4 successive cycles and again at a later age (all classes of factor *B*). Data are $\log_e$ (length of cycle at one age).

Diet (*A*)	Rat	(*B*) 1	2	3	4	5	Total
1	1	1.46	1.39	1.39	1.67	1.87	7.78
	2	1.54	1.61	1.56	1.39	1.47	7.57
	3	1.47	1.45	1.39	1.39	1.57	7.27
	4	1.61	1.54	1.39	1.49	2.20	8.23
	5	1.79	1.95	1.66	1.39	2.23	9.02
	6	1.61	1.61	1.79	1.46	1.65	8.12
	Total	9.48	9.55	9.18	8.79	10.99	47.99
	Mean	1.580	1.592	1.530	1.465	1.832	1.600
	Variance	0.0148	0.0385	0.0288	0.0119	0.1056	
2	7	1.61	1.61	2.14	1.57	1.61	8.54
	8	1.47	1.74	1.50	1.79	1.50	8.00
	9	1.39	1.61	1.66	1.61	1.79	8.06
	10	1.61	1.67	1.61	1.76	1.91	8.56
	11	1.47	1.39	1.39	1.57	1.72	7.54
	12	1.50	1.74	1.66	2.05	1.95	8.90
	Total	9.05	9.76	9.96	10.35	10.48	49.60
	Mean	1.508	1.627	1.660	1.725	1.747	1.653
	Variance	0.0075	0.0168	0.0663	0.0345	0.0300	
3	13	1.50	1.39	1.45	1.64	1.91	7.89
	14	1.47	1.54	1.66	1.76	1.91	8.34
	15	1.39	1.61	1.61	1.64	2.64	8.89
	16	1.61	1.61	1.61	1.57	2.23	8.63
	17	1.39	1.61	1.61	1.39	2.56	8.56
	18	1.61	1.39	1.61	1.82	2.11	8.54
	Total	8.97	9.15	9.55	9.82	13.36	50.85
	Mean	1.495	1.525	1.592	1.637	2.227	1.695
	Variance	0.0098	0.0117	0.0052	0.0228	0.0992	
Total		27.50	28.46	28.69	28.96	34.83	148.44
Mean		1.528	1.581	1.594	1.609	1.935	1.649

(a) Compute the analysis of variance and discuss validity of f tests.
(b) Compare all possible pairs of diets.
(c) Compare cycles 1-4 collectively with the older age (5).

8.16. Serum immune globulin levels were studied in dairy calves (1) nursed by the cow or (2) hand-fed colostrum (factor *A*). Each calf was measured at 1, 2, 3, and 14 days after birth (factor *B*). Data are turbidimetric units (relative to barium sulphate standard = 20).

Trt. (A)	Calf	1	2	3	14	Total
1	1	16.84	20.00	17.11	12.84	66.79
	2	15.02	18.56	19.37	11.70	64.65
	3	11.01	14.35	14.55	8.87	48.78
	4	11.45	12.55	12.64	9.43	46.07
	5	13.71	15.36	16.48	10.51	56.06
	6	8.97	10.11	10.30	7.35	36.73
	7	11.66	14.32	14.36	9.97	50.31
	8	6.36	9.59	9.74	6.54	32.23
	9	7.67	8.35	9.53	4.86	30.41
	Total	102.69	123.19	124.08	82.07	432.03
	Mean	11.41	13.69	13.79	9.12	12.00
	Variance	11.68	15.82	12.29	6.41	
2	10	13.48	14.63	16.41	9.29	53.81
	11	2.35	4.38	2.05	2.53	11.31
	12	7.15	9.28	7.72	4.94	29.09
	13	7.70	4.55	7.33	5.62	25.20
	14	5.63	4.41	5.62	3.11	18.77
	15	7.91	7.34	7.62	6.82	29.69
	16	13.95	13.73	15.27	9.91	52.86
	17	12.22	12.17	12.63	8.58	45.60
	18	9.17	8.08	8.62	7.76	33.63
	Total	79.56	78.57	83.27	58.56	299.96
	Mean	8.84	8.73	9.25	6.51	8.33
	Variance	14.58	16.17	21.63	6.99	
Total		182.25	201.76	207.35	140.63	731.99
Mean		10.12	11.21	11.52	7.81	10.17

(a) Compute the analysis of variance and discuss validity of f tests.
(b) Compare the first 3 days collectively with results for 14 days.
(c) Test orthogonal polynomial contrasts for the first 3 days.

8.17. A split-plot experiment with chickens was designed to discover if thymectomy (A) must be followed by irradiation (B) to ensure that no immunologically competent cells exist. Nine chicks were assigned to each of the 4 combinations (A_1B_1 = no surgery or X ray, A_1B_2 = no surgery with X ray, A_2B_1 = surgery without X ray, A_2B_2 = surgery with X ray). Surgery was performed within 1 day after hatching; irradiation was given 3 days after hatching. Lymphocytes were counted at 1/2, 1, 2, 3, and 5 months of age (C) for each bird. Data were recorded as y =

lymphocytes × 10^{-3}. Totals for various combinations are given in the following table.

Treatment (*AB*)		1/2	1	2	3	5	Total	Mean
A_1B_1	Total	467	387	368	676	672	2570	
	Mean	51.9	43.0	40.9	75.1	74.7		57.1
	Variance	653	76	161	64	602		
A_1B_2	Total	261	456	385	612	843	2557	
	Mean	29.0	50.7	42.8	68.0	93.7		56.8
	Variance	59	372	128	206	374		
A_2B_1	Total	354	490	294	363	596	2097	
	Mean	39.3	54.4	32.7	40.3	66.2		46.6
	Variance	77	82	158	253	210		
A_2B_2	Total	329	235	296	558	628	2046	
	Mean	36.6	26.1	32.9	62.0	69.8		45.5
	Variance	197	72	81	340	166		
Total		1411	1568	1343	2209	2739	9270	
Mean		39.2	43.6	37.3	61.4	76.1		51.5

Birds: $ss_{D/AB} = 8718.65$; total: $ss_y = 92{,}483.86$

(a) Compute the analysis of variance and discuss validity of *f* tests.
(b) Compare responses at 3 and 5 months collectively with results at earlier ages.

8.18. A split-plot experiment was designed to study uptake of ^{131}I by the thyroid of coturnix quail in (1) hypothyroid, (2) euthyroid, or (3) hyperthyroid state (*A*) at 20°C or 35°C (*B*). Six birds were used for each of the 6 combinations. Responses (cpm × 10^{-4}) were recorded for each bird at 24, 48, and 96 hours (factor *C*).

A	Bird	20°C 24	20°C 48	20°C 96	Total	Bird	35°C 24	35°C 48	35°C 96	Total	
1	1	802	520	394	1716	19	465	411	255	1131	
	2	602	363	177	1142	20	711	740	510	1961	
	3	552	408	592	1552	21	568	472	351	1391	
	4	930	619	425	1974	22	545	466	228	1239	
	5	779	408	297	1484	23	615	486	258	1359	
	6	1060	322	209	1591	24	418	431	332	1181	
	Total	4725	2640	2094	9459		3322	3006	1934	8262	17721

		20°C					35°C				
A	Bird	24	48	96	Total	Bird	24	48	96	Total	
	Mean	788	440	349	525		554	501	322	459	492
	Variance	36943	12068	23748			11013	14482	10741		
2	7	661	467	495	1623	25	586	829	312	1727	
	8	377	368	287	1032	26	967	766	670	2403	
	9	518	397	313	1228	27	742	631	560	1933	
	10	412	192	155	759	28	263	182	110	555	
	11	454	380	344	1178	29	354	305	246	905	
	12	310	386	289	985	30	386	315	264	965	
	Total	2732	2190	1883	6805		3298	3028	2162	8488	15293
	Mean	455	365	314	378		550	505	360	472	425
	Variance	15073	8406	12059			71946	73880	44623		
3	13	707	654	671	2032	31	405	435	332	1172	
	14	576	449	334	1359	32	597	639	567	1803	
	15	927	976	941	2844	33	493	491	360	1344	
	16	722	614	560	1896	34	795	726	603	2124	
	17	534	629	475	1638	35	611	571	433	1615	
	18	613	611	611	1835	36	538	499	407	1444	
	Total	4079	3933	3592	11604		3439	3361	2702	9502	21106
	Mean	680	656	599	645		573	560	450	528	586
	Variance	20025	29954	41883			17431	11605	12252		
Total		11536	8763	7569	27868		10059	9395	6798	26252	54120
Mean		641	487	420	516		559	522	378	486	501

(a) Compute the analysis of variance and discuss validity of f tests.
(b) Compare euthyroid results separately with hypothyroid and hyperthyroid results.
(c) Is response at 96 hours less than at earlier measurements?

8.19. A double-split-plot experiment involved 8 rats. Four were control rats (A_1) and 4 were given an oral dose of a contraceptive (A_2). The rats were measured at two ages (B). At each age, serum glucose was measured at 0, 20, and 40 minutes after feeding (C).

		B_1			Sub-	B_2			Sub-	
A	Rat	0	20	40	total	0	20	40	total	Total
1	1	38.6	44.8	29.3	112.7	38.2	73.8	18.6	130.6	243.3
	2	14.8	50.1	14.5	79.4	21.2	74.1	9.7	105.0	184.4

A	Rat	B_1 0	20	40	Sub-total	B_2 0	20	40	Sub-total	Total
	3	22.0	43.8	18.9	84.7	25.3	68.7	7.9	101.9	186.6
	4	31.3	54.7	28.1	114.1	34.2	80.6	21.3	136.1	250.2
	Total	106.7	193.4	90.8	390.9	118.9	297.2	57.5	473.6	864.5
	Mean	26.68	48.35	22.70	32.58	29.72	74.30	14.38	39.47	36.02
	Variance	108.8	25.6	51.5		61.4	23.8	43.2		
2	5	32.7	22.1	22.6	77.4	24.6	53.3	13.6	91.5	168.9
	6	33.1	32.3	18.4	83.8	27.1	13.6	15.4	56.1	139.9
	7	19.9	14.4	7.9	42.2	12.5	20.8	6.7	40.0	82.2
	8	24.6	20.1	13.8	58.5	18.3	25.9	9.3	53.5	112.0
	Total	110.3	88.9	62.7	261.9	82.5	113.6	45.0	241.1	503.0
	Mean	27.58	22.22	15.68	21.82	20.62	28.40	11.25	20.09	20.96
	Variance	41.5	55.8	39.8		43.0	301.0	15.8		
Total		217.0	282.3	153.5	652.8	201.4	410.8	102.5	714.7	1367.5
Mean		27.12	35.29	19.19	27.2	25.18	51.35	12.81	29.78	28.49

(a) Write a proper model for the analysis of variance and complete the analysis.

(b) Compare results at 0 minutes separately with results at 20 and 40 minutes after feeding.

8.20. A changeover experiment was designed in 3 Latin squares to study the effects of intramuscular injection of A = 200, B = 400, or C = 800 μg gonadotropin-releasing hormone on serum luteinizing hormone (ng/mL) in bull calves. Three calves each of 2, 4, and 6 months of age were used, calves of the same age forming the 3 rows of one square. The days (periods) of injection were the columns of the squares (not nested), as each calf received all 3 doses in sequence. Measurements were taken on each day at 1/2, 1, 2, and 4 hours after injection.

Age	Calf		Period 1 1/2	1	2	4		Period 2 1/2	1	2	4		Period 3 1/2	1	2	4
2	1	C:	32	25	23	24	B:	20	15	14	2	A:	16	12	8	0
	2	B:	30	19	20	3	A:	16	10	4	0	C:	23	22	23	10
	3	A:	31	18	15	5	C:	34	30	30	19	B:	18	12	14	0
4	4	B:	32	19	20	6	A:	21	15	12	2	C:	30	29	32	6
	5	A:	25	12	7	0	C:	30	28	28	16	B:	18	13	13	2
	6	C:	34	28	27	15	B:	15	9	9	0	A:	16	7	6	0
6	7	A:	31	19	14	5	C:	30	28	29	16	B:	23	20	20	6
	8	C:	33	38	27	14	B:	22	15	16	2	A:	17	11	9	0
	9	B:	22	11	13	0	A:	16	10	8	0	C:	21	19	20	8

(a) Write a model and compute the analysis of variance.
(b) Plot the interaction of dose and hours after injection and perform specific comparisons.

8.21. A study was made of 6 treatments involving animal wastes and nonprotein nitrogen in calf diets. Twelve blocks of 6 calves each were formed as calves reached 30 days of age. Body weights were recorded at 0, 2, 6, and 10 weeks after beginning treatment.
(a) Write a model for the analysis and define terms used.
(b) List sources of variation and degrees of freedom, and describe proper procedures for testing pertinent hypotheses.

8.22. Ten cows were used in a study to compare 4 doses of prostaglandin $F_{2\alpha}$ injected into the carotid artery at 48-hour intervals in random order. Period effects were assumed to be negligible (no nuisance trend). However, jugular blood samples were drawn from each cow at 10, 20, 30, and 60 minutes after injection on each treatment day.
(a) Write a model for the analysis and define terms used.
(b) List sources of variation and degrees of freedom, and describe proper procedures for testing pertinent hypotheses.

8.23. An experiment was designed to test the effect of supplementary dietary zinc (some or none) on sperm count in boars. Eight boars were randomly assigned, 4 to each diet. Within each diet, the 4 boars were paired by age for assignment to changeover sequences in 2 × 2 Latin squares. The treatment sequence consisted of ejaculatory frequencies, 7 or 14 times per week. Measurements were taken at the end of each period of treatment. For one week prior to each week of treatment, frequency was standardized at 3 times per week (to avoid residual effects).
(a) Write a model for the analysis and define terms used.
(b) List sources of variation and degrees of freedom, and describe proper procedures for testing pertinent hypotheses.

8.24. Pheasant behavior in responding to recorded calls is to be measured by the time required for a bird to move a specified distance in response to a call. Four types of calls are to be used and 16 birds are available. Each bird is to be measured in each of 4 trials (periods). Four designs that may be feasible are: (1) simple crossover, (2) Latin square crossover, (3) complete blocks with repeated measure, and (4) split plot with repeated measure. Discuss for each design (a) objectives and (b) conditions, relevant to the proposed experiment, that would lead you to select that design in preference to the others.

8.25. A calf-feeding experiment involved $A = 10\%$ or 15% dry matter and B = ad libitum or 10% of body weight for feeding level. The machinery used restricted the automatic feeding to one dry matter level at one time. Therefore, 24 calves were assigned levels of B during a three-week trial at 10% dry matter; then another 24 calves were assigned at 15% dry matter. The design is a split plot (no repeated measurement). Average daily gains (kg/day) are shown below.

	10% Dry Matter		15% Dry Matter		
Replicate	ad lib	10% b.w.	ad lib	10% b.w.	Total
1	0.344	0.139	0.350	0.293	1.126
2	0.380	0.157	0.237	0.325	1.099
3	0.259	0.263	0.429	0.115	1.066
4	0.364	0.380	0.256	0.333	1.333

Replicate	10% Dry Matter ad lib	10% Dry Matter 10% b.w.	15% Dry Matter ad lib	15% Dry Matter 10% b.w.	Total
5	0.592	0.161	0.188	0.179	1.080
6	0.260	0.194	0.187	0.103	0.744
7	0.368	0.416	0.339	0.454	1.577
8	0.352	0.205	0.297	0.210	1.064
9	0.442	0.209	0.354	0.308	1.313
10	0.356	0.201	0.248	0.434	1.239
11	0.227	0.193	0.270	0.218	0.908
12	0.301	0.203	0.423	0.247	1.174
Total	4.205	2.721	3.578	3.219	13.723
Mean	0.350	0.227	0.298	0.268	0.286
Total	6.926		6.797		
Mean	0.289		0.283		

(a) Compute the analysis of variance and interpret.
(b) Compute standard errors for (1) the mean difference between ad lib and feeding by 10% of body weight at 10% dry matter and (2) the mean difference between 10% and 15% dry matter with ad lib feeding.

8.26. A study was conducted to assess the effects of A = methionine hydroxy analogue (0 or 40 g/day) and B = protein (14% or 22%) in grain mixture on milk production in the first 8 weeks of lactation. The main effect of B was well known a priori; but interaction was of interest, so the experiment was conducted as a split plot (nonrepeat measure), confounding B with pair differences. As each replicate of 4 cows came into production, they were paired by prior production; one pair was assigned 14% and the other 22% protein. Within each pair the cows were randomly assigned 0 or 40 g MHA. Data are production in kg/day.

Replicate	14% Protein Control	14% Protein MHA	Total	22% Protein Control	22% Protein MHA	Total	Rep. Total
1	37.8	38.2	76.0	36.2	39.0	75.2	151.2
2	33.0	34.2	67.2	35.2	36.9	72.1	139.3
3	32.2	34.4	66.6	31.8	34.3	66.1	132.7
4	33.4	33.7	67.1	37.3	39.1	76.4	143.5
5	26.1	28.6	54.7	34.2	35.6	69.8	124.5
6	33.6	37.3	70.9	33.8	34.8	68.6	139.5
7	27.2	26.3	53.5	29.1	30.5	59.6	113.1
8	26.4	30.0	56.4	32.2	35.9	68.1	124.5
9	35.3	34.8	70.1	26.0	27.0	53.0	123.1

	14% Protein			22% Protein			
Replicate	Control	MHA	Total	Control	MHA	Total	Rep. Total
10	34.0	36.4	70.4	34.6	37.1	71.7	142.1
Total	319.0	333.9	652.9	330.4	350.2	680.6	1333.5
Mean	31.90	33.39	32.64	33.04	35.02	34.03	33.34

Control total: 649.4; Mean: 32.47

MHA total: 684.1; Mean: 34.20

(a) Compute the analysis of variance and interpret.
(b) Compute standard errors for (1) the mean difference between control and MHA at the same protein level and (2) the mean difference between MHA at 14% and MHA at 22%.

8.27. Exercise 8.12 contains data related to the effect of estrous cycle (repeated measure in 6 periods) on response to TRH injection in 6 dairy heifers. For this exercise, ignore the data in periods 2, 4, and 6. The remaining means are 45.12, 31.95, and 53.87. The relevant sample variance-covariance matrix is

$$S = \begin{bmatrix} 1517 & 541 & -94 \\ 541 & 262 & -107 \\ -94 & -107 & 341 \end{bmatrix}$$

(a) Test for homogeneity of the variance-covariance matrix.
(b) Test the hypothesis of equal period means, and compute 95% simultaneous confidence intervals for the mean differences of periods 3 and 5 from period 1.

8.28. Exercise 8.15 contains data from a split-plot, repeat-measure experiment with 3 treatments and 5 periods.
(a) Test for equality of variance-covariance matrices across treatments; test homogeneity of pooled variance-covariance.
(b) Use the multivariate procedures of Sec. 4.6.2 to test treatments, periods, and interaction.

8.29. Exercise 8.16 contains data from a split-plot, repeat-measure experiment with 2 treatments and 4 periods.
(a) Test for equality of the two variance-covariance matrices.
(b) Use the multivariate procedures of Sec. 4.6.2 to test treatments, periods, and interaction.

CHAPTER 9
Response Surface Designs

Experiments in which the primary objective is to find the combination of quantitative levels of 2 or more factors that will produce an optimum response (usually maximum or minimum) may be classified as response surface problems. Developments in response surface methodology since the initiation of the basic ideas by Box and Wilson (1951) have been aimed at determination of optimum operating conditions or specification of a region of treatment combinations in which certain practical requirements are met. The procedures may (or may not) aid the experimenter in understanding underlying natural phenomena.

Response surface designs may involve augmentation of ordinary factorial designs (Secs. 2.4, 5.8, 6.3, 6.4) or may be represented by relatively simple geometrical patterns of *design points* (combinations of levels of the treatment factors involved). A proper appreciation of the available designs requires some understanding of the analytical procedures used to characterize the nature of the response surface. Therefore, the analysis is presented first (Sec. 9.1) and the designs later (Secs. 9.2-9.4). Additional material on the subject is found in a review by Hill and Hunter (1966) and a text by Myers (1971).

9.1. ESTIMATION OF OPTIMUM COMBINATIONS

The satisfactory determination of treatment levels permitting maximum (or minimum) response often requires a sequential process involving several small experiments. One or more of the following steps may be needed:

1. Preliminary factor screening, using 2-level fractional experiments with many factors (Sec. 6.3.4)
2. Small 2-level factorial experiments for nearly linear response over various *local* regions (Sec. 2.4)
3. Additional experiments along the path of steepest ascent or descent (Myers, Chap. 5)
4. Second order (3-level) factorial experiments to locate and interpret the optimum conditions
5. Additional experiments to increase precision of estimates and confirm details

The first three steps utilize first order (linear) designs, which may or may not fit well. If the intervals between chosen treatment levels are too wide, the linear approximations almost inevitably will be poor; if the intervals are too narrow, real differences may be obscured by experimental error. In a series of experiments along a path of steepest change in response, a relatively large shift in factor levels may be made when estimates of parameters are large or a small shift may be used if estimates are small. The direction of steepest change is perpendicular to lines of constant predicted response (see 4.14). However, note that the direction is not invariant under linear transformation of scale (different units or coding of the levels of a factor). When one reaches the stage of second order (curvilinear) surfaces, the optimum combination of factor levels (*stationary point*) may be estimated and *canonical analysis* may be used to reduce the number of parameters and aid in interpretation of the nature of the shape of the response surface.

9.1.1. Parameter Estimation

In this section second order estimation is explained by using a 2-factor example. An experimenter wished to develop a certain sterilized, concentrated milk product that would exhibit minimum viscosity after 6 months storage at room temperature. Early experiments showed that 2 factors had substantial influence on viscosity when employed over certain ranges of values. The best temperature at which the milk solids and an additive should be warmed for 30 minutes was established to be in the range 80°-90°C. An adequate level of the additive, sodium polyphosphate, was determined to be in the range 0.04-0.12%. The initial second order experiment was performed at 80°, 85°, and 90° (coded $x_1 = -1, 0, +1$) and levels of additive 0.04, 0.08, and 0.12 (coded $x_2 = -1, 0, +1$). Response data (Y = centipoises on a Brookfield viscometer) for the 9 combinations are shown below

Y:	645	255	285	65	45	70	160	15	315
x_1:	-1	0	+1	-1	0	+1	-1	0	+1
x_2:	-1	-1	-1	0	0	0	+1	+1	+1

A model for the response is

$$Y = \beta_0 x_0 + \beta_1 x_1 + \beta_2 x_2 + \beta_{11} x_1^2 + \beta_{22} x_2^2 + \beta_{12} x_1 x_2 + E \tag{9.1}$$

where $x_0 = 1$ for each experimental unit. The model may be written in matrix notation (as for general multiple regression, Sec. 4.2) as

$$\mathbf{y} = \mathbf{X}\boldsymbol{\beta} + \mathbf{e} \tag{9.2}$$

where $\mathbf{y}$ is a 9×1 column vector of responses (viscosity); $\mathbf{X}$ is a 9×6 matrix including x_0, the coded levels, their squares, and their product; $\boldsymbol{\beta}$ is a 6×1 column vector of parameters (origin, linear$_1$, linear$_2$, quadratic$_1$, quadratic$_2$, and interaction); and $\mathbf{e}$ is a 9×1 column vector of experimental errors (which may include substantial nonrandom bias caused by lack of fit of the model). It is convenient, for further developments, to express each independent variable as a deviation from its mean, where $\overline{x}_1 = 0$, $\overline{x}_2 = 0$, $\overline{(x_1^2)} = 2/3$, $\overline{(x_2^2)} = 2/3$, and $\overline{(x_1 x_2)} = 0$. Then, model (9.1) may be written

$$Y = \beta_0' x_0 + \beta_1 x_1 + \beta_2 x_2 + \beta_{11}(x_1^2 - 2/3) + \beta_{22}(x_2^2 - 2/3) + \beta_{12} x_1 x_2 + E \tag{9.3}$$

where $\beta_0' = \beta_0 + (2/3)\beta_{11} + (2/3)\beta_{22}$, the mean or expected response given the array of *x*s. This model also may be written $\mathbf{y} = \mathbf{X}\beta + \mathbf{e}$. The second and third columns of $\mathbf{X}$ comprise the *design matrix*. The normal equations are $(\mathbf{X}'\mathbf{X})\hat{\beta} = (\mathbf{X}'\mathbf{y})$ as generally used for multiple regression or other linear models (see Chap. 4). The data, model, and normal equations are fully displayed in Table 9.1. Because the columns of $\mathbf{X}$ are mutually orthogonal (product of any two

Table 9.1. Data, Model, and Normal Equations for Estimation of Parameters for Second Order Response Surface

$$
\underset{\mathbf{y}}{\begin{bmatrix} 645 \\ 255 \\ 285 \\ 65 \\ 45 \\ 70 \\ 160 \\ 15 \\ 315 \end{bmatrix}}
=
\underset{\mathbf{X}}{\begin{bmatrix}
1 & -1 & -1 & 1/3 & 1/3 & 1 \\
1 & 0 & -1 & -2/3 & 1/3 & 0 \\
1 & 1 & -1 & 1/3 & 1/3 & -1 \\
1 & -1 & 0 & 1/3 & -2/3 & 0 \\
1 & 0 & 0 & -2/3 & -2/3 & 0 \\
1 & 1 & 0 & 1/3 & -2/3 & 0 \\
1 & -1 & 1 & 1/3 & 1/3 & -1 \\
1 & 0 & 1 & -2/3 & 1/3 & 0 \\
1 & 1 & 1 & 1/3 & 1/3 & 1
\end{bmatrix}}
\underset{\beta}{\begin{bmatrix} \beta_0' \\ \beta_1 \\ \beta_2 \\ \beta_{11} \\ \beta_{22} \\ \beta_{12} \end{bmatrix}}
+
\underset{e}{\begin{bmatrix} E_1 \\ E_2 \\ E_3 \\ E_4 \\ E_5 \\ E_6 \\ E_7 \\ E_8 \\ E_9 \end{bmatrix}}
$$

$$
\underset{(\mathbf{X}'\mathbf{X})}{\begin{bmatrix}
9 & 0 & 0 & 0 & 0 & 0 \\
0 & 6 & 0 & 0 & 0 & 0 \\
0 & 0 & 6 & 0 & 0 & 0 \\
0 & 0 & 0 & 2 & 0 & 0 \\
0 & 0 & 0 & 0 & 2 & 0 \\
0 & 0 & 0 & 0 & 0 & 4
\end{bmatrix}}
\underset{\hat{\beta}}{\begin{bmatrix} \hat{\beta}_0' \\ \hat{\beta}_1 \\ \hat{\beta}_2 \\ \hat{\beta}_{11} \\ \hat{\beta}_{22} \\ \hat{\beta}_{12} \end{bmatrix}}
=
\underset{(\mathbf{X}'\mathbf{y})}{\begin{bmatrix} 1855 \\ -200 \\ -695 \\ 303.3 \\ 438.3 \\ 515 \end{bmatrix}}
$$

columns is zero), the six terms of the model are linearly independent in contrast to the usual variables of multiple regression. As a consequence of orthogonality, $(\mathbf{X}'\mathbf{X})$ is a diagonal matrix and its inverse may be obtained simply by replacing each diagonal element by its reciprocal. Thus the usual solutions to the normal equations, $\hat{\beta} = (\mathbf{X}'\mathbf{X})^{-1}(\mathbf{X}'\mathbf{y})$, are quite easy to compute, each parameter being estimated (as in simple linear regression, Sec. 1.8) by the ratio of a sum of products (element of $\mathbf{X}'\mathbf{y}$) to a sum of squares (diagonal element of $\mathbf{X}'\mathbf{X}$). For the example, one obtains $\hat{\beta}_0' = 1855/9 = 206.1$, $\hat{\beta}_1 = -200/$

$6 = -33.3$, $\hat{\beta}_2 = -695/6 = -115.8$, $\hat{\beta}_{11} = 303.3/2 = 151.7$, $\hat{\beta}_{22} = 438.3/2 = 219.2$, and $\hat{\beta}_{12} = 515/4 = 128.8$. Also, note that the origin is $\hat{\beta}_0 = 206.1 - 0.67 \cdot (151.7) - 0.67(219.2) = -41.2$ [see definition following (9.3)]. Predictions of response in viscosity may be based either on the mean,

$$\hat{Y} = 206.1 - 33.3x_1 - 115.8x_2 + 151.7(x_1^2 - 2/3) + 219.2(x_2^2 - 2/3) + 128.8x_1x_2$$

or the origin,

$$\hat{Y} = -41.2 - 33.3x_1 - 115.8x_2 + 151.7x_1^2 + 219.2x_2^2 + 128.8x_1x_2$$

The sum of squares in Y associated with each x is simply $\hat{\beta}_i(sp_{x_i y})$ because of orthogonality ($sp_{x_i y}$ is an element of the vector of sums of products of xs with Y, $\mathbf{X'y}$). The residual or error sum of squares in Y is

$$\mathbf{y'y} - \hat{\beta}'(\mathbf{X'y}) = \Sigma y^2 - \Sigma\hat{\beta}_i(sp_{x_i y}) \qquad (9.4)$$

and the associated number of degrees of freedom is $n - p$ when p parameters are estimated from data on n experimental units. For the example, the error sum of squares is

$$ss_E = 698,475 - [(206.1)(1855) + (-33.3)(-200) + \dots + (128.8)(515)] = 20,601$$

and mean square error is $ms_E = 20,601/(9 - 6) = 6867$. If bias is trivial, ms_E is a reasonable estimate of variance caused by random error (σ^2). The variance-covariance matrix for sample estimates of parameters is $\hat{\sigma}^2(\mathbf{X'X})^{-1}$, as usual (see Sec. 4.2). However, because of orthogonality, all covariances are zero and the variance of each estimate is merely $\hat{\sigma}^2$ times the corresponding diagonal element of $(\mathbf{X'X})^{-1}$. Square roots of the variances provide standard errors of the estimated parameters: $\sqrt{6867/9} = \sqrt{763} = 27.6$ for $\hat{\beta}'_0$, $\sqrt{6867/6} = \sqrt{1144.5} = 33.8$ for $\hat{\beta}_1$ and $\hat{\beta}_2$, $\sqrt{6867/2} = \sqrt{3433.5} = 58.6$ for $\hat{\beta}_{11}$ and $\hat{\beta}_{22}$, and $\sqrt{6867/4} = \sqrt{1716.75} = 41.4$ for $\hat{\beta}_{12}$. Because

$$\hat{\beta}_0 = \hat{\beta}'_0 - (2/3)(\hat{\beta}_{11}) - (2/3)(\hat{\beta}_{22})$$

the standard error of $\hat{\beta}_0$ is

$$\sqrt{\hat{V}[\hat{\beta}'_0] + (2/3)^2\hat{V}[\hat{\beta}_{11}] + (2/3)^2\hat{V}[\hat{\beta}_{22}]} = 61.7$$

Note the unequal precision; linear effects are estimated more precisely than interaction and interaction more precisely than quadratic effects.

The statistical significance of each parameter may be evaluated by a t test:

$$t = \hat{\beta}_i / s_{\hat{\beta}_i} \tag{9.5}$$

where $s_{\hat{\beta}_i}$ is the standard error versus critical values $\pm t_{\alpha/2,n-p}$. Despite low power for the example ($n - p$ is only 3), four of the five treatment parameters are significant. Lack of significance of β_1 (linear parameter for temperature) may be merely a reflection of the fact that the optimum temperature is near the center of the design used for this experiment (85°). Significance of the quadratic parameter for temperature (β_{11}) suggests that a reasonably local optimum temperature exists. Significance of linear as well as quadratic parameter for sodium polyphosphate suggests that the optimum level may not be near the center of the design (0.08%) but it is not likely to be far away. Interaction can be illustrated by plotting response to change in x_1 (temperature) for each level of x_2 (sodium polyphosphate). The locally optimum temperature is more narrowly defined for $x_2 = +1$ (0.12%) than for $x_2 = 0$ (0.08%), a result that should be expected if the optimum temperature is near the center of the design.

The general reliability of predictions depends on the precision with which the parameters are estimated and also on the design points (levels of the factors) selected for predicting. For example, the variance of predicted viscosity for 90° ($x_1 = +1$) and 0.12% sodium polyphosphate ($x_2 = +1$) is

$$\begin{aligned}\hat{V}[\hat{Y}] &= \hat{V}[\hat{\beta}_0'] + (+1)^2\hat{V}[\hat{\beta}_1] + (+1)^2\hat{V}[\hat{\beta}_2] + [(+1)^2 - 2/3]^2\hat{V}[\hat{\beta}_{11}] \\ &\quad + [(+1)^2 - 2/3]^2\hat{V}[\hat{\beta}_{22}] + [(+1)(+1)]^2\hat{V}[\hat{\beta}_{12}] \\ &= \hat{V}[\hat{\beta}_0'] + 2\hat{V}[\hat{\beta}_1] + (2/9)\hat{V}[\hat{\beta}_{11}] + \hat{V}[\hat{\beta}_{12}] \\ &= 763 + (2)(1144.5) + (0.222)(3433.5) + 176.75 = 5531\end{aligned}$$

The predicted value and its standard error are $309 \pm \sqrt{5531} = 309 \pm 74.4$. Similarly, the prediction for ($x_1 = \sqrt{2}$, $x_2 = 0$), i.e., for 92° and 0.08%, is 215 ± 95.7. The center of the design is $x_1 = 0$, $x_2 = 0$; therefore the coordinates (+1, +1) and ($\sqrt{2}$, 0) are equidistant from the center of the design but do not provide equal reliability for prediction. That is, the orthogonal 3^2 design is direction dependent with respect to precision of prediction. Box and Hunter (1957) introduced "rotatable" designs to eliminate the influence of direction on reliability (Sec. 9.3).

9.1.2. Interpretation of Fitted Response Surface

A *stationary point* is the factor coordinate at which change in response is zero simultaneously for infinitesimally small changes in levels of factors involved. Practically speaking, for 2 factors, the stationary point may be a double maximum ("peak" of response), a double minimum (a "sink-hole"), or a combination of maximum for one factor and minimum for the other (a "saddle" or "mountain pass"). If the sample responses miss one maximum or minimum because of poor choice of levels for a factor, the resulting response surface may be a "rising ridge" or a "sinking valley." One should not be so unlucky as to miss both maxima or minima with the design region selected if sufficient early experimentation has been carried out carefully. Mathematical determination of the location of an estimated stationary point requires elementary calculus (i.e., partial differentiation of the prediction equation with respect to each factor involved), setting the results equal to zero, and solving the set of equations simultaneously. For the example at hand,

$$\partial\hat{Y}/\partial x_1 = -33.3 + 2(151.7)x_{1s} + 128.8x_{2s} = 0$$

$$\partial\hat{Y}/\partial x_2 = -115.8 + 128.8x_{1s} + 2(219.2)x_{2s} = 0$$

Solutions to the equations are $x_{1s} = -0.00274$ and $x_{2s} = +0.265$. Translated, by equating 1 x unit to 5 degree units and 0.04 percentage units, these results indicate an optimum combination of 84.99° and 0.0906% sodium polyphosphate. The optimum appears to be close to the center of the design (85°, 0.08%). The predicted response at the stationary point is

$$\hat{Y}_s = -41.2 - 33.3(-0.00274) - 115.8(+0.265) + 151.7(-0.00274)^2$$

$$+ 219.2(+0.265)^2 + 128.8(-0.00274)(+0.265) = -56.5$$

with standard error ±59.9. Box and Hunter (1954) described a procedure for computing a confidence region for the location of a stationary point.

The first equation used to estimate the stationary point leads to the result $x_{1s} = (33.3 - 128.8x_2)/[2(151.7)]$. For design points $x_2 = -1$, 0, and +1, one obtains $x_{1s} = +0.534$, +0.110, and -0.315 (87.7°, 85.6°, and 83.4°), the local stationary points of x_1 response profiles for different design values of x_2. Similarly, for design points $x_1 = -1$, 0, and +1, the second equation leads to $x_{2s} = (115.8 - 128.8x_1)/[2(219.2)] = +0.558$, 0.264, and -0.030 (0.102%, 0.091%, and 0.079%), the local stationary points of x_2 response profiles for different design values of x_1. When plotted on the (x_1, x_2) design grid, a line connecting the local x_1 stationary points intersects a line connecting the local x_2 stationary points at the joint stationary point (-0.00274, +0.265).

The second partial derivative of the prediction equation with respect to either x describes curvature, i.e., $\partial^2\hat{Y}/\partial x_1^2 = 2\hat{\beta}_{11}$ and $\partial^2\hat{Y}/\partial x_2^2 = 2\hat{\beta}_{22}$. The mixed second partial derivative, $\partial^2\hat{Y}/(\partial x_1)(\partial x_2) = \hat{\beta}_{12}$, describes the slope of the axes of any elliptical contour of estimated equal response, i.e., the interaction of the 2 factors.

Canonical analysis is a term applied to the transformation of a second order model in variables $\{x_1, x_2, \ldots, x_k\}$ to a canonical form involving only the squares of new variables $\{w_1, w_2, \ldots, w_k\}$, which are linear combinations of the xs. First order terms and interaction term(s) are "eliminated" so that interpretation of the response surface may be simplified, regardless of the number of factors involved. The procedure translates the origin from $\{x_1, x_2, \ldots, x_k\} = \{0, 0, \ldots, 0\}$ to the stationary point $\{x_{1s}, x_{2s}, \ldots, x_{ks}\}$ and rotates the axes from $\{x_i\}$ to the principal axes $\{w_i\}$ of the response pattern. For 2 factors, the stationary point is the center of elliptical (or hyperbolic) contours of equal response. The canonical equation for 2 factors is

$$\hat{Y} = \hat{Y}_s + \lambda_1 w_1^2 + \lambda_2 w_2^2 \tag{9.6}$$

where $\hat{Y}_s$ is the estimated stationary point, w_1 and w_2 are linear combinations of x_1 and x_2, and λ_1 and λ_2 are the *characteristic* (*latent*) *roots* or *eigenvalues* [see (4.5)] of a matrix involving estimates $\hat{\beta}_{11}$, $\hat{\beta}_{22}$, and $\hat{\beta}_{12}$ of the second order prediction equation. If both characteristic roots are negative, $\hat{Y}_s$ is a double maximum and contours of equal response are ellipses, elongated in the direction of the variable associated with the root of smaller absolute value. If both roots are positive, $\hat{Y}_s$ is a double minimum. If one root is positive and the other is negative, $\hat{Y}_s$ is a saddle and the response contours are hyperbolas. If one root is near zero, the change in response is small for changes in level of the corresponding factor and the surface is a slowly rising ridge (or slowly sinking valley). See Myers (1971, Chap. 5) for further ridge analysis. Of course, generalizations of these interpretations are required when more than 2 factors are involved.

The general transformation to canonical form involves separation of the linear terms from the second order and interaction terms of the prediction equation. It is convenient to represent an interaction term, such as $\hat{\beta}_{12}x_1x_2$, by $(1/2)\hat{\beta}_{12}x_1x_2 + (1/2)\hat{\beta}_{21}x_2x_1$. For k factors, the prediction equation may be written

$$\hat{Y} = \hat{\beta}_0 + \mathbf{x}'\hat{\beta} + \mathbf{x}'\hat{\mathbf{B}}\mathbf{x} \tag{9.7}$$

where $\mathbf{x}' = [x_1 x_2 \ldots x_k]$, $\hat{\beta}' = [\hat{\beta}_1 \hat{\beta}_2 \ldots \hat{\beta}_k]$, and

$$\hat{\mathbf{B}} = \begin{bmatrix} \hat{\beta}_{11} & (1/2)\hat{\beta}_{12} & \cdots & (1/2)\hat{\beta}_{1k} \\ (1/2)\hat{\beta}_{21} & \hat{\beta}_{22} & \cdots & (1/2)\hat{\beta}_{2k} \\ \cdot & \cdot & \cdots & \cdot \\ (1/2)\hat{\beta}_{k1} & (1/2)\hat{\beta}_{k2} & \cdots & \hat{\beta}_{kk} \end{bmatrix}$$

The set of partial derivatives of $\hat{Y}$ with respect to $\mathbf{x}$ may be equated to a zero vector and solved to obtain coordinates of the stationary point, i.e.,

$$\{\partial\hat{Y}/\partial\mathbf{x}\} = \hat{\beta} + 2\hat{\mathbf{B}}\mathbf{x}_s = 0 \tag{9.8}$$

$$\mathbf{x}_s = -(1/2)\hat{\mathbf{B}}^{-1}\hat{\boldsymbol{\beta}} \tag{9.9}$$

Letting $\mathbf{z} = \mathbf{x} - \mathbf{x}_s$, one obtains $\mathbf{x} = \mathbf{z} + \mathbf{x}_s$. Therefore,

$$\begin{aligned}\hat{Y} &= \hat{\beta}_0 + (\mathbf{z}' + \mathbf{x}_s')\hat{\boldsymbol{\beta}} + (\mathbf{z}' + \mathbf{x}_s')\hat{\mathbf{B}}(\mathbf{z} + \mathbf{x}_s) \\ &= [\hat{\beta}_0 + \mathbf{x}_s\hat{\boldsymbol{\beta}} + \mathbf{x}_s'\hat{\mathbf{B}}\mathbf{x}_s] + [\mathbf{z}'\hat{\boldsymbol{\beta}} + \mathbf{z}'\hat{\mathbf{B}}\mathbf{x}_s + \mathbf{x}_s'\hat{\mathbf{B}}\mathbf{z}] + [\mathbf{z}'\hat{\mathbf{B}}\mathbf{z}] \\ &= [\hat{\beta}_0 + \mathbf{x}_s'\hat{\boldsymbol{\beta}} + \mathbf{x}_s'(-\hat{\boldsymbol{\beta}}/2)] + [\mathbf{z}'(\hat{\boldsymbol{\beta}} + 2\hat{\mathbf{B}}\mathbf{x}_s)] + [\mathbf{z}'\hat{\mathbf{B}}\mathbf{z}]\end{aligned} \tag{9.10}$$

Myers (p. 73) has shown that the first term of (9.10) is equal to the predicted response at the stationary point $\hat{Y}_s$. Also, the second term vanishes because

$$\hat{\boldsymbol{\beta}} + 2\hat{\mathbf{B}}\mathbf{x}_s = \hat{\boldsymbol{\beta}} + 2\hat{\mathbf{B}}(-1/2\hat{\mathbf{B}}^{-1}\hat{\boldsymbol{\beta}}) = \hat{\boldsymbol{\beta}} - \mathbf{I}\hat{\boldsymbol{\beta}} = 0$$

Therefore,

$$\hat{Y} = \hat{Y}_s + \mathbf{z}'\hat{\mathbf{B}}\mathbf{z} \tag{9.11}$$

and the response is predictable in terms of response expected at the stationary point and the second order regression coefficients (including interaction). In moving from (9.7) to (9.11) the origin has been changed from the center of the design $\{x_1 = 0, x_2 = 0, \ldots, x_k = 0\}$ to the stationary point or "center of response" $\{x_{1s}, x_{2s}, \ldots, x_{ks}\}$. Next, the axes are to be changed from the design orientation $\{x_1, x_2, \ldots, x_k\}$ to the response orientation $\{w_1, w_2, \ldots, w_k\}$. Let $\mathbf{z} = \mathbf{M}\mathbf{w}$, where matrix $\mathbf{M}$ is to be determined. The second term of (9.11) becomes

$$\mathbf{z}'\hat{\mathbf{B}}\mathbf{z} = \mathbf{w}'(\mathbf{M}'\hat{\mathbf{B}}\mathbf{M})\mathbf{w} \tag{9.12}$$

Searle (1966, Chap. 7) has shown that for a symmetrical matrix such as $\hat{\mathbf{B}}$, there exists an orthogonal matrix $\mathbf{M}$ (one for which $\mathbf{M}' = \mathbf{M}^{-1}$, so $\mathbf{M}\mathbf{M}' = \mathbf{I}$) such that $\mathbf{M}'\hat{\mathbf{B}}\mathbf{M} = \Lambda$, where Λ is a diagonal matrix whose diagonal elements λ_i are the characteristic roots of $\hat{\mathbf{B}}$. Thus,

$$\hat{Y} = [\hat{Y}_s + \mathbf{z}'\hat{\mathbf{B}}\mathbf{z}] = [\hat{Y}_s + \mathbf{w}'\Lambda\mathbf{w}] = \hat{Y}_s + \lambda_1 w_1^2 + \lambda_2 w_2^2 + \ldots + \lambda_k w_k^2 \tag{9.13}$$

the desired canonical form. Let the orthogonal matrix $\mathbf{M}$ be partitioned into k column vectors,

$$\mathbf{M} = [\mathbf{m}_1\mathbf{m}_2\ldots\mathbf{m}_k]$$

where $\mathbf{m}_1$ contains elements m_{11} through m_{1k}, $\mathbf{m}_2$ contains m_{21} through m_{2k}, etc. The matrix $\mathbf{M}$ is orthogonal: the product of any two columns $(\mathbf{m}_i)$ is zero. One can force $\mathbf{m}_i'\mathbf{m}_i = 1$ by "normalizing" the elements. For example, an arbitrary element in $\mathbf{m}_1$, say m_{1j}, becomes $m_{1j}/\sqrt{\sum_{j=1}^{k} m_{1j}^2}$ in normalized form. Because $\mathbf{M}'\mathbf{M} = \mathbf{I}$ (identity matrix, Sec. 4.1) and Λ is diagonal,

$$\mathbf{M}(\mathbf{M}'\hat{\mathbf{B}}\mathbf{M}) = \mathbf{M}\Lambda \quad \text{or} \quad \hat{\mathbf{B}}\mathbf{M} = \Lambda\mathbf{M}$$

Thus $\hat{\mathbf{B}}\mathbf{M} - \Lambda\mathbf{M} = 0$, or

$$\hat{\mathbf{B}}[\mathbf{m}_1\mathbf{m}_2\ldots\mathbf{m}_k] - \Lambda[\mathbf{m}_1\mathbf{m}_2\ldots\mathbf{m}_k] = 0$$

For each $\mathbf{m}_i$ vector, $\hat{\mathbf{B}}\mathbf{m}_i - (\lambda_i\mathbf{I})\mathbf{m}_i = 0$, or

$$(\hat{\mathbf{B}} - \lambda_i\mathbf{I})\mathbf{m}_i = 0 \tag{9.14}$$

The linear homogeneous equations of (9.14) have solutions only if the determinant of $[\hat{\mathbf{B}} - \lambda\mathbf{I}]$ is zero. On setting the determinant equal to zero, the solutions are the characteristic roots λ_i of the matrix $\hat{\mathbf{B}}$ [see (4.5)].

In practice one begins the computations for a canonical analysis at this point. For the example with 2 factors,

$$\begin{vmatrix} (\hat{\beta}_{11} - \lambda) & 1/(2\hat{\beta}_{12}) \\ 1/(2\hat{\beta}_{12}) & (\hat{\beta}_{22} - \lambda) \end{vmatrix} = \begin{vmatrix} (151.7 - \lambda) & 64.4 \\ 64.4 & (219.2 - \lambda) \end{vmatrix} = 0$$

or $(151.7 - \lambda)(219.2 - \lambda) - (64.4^2) = 0$, becomes a quadratic equation (in general, a kth degree polynomial for k factors),

$$\lambda^2 - 370.9\lambda + 29{,}105.3 = 0 \tag{9.15}$$

The roots of (9.15) are

$$\lambda_i = [370.9 \pm \sqrt{(-370.9)^2 - 4(29{,}105.3)}]/2 = 258.1 \text{ and } 112.8$$

Subscripts are assigned to the λ_i according to the ranked magnitude (sign considered) of the quadratic regression coefficients $\hat{\beta}_{ii}$. In the example, $\hat{\beta}_{11} < \hat{\beta}_{22}$, so $\lambda_1 = 112.8$ and $\lambda_2 = 258.1$. Note that both roots are positive, indicating a double minimum of viscosity as desired. Contours of equal response are ellipses, elongated in the direction of temperature (variable 1) because $\lambda_1 < \lambda_2$. Practically speaking, such a result indicates that change in viscosity is less sensitive to a 5° change in temperature (one unit of x_1) than to a change of 0.04% in sodium polyphosphate (one unit of x_2).

The canonical form for the 2-factor example is

$$\hat{Y} = -56.5 + 112.8w_1^2 + 258.1w_2^2$$

but the relation of the old variables (x_i) to the new ones (w_i) is yet to be determined. Examining (9.14) for rotated axes, one may begin with λ_1 and $\mathbf{m}_1$:

$$\begin{bmatrix} (151.7 - 112.8) & 64.4 \\ 64.4 & (219.2 - 112.8) \end{bmatrix} \begin{bmatrix} m_{11} \\ m_{12} \end{bmatrix} = \begin{bmatrix} 0 \\ 0 \end{bmatrix}$$

Because the equations are homogeneous, the easiest procedure is to obtain relative values for m_{11} and m_{12} (to be normalized later) by taking an arbitrary value for one of them, say $m_{11} = 1$. Given $m_{11} = 1$, one obtains $m_{12} = -0.6040$ or -0.6052 from the two equations, with average value $m_{12} = -0.6046$. If the two results for m_{12} do not agree rather closely, it is likely that $\hat{\beta}_{11}$ and $\hat{\beta}_{22}$ were computed incorrectly. Similarly, using λ_2 and m_2, one obtains

$$\begin{bmatrix} (151.7 - 258.1) & 64.4 \\ 64.4 & (219.2 - 258.1) \end{bmatrix} \begin{bmatrix} m_{21} \\ m_{22} \end{bmatrix} = \begin{bmatrix} 0 \\ 0 \end{bmatrix}$$

Arbitrarily choosing $m_{21} = 1$, one obtains average $m_{22} \simeq 1.654$. For vector $\mathbf{m}_1$, $\sqrt{\Sigma m_{1j}^2} = 1.169$, and for vector $\mathbf{m}_2$, $\sqrt{\Sigma m_{2j}^2} = 1.933$. Therefore, the normalized elements are $m_{11} = 1/1.169 = +0.855$, $m_{12} = -0.6046/1.169 = -0.517$, $m_{21} = 1/1.933 = +0.517$, and $m_{22} = 1.654/1.933 = +0.855$, so

$$\mathbf{M} = \begin{bmatrix} +0.855 & +0.517 \\ -0.517 & +0.855 \end{bmatrix} \qquad \mathbf{M}' = \begin{bmatrix} +0.855 & -0.517 \\ +0.517 & +0.855 \end{bmatrix}$$

Because $\mathbf{z} = \mathbf{Mw}$, then $\mathbf{M'z} = \mathbf{M'Mw} = \mathbf{Iw} = \mathbf{w}$. Therefore,

$$\mathbf{w} = \mathbf{M'z} = \mathbf{M'}(\mathbf{x} - \mathbf{x}_s) \tag{9.16}$$

indicates the relation between new and old variables ($\mathbf{w}$ and $\mathbf{x}$). For the 2-factor example, one obtains

$$\begin{bmatrix} w_1 \\ w_2 \end{bmatrix} = \begin{bmatrix} +0.855 & -0.517 \\ +0.517 & +0.855 \end{bmatrix} \begin{bmatrix} (x_1 + 0.00274) \\ (x_2 - 0.265) \end{bmatrix}$$

or $w_1 = 0.855x_1 - 0.517x_2 + 0.139$ and $w_2 = 0.517x_1 + 0.855x_2 - 0.225$. By alternately setting w_2 and w_1 equal to zero, one obtains equations for the w_1 and w_2 axes, respectively,

$$0.517x_1 + 0.855x_2 = 0.225 \qquad 0.855x_1 - 0.517x_2 = -0.139 \tag{9.17}$$

For k factors, a particular w_i axis is determined by setting the remaining k -

1 w equations equal to zero simultaneously. Equations like (9.17) are especially important for problems in which the surface is a slowly rising ridge or sinking valley (one λ near zero). The axis corresponding to a near-zero λ is the equation of the ridge (or valley) and should be followed to locate the next design. Simultaneous solution of (9.17) produces values $x_{1s} = -0.00252$ and $x_{2s} = +0.264$, values within rounding error of coordinates of the stationary point determined earlier. The canonical axes may be plotted as follows: For the w_1 axis with $x_1 = +1$, one obtains $x_2 = (0.225 - 0.517)/0.855 = -0.342$; with $x_1 = -1$, $x_2 = (0.225 + 0.517)/0.855 = +0.868$. For the w_2 axis with $x_2 = +1$, one obtains $x_1 = (-0.139 + 0.517)/0.855 = +0.442$; with $x_2 = -1$, $x_1 = (-0.139 - 0.517)/0.855 = -0.767$. See Fig. 9.1. Contour(s) of expected equal

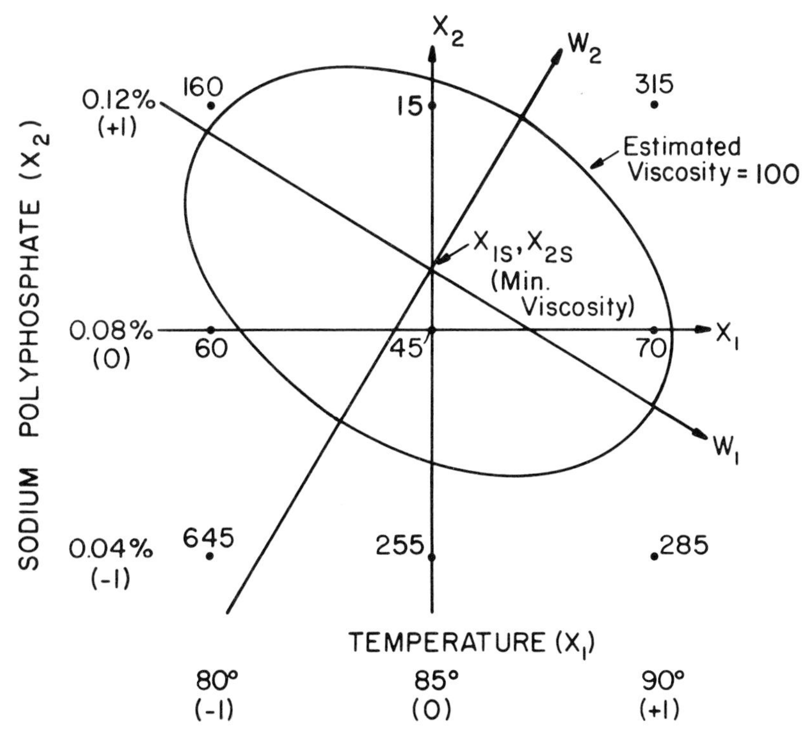

Fig. 9.1. Combinations of temperature and sodium polyphosphate for viscosity of 100 centipoises or less.

response can be constructed by setting the prediction equation ($\hat{Y}$) equal to the selected response value(s) and computing values of x_2 for several selected values of x_1. A response contour for $\hat{Y} = 100$ is shown in Fig. 9.1. For example, let $x_1 = +1$ and one obtains

$$\hat{Y} = 100 = -41.2 - 33.3(+1) - 115.8x_2 + 151.7(+1)^2 + 219.2x_2^2 + 128.8(+1)x_2$$

$$= 219.2x_2^2 + 13x_2 - 22.8 = 0$$

$$x_2 = [-13 \pm \sqrt{13^2 + 4(219.2)(22.8)}]/2(219.2) = +0.29 \text{ and } -0.35$$

For $x_1 = -1$, $x_2 = +0.89$ and $+0.22$; for $x_1 = 0$, $x_2 = +1.11$ and -0.58, etc. Any combination of temperature and sodium polyphosphate inside the ellipse is predicted to produce viscosity lower than 100. The w_1 axis (temperature) is longer, indicating less sensitivity of viscosity to changes of 5° in temperature than to changes of 0.04% in sodium polyphosphate.

Any further experimentation should be devoted to confirming details of the second order model. Estimates of parameters have large standard errors; more precise estimates may be needed if the true stationary point is to be estimated almost exactly. Note that the design point nearest the estimated minimum ($x_1 = 0$, $x_2 = 0$) produced a higher sample value ($y = 45$) than produced by the ($x_1 = 0$, $x_2 = +1$) point ($y = 15$). Also the sample value at point ($x_1 = -1$, $x_2 = 0$) falls outside the contour for $\hat{Y} = 100$ although the response was lower ($y = 65$). Such features accentuate the lack of precision but may point to real irregularities in the response that cannot be represented adequately by a second order polynomial model.

For problems involving 3 factors, one may construct a contour for x_1 and x_2 at each of the levels of x_3. For more than 3 factors the value of the geometry diminishes as a visual aid for interpretation, but the canonical analysis should be completed, nevertheless. In Davies (1956, p. 532), procedures are discussed for a response surface involving more than 2 factors. See Nielsen et al. (1973) for an example with 4 factors.

9.2. DESIGNS FOR LINEAR RESPONSE SURFACES

In the early stages of exploration of a population response surface, it may be expedient to rely on first order or linear surface models as approximations to the true surface in small regions. In such cases the classical multiple regression model,

$$Y = \beta_0 + \beta_1 x_1 + \beta_2 x_2 + \dots + \beta_k x_k + E \quad \text{or} \quad \mathbf{y} = \mathbf{X}\boldsymbol{\beta} + \mathbf{e} \qquad (9.18)$$

may be employed. The first column of $\mathbf{X}$ consists entirely of elements 1, as usual. The remaining k columns each contain the levels (sometimes replicated) of one factor and collectively make up the *design matrix*. The method of specification of the levels comprising the design matrix is the primary aspect of response surface design. Many designs are orthogonal because the design matrix specified leads to an $[\mathbf{X}'\mathbf{X}]$ matrix that is diagonal, i.e., the off-diagonal "covariance" elements are zero. The 2^k symmetrical factorial experiments described in Chap. 2 and the fractionally replicated designs for 2^k factorials of Sec. 6.3.4 are primary examples of orthogonal first order designs. Myers (1971, p. 109) shows that the design matrix minimizing the variances of the estimates of the parameters (β) is necessarily orthogonal--an excellent reason for selecting an orthogonal design.

The main-effect parameter estimates ($\hat{\beta}_1$, $\hat{\beta}_2$, ..., $\hat{\beta}_k$) remain unbiased if a complete factorial is used with a first order model even if second order terms ($\hat{\beta}_{ii}$, $\hat{\beta}_{ij}$) are not trivial. However, $\hat{\beta}_0$ would be biased by $\sum_{i=1}^{k} \beta_{ii}$. When fractional factorials are used, each main-effect estimate is biased by any nontrivial interaction(s) aliased with the main effect (Sec. 6.3.4). Unreplicated or fractionally replicated $2k$ designs provide no estimate of "pure" error. Such an estimate may be obtained by augmenting the design with several experimental units in the center of the design, i.e., at the design point (x_1, x_2, ..., x_k) = (0, 0, ..., 0) in coded units; the ordinary 2^k points involve x_i = +1 or -1 in coded units. Such a procedure not only provides an estimate of pure error but also permits one to test H: $\sum_{i=1}^{k} \beta_{ii} = 0$ (zero quadratic terms, collectively). Let the mean of the r_1 "ordinary" points be $\bar{y}_1$ and the mean of r_2 center points be $\bar{y}_2$. Then the test statistic

$$f = (\bar{y}_1 - \bar{y}_2)^2/[ms_E(1/r_1 + 1/r_2)] \quad (9.19)$$

may be compared with critical value $f_{\alpha,1,r_2-1}$ (App. A.5).

Orthogonal first order (*simplex*) designs may be useful if one wishes to estimate parameters with a minimum of experimental units ($n = k + 1$ for k factors). The design points of a simplex are the vertices of a k-dimensional, regular-sided figure. For $k = 2$, the 3 points are represented by an equilateral triangle; for $k = 3$, the 4 points form a tetrahedron; etc. For a given number of factors (k) more than one design matrix can be used, depending on the geometrical orientation of the simplex with respect to the coordinates (x_1, x_2, ..., x_k). Because of complex geometry for $k \geq 3$, it is easier to construct simplex designs by a matrix device. The $\mathbf{X}$ matrix of any first order design consists of a first column of n elements of 1 plus k columns of n elements each, describing design points for the k factors (the design matrix). For simplex designs, $n = k + 1$, so $\mathbf{X}$ is simply obtained by devising any orthogonal matrix of size $(k + 1) \times (k + 1)$ such that the elements of the first column are 1 and multiplying its elements by $\sqrt{n}$. An easy way to find an orthogonal matrix is to use the orthogonal polynomial coefficients of App. A.7.1, where $n = k + 1 = t$ of the table. Simplex designs ($\mathbf{X}$ matrices) of such construction, for k = 2, 3, and 4 factors, are given in Table 9.2. One can find simpler orthogonal matrices in some cases. For example, with $k = 3$, the second, third, and fourth columns could have elements (1, 1, -1, -1), (1, -1, 1, -1), and (1, -1, -1, 1), respectively. As Myers (p. 121) has pointed out, the biases of first order regression coefficients in the presence of a true quadratic response surface depend on which simplex orientation is used. When lack of fit of first order models is likely to be a problem and especially if simplex designs have been used, one should not rely heavily on the results of first order experiments. Simplex designs permit no assessment whatever of lack of fit.

9.3. DESIGNS FOR CURVILINEAR RESPONSE SURFACES

Polynomial functions do not necessarily represent the exact shape of many curvilinear population response surfaces, but they are useful approximations for the practical business of estimating optimal or near-optimal conditions.

Table 9.2. Simplex Designs for Linear Response Surfaces Involving k=2, 3, and 4 Factors

$$X=\sqrt{3}\begin{bmatrix} 1 & -1 & 1 \\ 1 & 0 & -2 \\ 1 & 1 & 1 \end{bmatrix} \quad X=2\begin{bmatrix} 1 & -3 & 1 & -1 \\ 1 & -1 & -1 & 3 \\ 1 & 1 & -1 & -3 \\ 1 & 3 & 1 & 1 \end{bmatrix} \quad X=\sqrt{5}\begin{bmatrix} 1 & -2 & 2 & -1 & 1 \\ 1 & -1 & -1 & 2 & -4 \\ 1 & 0 & -2 & 0 & 6 \\ 1 & 1 & -1 & -2 & -4 \\ 1 & 2 & 2 & 1 & 1 \end{bmatrix}$$

Three or more levels of each factor must be used to examine response with second (or higher) order models. Most response surface practice has centered around second order (3-level) experiments. Three basic categories of designs for such experiments are: (1) the orthogonal 3^k factorial system (Secs. 2.4, 6.4), (2) central composite designs (2^k or fractional 2^k factorials augmented with extra design points), and (3) rotatable designs (for which the variance of predicted response is a function of distance but not direction from the center of the design. Orthogonal 3^k factorials often require more design points than seem warranted for the number of parameters to be estimated. For example, k = 4 factors require 3^4 = 81 design points and the model

$$Y = \beta_0' + \sum_{i=1}^{4} \beta_i x_i + \sum_{i=1}^{4} \beta_{ii}(x_i^2 - \overline{x_i^2}) + \sum_{i<i'}\sum \beta_{ii'}(x_i x_{i'}) + E$$

has only 15 parameters (ignoring interactions among 3 or more factors simultaneously).

9.3.1. Central Composite Designs

Box and Wilson (1951) devised composite designs that require only $2^k + 2k + r_2$ design points (or fewer for fractions of 2^k), where $r_2 \geq 1$ experimental unit at the center of the design. With $r_2 = 1$, $2^k + 2k + 1 = 25$ units for k = 4 factors in contrast to $3^4 = 81$. Or for only k = 3 factors, $2^3 + 2(3) + 1$ = 15 units (instead of 3^3 = 27) to estimate 10 parameters. The first 2^k points (or fraction thereof) of a composite design are the usual combinations of factorial levels (coded -1, +1) in full or fractional 2^k designs. The next $2k$ points are called *axial* points because each involves coded level zero for all but one of the factors. Two nonzero levels are used for each factor, say $\pm m$. If $m \neq 1$, each factor actually is studied at 5 levels, but the number of combinations is far fewer than 5^k. The final point (replicated r_2 times) is at the center of the design $(x_1, x_2, \ldots, x_k) = (0, 0, \ldots, 0)$. For example, the X' matrix for a 3-factor composite design with m = 2 and r_2 = 1 is

$$X' = \begin{bmatrix} 1 & 1 & 1 & 1 & 1 & 1 & 1 & 1 & 1 & 1 & 1 & 1 & 1 & 1 & 1 \\ -1 & -1 & -1 & -1 & 1 & 1 & 1 & 1 & -2 & 2 & 0 & 0 & 0 & 0 & 0 \\ -1 & -1 & 1 & 1 & -1 & -1 & 1 & 1 & 0 & 0 & -2 & 2 & 0 & 0 & 0 \\ -1 & 1 & -1 & 1 & -1 & 1 & -1 & 1 & 0 & 0 & 0 & 0 & -2 & 2 & 0 \end{bmatrix}$$

where the last 3 elements of each column are the treatment combination $(x_1 x_2 x_3)$ for one experimental unit. For $k = 3$, $m = 1$, and $r_2 = 1$, the design can be represented as a cube with 8 vertices, 6 points in the centers of the faces, and 1 point in the center of the cube. In general, X'X is not diagonal and therefore not orthogonal except for certain values of m. For a full 2^k factorial, the $k \times k$ submatrix of X'X that pertains to $\beta_{11}, \beta_{22}, \ldots, \beta_{kk}$ (rows and columns) has off-diagonal elements all equal to

$$a_{ii} = [2^k(2k + r_2) - 4(2^k)(m^2) - 4m^4]/(2^k + 2k + r_2) \qquad (9.20)$$

For the full 2^k factorial plus $2k$ axial points and $r_2 = 1$ center point, the off-diagonal elements can be made to equal zero (and thus achieve orthogonality) by choosing $m = 1$, 1.216, 1.414, 1.596, 1.761, 1.910, 2.045 for $k = 2$, 3, ..., 8. Note that for $k = 2$, orthogonality is achieved for $m = 1$, implying the standard 3^2 factorial arrangement of 9 design points.

Orthogonality is not the only criterion for selecting a value of m, however. The efficiency of a design with respect to precision of estimated parameters and number of experimental units required also is important. In general, for central composite designs constructed around a full 2^k factorial, variances of the estimates are $V[\hat{\beta}'_0] = \sigma^2/n$; $V[\hat{\beta}_i] = \sigma^2/(2^k + 2m^2)$; $V[\hat{\beta}_{ii'}] = \sigma^2/2^k$ $(i \neq i')$; and

$$V[\hat{\beta}_{ii}] = \sigma^2/(2m^4\{1 + a_{ii}/[2m^4 + (k - 1)a_{ii}]\})$$

where a_{ii} is defined by (9.20) and is equal to zero for an orthogonal design. For fractions of 2^k designs, substitute the number of "ordinary" design points for 2^k in the variances of first order and interaction terms. All covariances are zero, excepting

$$\text{Cov}[\hat{\beta}_{ii}, \hat{\beta}_{i'i'}] = -a_{ii}\sigma^2/[2m^4(2m^4 + ka_{ii})] \qquad (i \neq i')$$

which is also zero for an orthogonal design. The orthogonal central composite designs are as or sometimes slightly more efficient than the standard 3^k factorial designs; but nonorthogonal composite designs often incur relative efficiencies lower than 1.0 (especially with respect to precision of interaction terms), the loss in efficiency being somewhat proportional to the deviation of

m from its orthogonal value. See Myers (1971, Chap. 9) for discussion of more complex criteria for choosing a design.

9.3.2. Rotatable Designs

Box and Hunter (1957) defined *rotatable* designs as those for which the variance of predicted response ($\hat{Y}$) is a function only of the distance (not direction) of a point of interest ($x_1, x_2, \ldots, x_k$) from the center of the design (0, 0, ..., 0). The quality of rotatability is important because an investigator may have little knowledge a priori of the manner in which the axes of the response contours will be oriented to the axes of the design (e.g., Fig. 9.1).

Although the concept of rotatability is general, most practical designs involve second order polynomial models. The property of rotatability depends on the coding system used to describe the levels of factors involved. A design that is rotatable under one coding may not be rotatable under a second coding for the same problem. For example, if 0.04%, 0.08%, and 0.12% of an additive are coded (-1, 0, +1), a rotatable design developed for the problem is based on one unit equal to 0.04%. If the investigator later decides to equate one unit to only 0.02%, the levels are (-2, 0, +2) and the initial design may not have the quality of rotatability. Any first order design (Sec. 9.2) that is orthogonal (has diagonal $\mathbf{X'X}$) also is rotatable. The necessary conditions for rotatability of a second order design are

1. All terms in $\mathbf{X'X}$ similar to $\sum_{j=1}^{n} x_{ij}^{g}$ or $\sum_{j=1}^{n} x_{ij}^{g} x_{i'j}^{h}$ ($i \neq i'$) such that g or h are odd powers ($g + h \leq 4$) must be zero.
2. $\sum_{j=1}^{n} x_{ij}^{4} = 3 \sum_{j=1}^{n} x_{ij}^{2} x_{i'j}^{2}$ ($i \neq i'$) for each $i = 1, 2, \ldots, k$ factors.

An example of a rotatable design is a central composite design with $m = \sqrt{2}$, where the 9 experimental units have factor levels

x_1:	-1	-1	+1	+1	$\sqrt{2}$	$-\sqrt{2}$	0	0	0
x_2:	-1	+1	-1	+1	0	0	$\sqrt{2}$	$-\sqrt{2}$	0

Note that $\Sigma x_1 = \Sigma x_2 = \Sigma x_1 x_2 = \Sigma x_1^2 x_2 = \Sigma x_1 x_2^2 = \Sigma x_1^3 = \Sigma x_2^3 = \Sigma x_1^3 x_2 = \Sigma x_1 x_2^3 = 0$, fulfilling condition 1, and $\Sigma x_1^4 = \Sigma x_2^4 = 12$, which is $3\Sigma x_1^2 x_2^2 = 3(4)$, thus meeting condition 2.

A critical feature of rotatable designs is the mixed second moment parameter $\lambda_4 = \sum_{j=1}^{n} x_{ij}^2 x_{i'j}^2$ ($i \neq i'$), where factor levels are scaled so that $\sum_{j=1}^{n} x_{ij} = 0$ and $\sum_{j=1}^{n} x_{ij}^2 = n$. Any design for which $\lambda_4 = 1$ is orthogonal. Box and Hunter described a useful class of rotatable designs, "uniform precision designs," that maintain equal variance of predicted response from the center of the design to any point no farther from the center than $\sqrt{\sum_{i=1}^{k} x_i^2} = 1$. The values of

λ_4 required for uniform precision with k = 2, 3, ..., 8 factors are 0.78, 0.84, 0.87, 0.89, 0.91, 0.92, 0.93, respectively. In practice, designs with λ_4 near these values often can be obtained by starting with a basic rotatable design and adding additional experimental units at the center of the design. Any rotatable design for which $\lambda_4 = k/(k + 2)$ will have singular matrix X'X (Sec. 4.1.2). This always occurs for rotatable designs consisting solely of points equidistant from the center (without center points). One or more experimental units may be added at the center point to avoid the problem. Myers (1971, p. 151) shows that a central composite design is rotatable if one selects $m = (\sqrt{2^k})^{1/2}$ for the axial points of the design. Or for half-replicates, use $m = (\sqrt{2^{k-1}})^{1/2}$. Basic rotatable central composite designs for k = 2, 3, ..., 8 factors with a single center point are shown in Table 9.3. The

Table 9.3. Basic Rotatable Central Composite Designs for 2 to 8 Factors (1 center point)

k Factors	2^k or 2^{k-1} Factorial Combinations	$2k+1$ Axillary Points	n Exp. Units	$\pm m$ Axillary Levels
2	4	5	9	1.414
3	8	7	15	1.682
4	16	9	25	2.000
5	32	11	43	2.378
5	16 (1/2 rep.)	11	27	2.000
6	64	13	77	2.828
6	32 (1/2 rep.)	13	45	2.378
7	128	15	143	3.364
7	64 (1/2 rep.)	15	79	2.828
8	256	17	273	4.000
8	128 (1/2 rep.)	17	145	3.364

basic designs can be changed to designs of near uniform precision by the addition of experimental units at the center point. The relationship between number of points and uniform precision is given by $\lambda_4 = n/(2^k + 4\sqrt{2^k} + 4)$, where $n = 2^k + 2k + r_2$ with r_2 units at the center (substitute 2^{k-1} for 2^k in a half-replicate). For full replicates the total number of units at the center for uniform precision is r_2 = 5, 6, 7, 10, 15 for k = 2, 3, ..., 6 factors, respectively. For half-replicates of k = 5, 6, 7, 8, use r_2 = 6, 9, 14, 20.

If one desires orthogonality ($\lambda_4 = 1$) rather than uniform precision, $r_2 = 8$, 9, 12, 17, 24 for $k = 2, 3, \ldots, 6$ factors (full replicate) or $r_2 = 10$, 15, 22, 33 for $k = 5, 6, 7, 8$ factors (half-replicate). For example, a design of near uniform precision for $k = 2$ factors consists of 13 experimental units,

$$x_1 = +1 \quad +1 \quad -1 \quad -1 \quad +\sqrt{2} \quad -\sqrt{2} \quad 0 \quad 0 \quad 0 \quad 0 \quad 0 \quad 0 \quad 0$$

$$x_2 = +1 \quad -1 \quad +1 \quad -1 \quad 0 \quad 0 \quad \sqrt{2} \quad -\sqrt{2} \quad 0 \quad 0 \quad 0 \quad 0 \quad 0$$

Addition of 3 more units at the center (0, 0) makes the design orthogonal.

Myers (Chap. 7) has discussed other rotatable designs, such as those with equiradial points, combinations of equiradial designs, and other geometric configurations.

9.4. BLOCK DESIGNS

For experiments in which the experimental units are not uniform (with respect to attributes likely to influence response in the primary variable) or in which the experimental conditions are not uniform (e.g., nuisance time trends), *blocking* (see Chap. 5) may be essential to reduce experimental error and avoid bias. With respect to response surface designs, the problem is to develop plans for blocking such that the effects of blocks are orthogonal to those of the treatment factors.

For first order designs (Sec. 9.2), the incomplete block designs for 2^k treatment combinations discussed in Sec. 6.3.1 are adequate. For analytical purposes, model (9.18) becomes

$$Y = \beta_0 + \sum_{i=1}^{k} \beta_i x_i + \sum_{m=1}^{b} \delta_m z_m + E \qquad (9.21)$$

where $x_i = +1$ or -1 (coded factor levels), $z_m = 1$ or 0 (depending on whether a particular experimental unit is in the mth block or not), and the δ_m represent effects of fixed blocks. All effects except defining contrast(s) are orthogonal to blocks. Ordinarily, model (9.21) is "centered" by substituting $z_m - \bar{z}_m$ for z_m and β_0' for β_0. The so-called X'X matrix is expanded to include as many extra rows and columns as there are blocks. Equal numbers of additional units at the center point may be added to each block to estimate pure error without loss of orthogonality.

Second order designs that permit orthogonal blocking make use of the model

$$Y = \beta_0' + \sum_{i=1}^{k} \beta_i x_i + \sum_{i=1}^{k} \beta_{ii}(x_i^2) + \sum_{i<i'}\sum \beta_{ii'} x_i x_{i'} + \sum_{m=1}^{b} \delta_m (z_m - \bar{z}_m) + E \qquad (9.22)$$

as in Sec. 9.3, with addition of terms for b blocks as for model (9.21). Three conditions must be met for a second order design to permit orthogonal blocking:

1. $\sum_j x_{ij} = 0$ for units in each block for each variable

2. $\sum_j x_{ij} x_{i'j} = 0$ for units in each block for each pair of variables ($i \neq i'$)

3. $\sum_{j=1}^{n} x_{ij}^2 z_{mj} = \sum_{j=1}^{n} x_{ij}^2 \bar{z}_m$ for each block for each variable

Conditions 1 and 2 imply that *each* block must contain factor combinations that fulfill first order orthogonality requirements. Condition 3 implies that the contribution of each block to $\sum_{j=1}^{n} x_{ij}^2$ must be proportional to the number of units in the block because $\bar{z}_m$ is the proportion of n units assigned to the mth block. The central composite designs (Sec. 9.3.1) can be modified easily to meet such conditions by adding an appropriate number of units at the center point to certain blocks. For example, consider a central composite design for $k = 3$ factors with $m = 1.633$:

$x_1 =$	+1	+1	-1	-1	+1	+1	-1	-1	+1.633	-1.633	0	0	0	0	0
$x_2 =$	+1	-1	+1	-1	+1	-1	+1	-1	0	0	+1.633	-1.633	0	0	0
$x_3 =$	+1	-1	-1	+1	-1	+1	+1	-1	0	0	0	0	+1.633	-1.633	0

The first four combinations make an orthogonal half-replicate of a 2^3 factorial, the next four make the other half-replicate and are orthogonal to the first four, and the last seven axial points are orthogonal to previous points. For condition 3, note that $\sum x_{ij}^2 = 4$, 4, and 16/3 for the 3 blocks of combinations (for any variable). Because 4 + 4 + 16/3 = 40/3, the relative proportions are 0.3, 0.3, and 0.4 for the 3 blocks. Those proportions when multiplied by the total number of experimental units ($n = 15$) do not produce integers. However, if one adds 2 units to each of the first two blocks and 1 unit to the third at the center point (0, 0, 0), the x statistics remain unchanged but the total number of units is 20; and 0.3, 0.3, and 0.4 of 20 units is 6, 6, and 8 units in blocks 1, 2, and 3, respectively, as specified by the additional points. Hence, the basic central composite design plus 5 additional center points permits orthogonal blocking in blocks of unequal size. A central composite design also may be divided into 2 blocks with 2^k factorial points plus r_1 center points in the first block and $2k$ axial points plus r_2 center points in the second block. For given numbers r_1 and r_2, orthogonal blocking is achieved by using axial treatment combinations in which the nonzero factor is at value

$$\pm m = \pm \sqrt{2^{k-1}(2k + r_2)/(2^k + r_1)} \tag{9.23}$$

For fractional factorials, substitute the number of factorial combinations for 2^k. Rotatability ($m = 2^{k/4}$) may not be obtainable in conjunction with orthogonal blocking; it is not obtainable for $k = 3$ factors. For $k = 4$, rotatability is achieved with $r_1 = 4$, $r_2 = 2$, and $m = 2.0$.

Table 9.4. Rotatable (or near-rotatable) Central Composite Designs in Blocks

	k=2	3	4	5	5	6	6	7	7
Factorial block(s),									
2^k or 2^{k-1} (half-rep.)	4	8	16	32	16	64	32	128	64
Number of blocks (b)	1	2	2	4	1	8	2	16	8
Number of added center points per block	3	2	2	2	6	1	4	1	1
Axial block									
$2k$	4	6	8	10	10	12	12	14	14
Number of added center points	3	2	2	4	1	6	2	11	4
Number of exp. units	14	20	30	54	33	90	54	169	80
Value of m									
For orthogonal blocking	1.414	1.633	2.000	2.366	2.000	2.828	2.366	3.364	2.828
For rotatability	1.414	1.682	2.000	2.378	2.000	2.828	2.378	3.333	2.828

Source: Box and Hunter, 1957, *Ann. Math. Stat.* 28:231. Reprinted by permission.

In Table 9.4 (Box and Hunter 1957) are listed several rotatable (or near rotatable) central composite designs that permit orthogonal blocking.

EXERCISES

9.1. Suppose an experimenter wishes to determine optimum combinations of methionine (x_1) and histidine (x_2) as supplements to fish protein concentrate to maximize the growth of dairy calves. Further, suppose that a 3^2 factorial experiment is carried out with each amino acid at levels 100%, 125%, and 150% of the amount in milk-replacer protein (coded -1, 0, +1), with y recorded as mean gain of 3 calves for each combination:

x_1 =	-1	0	+1	-1	0	+1	-1	0	+1
x_2 =	-1	-1	-1	0	0	0	+1	+1	+1
y =	71.7	79.2	80.1	75.2	81.5	79.1	76.3	80.2	75.8

(a) Estimate parameters of the full second order model,

$$Y = \beta_0 x_0 + \beta_1 x_1 + \beta_2 x_2 + \beta_{11} x_1^2 + \beta_{22} x_2^2 + \beta_{12} x_1 x_2 + E$$

(b) Use canonical analysis to compute the coordinate of the stationary point, the characteristic roots of the canonical form, and the transformation equations relating the original variables to the canonical variables.

9.2. Brown and Lassiter (1962) studied combinations of protein percentage (x_1) and protein:energy ratio (x_2) in calf starter diets. A 3^2 factorial experiment was performed with 14%, 16%, and 18% protein and 1:50, 1:48, 1:46 ratios (coded -1, 0, +1), with y recorded as mean feed efficiency (gain/estimated net energy) of 8 calves for each combination:

x_1 =	-1	0	+1	-1	0	+1	-1	0	+1
x_2 =	-1	-1	-1	0	0	0	+1	+1	+1
y =	0.42	0.28	0.31	0.42	0.40	0.38	0.53	0.41	0.36

(a) Estimate parameters of the full second order model.
(b) Compute the stationary point and characteristic roots of the canonical form; interpret the nature of the response surface.
(c) Compute transformation equations relating the original variables to the canonical variables.

9.3. Cragle et al. (1955) designed an experiment to estimate optimal conditions for storing bull semen to ensure maximum survival of spermatozoa. A central composite design with m = 2 was used. Each factor had coded levels (-2, -1, 0, +1, +2), where percent sodium citrate (x_1) was used (1.6, 2.3, 3.0, 3.7, 4.4), percent glycerol (x_2) was used (2, 5, 8, 11, 14), and equilibration time in hours (x_3) was (4, 10, 16, 22, 28), with y = % survival as an average of 2 ejaculates from each of 4 bulls:

x_1 =	-1	+1	-1	+1	-1	+1	-1	+1	-2	+2	0	0	0	0	0
x_2 =	-1	-1	+1	+1	-1	-1	+1	+1	0	0	-2	+2	0	0	0
x_3 =	-1	-1	-1	-1	+1	+1	+1	+1	0	0	0	0	-2	+2	0
y =	57	40	19	40	54	41	21	43	28	11	2	18	56	46	63

(a) Compute estimates for parameters of a full second order prediction equation (10 parameters, ignoring 3-factor interaction).
(b) Compute the stationary point, characteristic roots of the canonical form, and transformation equations relating the original variables to the canonical variables.

9.4. Compare the variance of estimated "pure" second order coefficients ($\hat{\beta}_{ii}$) from Ex. 9.3 with the variance expected if the axial points had been at levels that would make the design orthogonal. Assume σ^2 unknown but equal in both cases.

9.5. A scientist, who has 36 volunteer subjects available for a response surface study of the effects of either 4 or 5 factors on blood pressure, wishes to use only levels 0, ±1, and ±2 for each factor adopted. Find two alternative, orthogonal, rotatable, central composite designs that will permit the use of exactly 36 subjects and either 4 or 5 factors.

9.6. A scientist, who intends to design a central composite experiment for 5 factors, has 68 animals, which should be divided into only 2 blocks of equal size (34 operations per day). At what levels should the axial points be set to achieve orthogonal blocking?

Classification of Experimental Designs

TYPES OF DESIGNS

- I. Completely randomized designs
  - A. Nuisance variables controlled or unknown (Chap. 2)
    - 1. One-way classification
    - 2. Hierarchical (nested) classification
    - 3. Cross classification (factorial)
      - a. General designs
      - b. Response surface designs (Chap. 9)
    - 4. Mixed (partially hierarchical) classification
  - B. Nuisance variables used as covariates
    - 1. One covariate (Chap. 3)
    - 2. Multiple covariates (Chap. 4)
- II. Designs with restricted randomization (nuisance variables used for blocking)
  - A. Designs having subjects within blocks
    - 1. One blocking restriction
      - a. Complete blocks (number of subjects per block = number of treatments) (Chap. 5)
        - (I) Nonfactorial experiments
        - (II) Factorial experiments
          - (A) General designs
          - (B) Response surface designs (Chap. 9)
      - b. Incomplete blocks (number of subjects per block < number of treatments) (Chap. 6)
        - (I) Nonfactorial experiments
          - (A) One-dimensional designs
            - (1) Balanced designs (BIB)
            - (2) Partially balanced designs (PBIB)
              - (a) Designs with two associate classes
              - (b) Designs with more than two associate classes
          - (B) Multidimensional designs (incomplete blocks grouped into complete blocks by replications)
            - (1) Nonlattice designs
            - (2) Balanced lattices
            - (3) Partially balanced lattices
              - (a) Double (simple) lattices
              - (b) Triple lattices
              - (c) Rectangular lattices
              - (d) Cubic lattices

- (II) Factorial experiments
    - (A) Symmetrical designs (all factors having same number of levels)
        - (1) Replicated designs
        - (2) Unreplicated designs
        - (3) Fractionally replicated designs
            - (a) Regular fractions
                - (i) Resolution III
                - (ii) Resolution IV
                - (iii) Resolution V
                - (iv) Resolution VI or higher
            - (b) Irregular fractions
                - (i) Orthogonal plans
                - (ii) Nonorthogonal plans
            - (c) Sequences of fractions
    - (B) Asymmetrical designs (mixtures of levels)
        - (1) Unreplicated designs
        - (2) Fractionally replicated designs
            - (a) Orthogonal plans
            - (b) Nonorthogonal plans
    - (C) Split-plot designs (subject = subplot) (Chap. 8)

2. More than one blocking restriction (Chap. 7)
    - a. Complete blocks (Latin squares)
        - (I) Nonfactorial experiments
        - (II) Factorial experiments
    - b. Incomplete blocks (squares)
        - (I) Nonfactorial experiments
            - (A) One-dimensional designs
                - (1) Balanced Youden squares
                - (2) Partially balanced incomplete Latin squares
            - (B) Multidimensional designs
                - (1) Balanced lattice squares
                - (2) Semibalanced and incomplete lattice squares
        - (II) Factorial experiments (double confounding)

B. Designs having repeated measurement of subjects (subject = block) (Chap. 8)

1. Designs with more than one treatment per subject (balanced changeover designs)
    - a. Crossover designs
    - b. Latin square crossover designs
    - c. Reversal (switchback) designs
2. Designs to study trends following treatment of subjects (nonrandom sampling of subjects in time or space)
    - a. Complete block designs (subject = complete block)
    - b. Split-plot designs (subject = whole plot or incomplete block)

C. Sequential designs (Chap. 6)

KEY FOR SELECTION FROM MAJOR CLASSES OF DESIGNS

I. One measurement (or several *random* samples) per subject (per variable of interest)

Number of Nuisance Variables	Number of Treatment Factors	
	1	2 or more
0	CRD(2.1-2.3)*	CRD(2.3-2.4, 9.2-9.3)
1 Covariate:	CRD-AOC(3.2-3.6)	CRD-AOC(3.7)
Complete block:	RCBD(5.1-5.7)	RCBD(5.8)
Incomplete block:	BIBD(6.2)	RIBD(6.3-6.5, 9.4); SPD(8.3)
2 Complete block:	LSD(7.2-7.5)	LSD(7.6)
Incomplete block:	YSD, ILSD, LCSD(7.8)	QLSD(7.8.4)
1 Cov, complete:	RCBD-AOC(3.7, 5.1-5.7)	RCBD-AOC(3.7, 5.8)
1 Cov, incomplete:	BIBD-AOC(3.7, 6.2)	RIBD-AOC(3.7, 6.3-6.5)
2 Cov:	CRD-AOC(4.3-4.4)	CRD-AOC(4.3-4.4)

*Sections of the text.

II. Nonrandom repeated measurement on each subject (same variable)

Grouping of Subjects	Number of Treatments Applied to Each Subject	
	1 (trends)	2 or more (crossovers)
No	CBD(8.2.1)*	RCBD, BCBCD(8.1.2)
Yes	SPD(8.2.2-8.2.3)	LSCD, etc.(8.1.3-8.1.6)

*Sections of the text.

We shall not cease from exploration,
and the end of all our exploring
will be to arrive where we started
and know the place for the first time.

—T. S. ELIOT

BIBLIOGRAPHY

Pertinent sections of the text are listed at the end of each reference.

Addelman, S. 1962. Orthogonal main effect plans for asymmetrical factorial experiments. *Technometrics* 4:21-46. (6.5.2.2)

———. 1963. Techniques for constructing fractional replicate plans. *J. Am. Stat. Assoc.* 58:45-71. (6.5.2.2)

———. 1969. Sequences of two-level fractional factorial plans. *Technometrics* 11:477-509, (6.3.4.5, 6.3.4.6)

———. 1975. Systematic Latin squares. Mimeographed. Paper read at annual meeting of Am. Stat. Assoc., Aug. 1975, Atlanta, Ga. (7.1)

Alling, D. W. 1966. Closed sequential tests for binomial probabilities. *Biometrika* 53:73-84. (6.3.4.7)

Anderson, R. L. 1946. Missing-plot techniques. *Biom. Bull.* 2:41-47. (8.3)

Anderson, V. L. 1970. Restriction errors for linear models (an aid to develop models for designed experiments). *Biometrics* 26:255-68. (5.2)

Anscombe, F. J., and J. W. Tukey. 1963. The examination and analysis of residuals. *Technometrics* 5:141-60. (5.3.3)

Armitage, P. 1960. *Sequential medical trials.* Oxford: Oxford Univ. Press. (6.3.4.7)

———. 1971. *Statistical methods in medical research.* New York: Wiley. (6.3.4.7, Ex. 8.2)

Aroian, L. A., and D. Öksoy. 1970. New exact sequential methods for the comparison of two medical treatments. *Biometrics* 26:598. (6.3.4.7)

Atkinson, G. F. 1966. Designs for sequences of treatments with carry-over effects. *Biometrics* 22:292-309. (8.1.6)

Babcock, S. M. 1889. Variations in yield and quality of milk. Wis. Agr. Exp. Stn. Ann. Rep. 6:42-67. (8.1.6)

Bacharach, A. L., M. R. A. Chance, and T. R. Middleton. 1940. The biological assay of testicular diffusing factor. *Biochem. J.* 34:1464-71. (Ex. 8.6)

Balaam, L. N. 1968. A two-period design with t^2 experimental units. *Biometrics* 24:61-74. (8.1.6)

Berenblut, I. I. 1964. Change-over designs with complete balance for first residual effects. *Biometrics* 20:707-12. (8.1.6)

———. 1967. The analysis of change-over designs with complete balance for first residual effects. *Biometrics* 23:578-80. (8.1.6)

———. 1968. Change-over designs balanced for the linear component of first residual effects. *Biometrika* 55:297-303. (8.1.6)

———. 1970. Treatment sequences balanced for linear component of residual effects. *Biometrics* 26:154-56. (8.1.6)

Berenblut, I. I., and G. I. Webb. 1974. Experimental design in the presence of autocorrelated errors. *Biometrika* 61:427-37. (8.1.4)

Bliss, C. I. 1967. *Statistics in biology*. Vol. 1. New York: McGraw-Hill. (Ex. 8.7)

Bose, R. C. 1939. On the construction of balanced incomplete block designs. *Ann. Eugen.* 9:353-400. (6.2.1.2)

Bose, R. C., W. H. Clatworthy, and S. S. Shrikhande. 1954. Tables of partially balanced designs with two associate classes. N.C. Agr. Stn. Tech. Bull. 107. (6.2.2.1, 7.8.2)

Box, G. E. P. 1949. A general distribution theory for a class of likelihood criteria. *Biometrika* 36:317-46. (8.4.1, 8.4.2)

———. 1950. Problems in the analysis of growth and wear curves. *Biometrics* 6:362-89. (8.4.1, 8.4.2)

———. 1954a. Some theorems on quadratic forms applied in the study of analysis of variance problems. I. Effect of inequality of variance in the one-way classification. *Ann. Math. Stat.* 25:290-302. (8.2.2)

———. 1954b. Some theorems on quadratic forms applied in the study of analysis of variance problems. II. Effects of inequality of variance and of correlation of errors in the two-way classification. *Ann. Math. Stat.* 25:484-98. (5.3.3, 8.2.1)

Box, G. E. P., and J. S. Hunter. 1954. A confidence region for the solution of a set of simultaneous equations with an application to experimental design. *Biometrika* 41:190-99. (9.1.2)

———. 1957. Multifactor experimental designs for exploring response surfaces. *Ann. Math. Stat.* 28:195-241. (9.1.1, 9.3.2, 9.4)

Box, G. E. P., and K. B. Wilson. 1951. On the statistical attainment of optimum conditions. *J. R. Stat. Soc.* (B)13:1-45. (6.3.4.5, Chap. 9, 9.3.1)

Brandt, A. E. 1938. Tests of significance in reversal or switchback trials. Iowa Agr. Exp. Stn. Res. Bull. 234. (8.1.5)

Breslow, N. 1970. Sequential modification of the UMP test for binomial probabilities. *J. Am. Stat. Assoc.* 65:639-48. (6.3.4.7)

Bross, I. 1952. Sequential medical plans. *Biometrics* 8:188-205. (6.3.4.7)

Brown, L. D., and C. A. Lassiter. 1962. Protein-energy ratios for dairy calves. *J. Dairy Sci.* 45:1353-56. (Ex. 9.2)

Brownlee, K. A., C. S. Delves, M. Dorman, C. A. Green, E. Greenfell, J. D. A. Johnson, and N. Smith. 1948. The biological assay of streptomycin by a modified cylinder plate method. *J. Gen. Microbiol.* 2:40-53. (Ex. 7.7)

Carmichael, R. D. 1956. *Introduction to the theory of groups of finite order*. New York: Dover. (6.5.1.2)

Chen, K. K., C. I. Bliss, and E. B. Robbins. 1942. The digitalis-like principles of *Calotropis* compared with other cardiac substances. *J. Pharm. Exp. Ther.* 74:223-34. (Ex. 7.6)

Choi, S. C. 1968. Truncated sequential designs for clinical trials based on Markov chains. *Biometrics* 24:159-68. (6.3.4.7)

Chow, Y. S., H. Robbins, and B. J. D. Siegmund. 1971. *Great expectations: The theory of optimal stopping*. Boston: Houghton-Mifflin. (6.3.4.7)

Ciminera, J. L., and E. K. Wolfe. 1953. An example of the use of extended cross-over designs in the comparison of NPH insulin mixtures. *Biometrics* 9:431-46. (8.1.5)

Clatworthy, W. H., J. M. Cameron, and J. A. Speckman. 1973. Tables of two-associate-class partially balanced designs. Washington, D.C.: U.S. Nat. Bur. Stand. (6.2.2.1)

Cochran, W. G., and G. M. Cox. 1957. *Experimental designs*. New York: Wiley. (6.2.1.1, 6.2.1.2, 6.2.2.2, 6.2.2.4, 6.2.3.1, 6.2.3.3, 6.2.3.4,

6.3.4.3, 6.5.2.1, 7.8.1, 7.8.2, 7.8.3, 7.8.4)

Cole, J. W. L., and J. E. Grizzle. 1966. Application of multivariate analysis of variance to repeated measurements experiments. *Biometrics* 22: 810-28. (8.4.2)

Connor, W. S., and S. Young. 1961. Fractional factorial designs for experiments with factors at two and three levels. U.S. Nat. Bur. Stand. Appl. Math. Ser. 58. (6.5.2.2)

Connor, W. S., and M. Zelen. 1959. Fractional factorial experiment designs for factors at three levels. U.S. Nat. Bur. Stand. Appl. Math. Ser. 54. (6.4.3.3)

Cornell, J. A. 1974. More on extended complete block designs. *Biometrics* 30:179-86. (5.3.2)

Cox, C. P. 1958a. Experiments with two treatments per experimental unit in the presence of an individual covariate. *Biometrics* 14:499-512. (8.1.5)

———. 1958b. The analysis of Latin square designs with individual curvatures in one direction. *J. R. Stat. Soc.* (B)20:193-204. (8.1.5)

Cox, D. R. 1958. *Planning of experiments.* New York: Wiley. (8.3)

Cragle, R. G., R. M. Myers, R. K. Waugh, J. S. Hunter, and R. L. Anderson. 1955. The effects of various levels of sodium citrate, glycerol, and equilibrium time on survival of bovine spermatozoa after storage at -79°C. *J. Dairy Sci.* 38:508-14. (Ex. 9.3)

Daniel, C. 1956. Fractional replication in industrial research. *Proc. Third Berkeley Symp. Math. Stat. Probab.* 5:87-98. (6.3.4.6)

———. 1958. On varying one factor at a time. *Biometrics* 14:430-31. (6.3.4.6)

———. 1962. Sequences of fractional replicates in the 2^{p-q} series. *J. Am. Stat. Assoc.* 57:403-29. (6.3.4.6)

Daniel, C., and F. Wilcoxon. 1966. Factorial 2^{p-q} plans robust against linear and quadratic trends. *Technometrics* 8:259-78. (6.3.4.6)

Das, M. N., and P. W. Rao. 1967. Construction and analysis of some new series of confounded asymmetrical factorial designs. *Biometrics* 23:813-22. (6.5.2.1)

David, H. A. 1952. Upper 5 and 1% points of the maximum *F*-ratio. *Biometrika* 39:422-24. (A.6.1)

Davies, O. L., ed. 1956. *The design and analysis of industrial experiments*, 2d ed. London: Oliver & Boyd. (9.1.2)

Davies, O. L., and W. A. Hay. 1950. The construction and uses of fractional factorial designs in industrial research. *Biometrics* 6:233-49. (6.3.4.6)

DeGray, R. J. 1970. The sequential Youden square. *Technometrics* 12: 147-52. (7.8.1)

Denes, J., and A. D. Keedwell. 1974. *Latin squares and their applications.* New York: Academic Press. (Chap. 7)

Draper, N. R., and D. M. Stoneman. 1964. Estimating missing values in unreplicated two-level factorial and fractional factorial designs. *Biometrics* 20:443-58. (6.3.4.2)

———. 1968a. Response surface designs for factors at two and three levels. *Technometrics* 10:177-92. (6.5.2.2)

———. 1968b. Factor changes and linear trends in eight-run two-level factorial designs. *Technometrics* 10:301-12. (6.3.4.6)

Eaton, M. L. 1969. Some remarks on Scheffé's solution to the Behrens-Fisher problem. *J. Am. Stat. Assoc.* 64:1318-22. (8.4.2)

Elfring, G. L., and J. R. Schultz. 1973. Group sequential designs for clinical trials. *Biometrics* 29:471-78. (6.3.4.7)

Federer, W. T. 1955. *Experimental design*. New York: Macmillan. (6.2.2.2, 6.2.3.3, 6.2.3.4, 6.4.1.2, 7.7)

Federer, W. T., and G. F. Atkinson. 1964. Tied double change-over designs. *Biometrics* 20:168-81. (8.1.6)

Finney, D. J. 1945. The fractional replication of factorial arrangements. *Ann. Eugen.* 12:291-301. (6.3.4)

Fisher, R. A. 1925. Theory of statistical estimation. *Proc. Camb. Philos. Soc.* 22:700-725. (5.6)

———. 1935. *The design of experiments*. London: Oliver & Boyd. (5.6)

———. 1942. The theory of confounding in factorial experiments in relation to the theory of groups. *Ann. Eugen.* 11:341-53. (6.3.1.3)

Fisher, R. A., and F. Yates. 1963. *Statistical tables for biological, agricultural, and medical research*, 6th ed. London: Oliver & Boyd. (7.7)

Geisser, S., and S. W. Greenhouse. 1958. An extension of Box's results on the use of the F distribution in multivariate analysis. *Ann. Math. Stat.* 29:885-91. (8.2.1, 8.2.2)

Ghosh, B. K. 1970. *Sequential tests of statistical hypotheses*. Reading, Mass.: Addison-Wesley. (6.3.4.7)

Gill, J. L., and H. D. Hafs. 1971. Analysis of repeated measurements of animals. *J. Anim. Sci.* 33:331-36. (8.2.2)

Gill, J. L., and W. T. Magee. 1976. Balanced two-period changeover designs for several treatments. *J. Anim. Sci.* 42:775-77. (8.1.6)

Gravett, I. M. 1971. A two-thirds replicate of a 3^4 factorial. *Biometrics* 27:1043-56. (6.4.3.1)

Graybill, F. A. 1961. *An introduction to linear statistical models*. Vol. 1. New York: McGraw-Hill. (6.2.1.2, 6.2.1.3)

Grizzle, J. E. 1965. The two-period change-over design and its use in clinical trials. *Biometrics* 21:467-80. (8.1.4)

Gupta, S. D., and M. D. Perlman. 1974. Power of the noncentral F-test: Effect of additional variables on Hotelling's T^2-test. *J. Am. Stat. Assoc.* 69:174-80. (8.4.1)

Han, C. 1969. Testing the homogeneity of variances in a two-way classification. *Biometrics* 25:153-58. (5.3.3)

Harshbarger, B. 1947. Preliminary report on the rectangular lattices. *Biometrics* 2:115-19. (6.2.3.4)

———. 1949. Triple rectangular lattices. *Biometrics* 5:1-13. (6.2.3.4)

Harter, H. L. 1961. On the analysis of split-plot experiments. *Biometrics* 17:144-49. (8.3)

———. 1969. *Order statistics and their use in testing and estimation*. Vol. 1. *Tests based on range and studentized range of samples from a normal population*. Aerosp. Res. Lab., USAF. Washington, D.C.: USGPO. (5.3.3)

Haseman, J. K., and D. W. Gaylor. 1973. An algorithm for non-iterative estimation of multiple missing values for crossed classifications. *Technometrics* 15:631-36. (5.4, 7.2.3, 8.3)

Hill, W. J., and W. G. Hunter. 1966. A review of response surface methodology: A literature survey. *Technometrics* 8:571-90. (Chap. 9)

Hodson, A. Z. 1956. Vitamin B_6 in sterilized milk and other milk products. *J. Agr. Food Chem.* 4:876-81. (Ex. 5.7)

Hoke, A. T. 1974. Economical second-order designs based on irregular fractions of the 3^n factorial. *Technometrics* 16:375-84. (6.4.3.3)

Holms, A. G., and J. N. Berettoni. 1969. Chain-pooling ANOVA for two-level factorial replication-free experiments. *Technometrics* 11:725-46. (6.3.1.2)

Huntsberger, D. V., and P. E. Leaverton. 1970. *Statistical inference in the biomedical sciences.* Boston: Allyn & Bacon. (Ex. 8.1)

Huynh, H., and L. S. Feldt. 1970. Conditions under which mean square ratios in repeated measurements designs have exact *F*-distributions. *J. Am. Stat. Assoc.* 65:1582-89. (8.2.1)

Jackson, J. E. 1960. Bibliography on sequential analysis. *J. Am. Stat. Assoc.* 55:561-624. (6.3.4.7)

Jensen, D. R. 1972. Some simultaneous multivariate procedures using Hotelling's T^2 statistic. *Biometrics* 28:39-54. (8.4.2)

John, P. W. M. 1961. Three-quarter replicates of 2^4 and 2^5 designs. *Biometrics* 17:319-21. (6.3.4.5)

———. 1962. Three-quarter replicates of 2^n designs. *Biometrics* 18: 172-84. (6.3.4.5)

———. 1963. Extended complete block designs. *Austral. J. Stat.* 5: 147. (5.1)

———. 1970. A fraction of the 4^3 factorial design for estimating main effects. *Biometrics* 26:157-60. (6.5.1.2)

Johnson, D. E., and F. A. Graybill. 1972. An analysis of a two-way model with interaction and no replication. *J. Am. Stat. Assoc.* 67:862-68. (5.3.2)

Johnson, W. D., and J. E. Grizzle. 1971. Analysis of change-over designs when the response is ordinal. *Biometrics* 27:768. (8.1.3)

Kastenbaum, M. A., D. G. Hoel, and K. O. Bowman. 1970. Sample size requirements: Randomized block designs. *Biometrika* 57:573-77. (5.5)

Kempthorne, O. 1952. *Design and analysis of experiments.* New York: Wiley. (5.6, 6.2.2.1, 6.2.2.2, 6.2.3.2, 6.2.3.3, 6.2.3.4, 6.5.1.2, 6.5.2.1, 7.1, 7.8.3, 7.8.4)

Kendall, M. G., and A. Stuart. 1966. *The advanced theory of statistics.* Vol. 3. *Design and analysis of surveys and experiments.* London: Griffin. (6.2.2.2)

———. 1967. *The advanced theory of statistics.* Vol. 2. *Inference and relationship,* 2d ed. London: Griffin. (5.6)

Kiefer, J. 1958. On the non-randomized optimality and randomized non-optimality of symmetrical designs. *Ann. Math. Stat.* 29:675-99. (7.8.2)

Kirk, R. E. 1968. *Experimental design: Procedures for the behavioral sciences.* Belmont, Calif.: Wadsworth. (5.4, 6.4.1.2, 6.5.2.1, 7.8.1, 8.3)

Kirton, H. C. 1966. "Non-additivity in the analysis of variance." Master's thesis. Univ. Aberdeen, Aberdeen, Scotland. (5.3.2)

———. 1971. Detection of non-additivity in the analysis of variance. Mimeographed. Paper read at Australas. Stat. Conf., Aug. 1971, Sydney. (5.3.2)

Koch, G. G. 1970. The use of non-parametric methods in the statistical analysis of a complex split-plot experiment. *Biometrics* 26:105-28. (8.4.2)

———. 1972. The use of non-parametric methods in the statistical analysis of the two-period change-over design. *Biometrics* 28:577-84. (8.1.3)

Koch, G. G., and P. K. Sen. 1968. Some aspects of the statistical analysis of the mixed model. *Biometrics* 24:27-48. (Ex. 8.13)

Kosswig, W. 1970. Die Aufspaltung der Wechselwirkung, 1. Ordnung aus $n \times k$ Tafeln. *Biom. Z.* 12:137-51. (8.2.2)

———. 1971. Die Analyse der Wechselwirkung Tier × Zeit in ZeitWirkungkurven mit Korrelierten Beobachtungen. Personal communication. Katzenburgweg 2a, Bonn, W. Germany. (8.2.2)

Kramer, C. Y. 1972. *A first course in methods of multivariate analysis.* Va. Polytech. Inst. & State Univ., Dept. Stat., Blacksburg. (8.4.2)

Lucas, H. L. 1956. Switch-back trials for more than two treatments. *J. Dairy Sci.* 39:146-54. (8.1.5)

———. 1957. Extra-period Latin-square change-over designs. *J. Dairy Sci.* 40:225-39. (8.1.6)

McNee, R. C., and P. P. Crump. 1962. Analysis of repeated measurements with disproportionate samples for treatments in a two-way classification. USAF Sch. Aerosp. Med. Rep. SAM-TDR-62-149. (8.2.2)

Mandel, J. 1954. Chain block designs with two-way elimination of heterogeneity. *Biometrics* 10:251-72. (7.8.2)

———. 1961. Non-additivity in two-way analysis of variance. *J. Am. Stat. Assoc.* 56:878-88. (5.3.2)

Margolin, B. H. 1968. Orthogonal main-effect 2^n3^m designs and two-factor interaction aliasing. *Technometrics* 10:559-73. (6.5.2.2)

———. 1969a. Results on factorial designs of resolution IV for the 2^n and 2^n3^m series. *Technometrics* 11:431-44. (6.3.4.5)

———. 1969b. Resolution IV fractional factorial designs. *J. R. Stat. Soc.* (B)31:514-23. (6.5.2.2)

———. 1969c. Orthogonal main-effect plans permitting estimation of all two-factor interactions for the 2^n3^m series of designs. *Technometrics* 11: 747-62. (6.3.4.5, 6.5.2.2)

Mason, J. M., and K. Hinkelmann. 1971. Change-over designs for testing different treatment factors at several levels. *Biometrics* 27:430-35. (8.1.6)

Morrison, D. F. 1967. *Multivariate statistical methods.* New York: McGraw-Hill. (8.4.1)

Myers, R. H. 1971. *Response surface methodology.* Boston: Allyn & Bacon. (Chap. 9, 9.1, 9.1.2, 9.2, 9.3.1, 9.3.2)

Nielsen, M. A., S. T. Coulter, C. V. Morr, and J. R. Rosenau. 1973. Four factor response surface experimental design for evaluating the role of processing variable upon protein denaturation in heated whey systems. *J. Dairy Sci.* 56:76-83. (9.1.2)

Ogilvie, J. C. 1963. A simple method for the elimination of individual trends in the analysis of balanced sets of Latin squares. *Biometrics* 19:264-72. (8.1.5)

Paape, M. J., and H. A. Tucker. 1969. Mammary nucleic acid, hydroxyproline, and hexosamine of pregnant rats during lactation and post-lactational involution. *J. Dairy Sci.* 52:380-85. (8.2.2)

Patterson, H. D. 1952. The construction of balanced designs for experiments involving sequences of treatments. *Biometrika* 39:32-48. (8.1.6)

———. 1970. Nonadditivity in change-over designs for a quantitative factor at four levels. *Biometrika* 57:537-49. (8.1.6)

Patterson, H. D., and H. L. Lucas. 1959. Extra-period change-over designs. *Biometrics* 15:116-32. (8.1.6)

———. 1962. Change-over designs. N.C. Agr. Exp. Stn. Tech. Bull. 147. (8.1.6)

Peng, K. C. 1967. *The design and analysis of scientific experiments.* Reading, Mass.: Addison-Wesley. (5.6)

Pimentel-Gomes, F. 1970. An extension of the method of joint analysis of experiments in complete randomized blocks. *Biometrics* 26:332-36. (5.9)

Raktoe, B. L. 1969. Combining elements from distinct finite fields in mixed factorials. *Ann. Math. Stat.* 40:498-504. (6.5.2.1)

———. 1970. Generalized combining of elements from finite fields. *Ann. Math. Stat.* 41:1763-67. (6.5.2.1)

Raktoe, B. L., and W. T. Federer. 1972. Construction of confounded mixed factorial and mixed lattice designs. *Austral. J. Stat.* 14:25-36. (6.5.2.1)

Robinson, J. 1967. Incomplete split-plot designs. *Biometrics* 23:793-802. (8.3)

Rojas, B. A. 1973. On Tukey's test of additivity. *Biometrics* 29:45-52. (5.3.2, 7.2.2)

Ruiz, F., and E. Seiden. 1974. On construction of some families of generalized Youden designs. *Ann. Stat.* 2:503-19. (7.8.2)

Schefler, W. C. 1969. *Statistics for the biological sciences.* Reading, Mass.: Addison-Wesley. (Ex. 5.3)

Sclove, S. L. 1972. On missing value estimation in experimental design models. *Am. Stat.* 26(Apr.):25-26. (5.4)

Scott, C. C., and K. K. Chen. 1944. Comparison of the action of 1-ethyl-theobromine and caffeine in animals and man. *J. Pharm. Exp. Ther.* 82:89-97. (8.1.2)

Searle, S. R. 1966. *Matrix algebra for biologists.* New York: Wiley. (9.1.2)

Shukla, G. K. 1972. An invariant test for the homogeneity of variances in a two-way classification. *Biometrics* 28:1063-72. (5.3.3)

Snee, R. D. 1972. On the analysis of response curve data. *Technometrics* 14:47-62. (8.2.2)

Spicer, C. C. 1962. Some new closed sequential designs for clinical trials. *Biometrics* 18:203-11. (6.3.4.7)

Stefansky, W. 1972. Rejecting outliers in factorial designs. *Technometrics* 14:469-80. (5.3.3)

Stewart, I. 1972. *Galois theory.* New York: Wiley. (6.5.1.2)

Storm, L. E. 1962. Nested analysis of variance: Review of methods. *Metrika* 5:158-83. (8.3)

Taylor, J. 1948. Errors of treatment comparisons when observations are missing. *Nature* (London) 162:262-63. (5.4)

———. 1967. The value of orthogonal polynomials in the analysis of change-over trials with dairy cows. *Biometrics* 23:297-312. (8.1.5)

Taylor, W. B., and P. J. Armstrong. 1953. The efficiency of some experimental designs used in dairy husbandry experiments. *J. Agr. Sci.* 43:407-12. (8.1.5)

Trail, S. M., and D. L. Weeks. 1973. Extended complete block designs generated by BIBD. *Biometrics* 29:565-78. (5.1)

Tukey, J. W. 1949. One degree of freedom for non-additivity. *Biometrics* 5:232-42. (5.3.2, 7.2.2)

———. 1955. Answer to a query on non-additivity in Latin squares. *Biometrics* 11:111-13. (7.2.2)

U.S. National Bureau of Standards. 1957. Fractional factorial experimental designs for factors at two levels. Washington, D.C.: Appl. Math. Ser. (6.3.4.3)

Wald, A. 1947. *Sequential analysis.* New York: Wiley. (6.3.4.7)

Webb, S. R. 1965. Design, testing, and estimation in complex experimentation. 1. Expansible and contractible factorial designs and the application of linear programming to combinatorial problems. Aerosp. Res. Lab. Rep. 65-116 (part 1). (6.3.4, 6.3.4.5, 6.3.4.6)

———. 1968. Nonorthogonal designs of even resolution. *Technometrics* 10:535-50. (6.3.4.5)

———. 1971. Small incomplete factorial experimental designs for two- and three-level factors. *Technometrics* 13:243-56. (6.3.4.5, 6.4.3.3, 6.5.2.2)

Westlake, W. J. 1974. The use of balanced incomplete block designs in bioavailability trials. *Biometrics* 30:319-28. (8.2.3)

Wetherill, G. B. 1966. *Sequential methods in statistics*. New York: Wiley. (6.3.4.7)

White, D., and R. A. Hultquist. 1965. Construction of confounding plans for mixed factorial designs. *Ann. Math. Stat.* 36:1256-71. (6.5.2.1)

Williams, E. J. 1949. Experimental designs balanced for the estimation of residual effects of treatments. *Austral. J. Sci. Res.* (A)2:149-68. (8.1.4, 8.1.6)

———. 1950. Experimental designs balanced for pairs of residual effects. *Austral. J. Sci. Res.* (A)3:351-63. (8.1.4, 8.1.6)

———. 1970. Comparing means of correlated variates. *Biometrika* 57: 459-61. (8.4.1)

Winer, B. J. 1962. *Statistical principles in experimental design*. New York: McGraw-Hill. (6.2.3.3, 6.5.2.1, Ex. 6.7, 7.8.1, 7.8.3, 8.1.3, 8.2.3)

Worthley, R., and K. S. Banerjee. 1974. A general approach to confounding plans in mixed factorial experiments when the number of levels of a factor is any positive integer. *Ann. Stat.* 2:579-85. (6.5.2.1)

Yao, Y. 1965. An approximate degrees of freedom solution to the multivariate Behrens-Fisher problem. *Biometrika* 52:139-47. (8.4.2)

Yates, F. 1937. The design and analysis of factorial experiments. Imp. Bur. Soil Sci. Tech. Comm. 35. (7.8.4)

———. 1939. The recovery of inter-block information in variety trials arranged in three-dimensional lattices. *Ann. Eugen.* 9:136-56. (6.2.3.1)

———. 1940. The recovery of inter-block information in balanced incomplete block designs. *Ann. Eugen.* 10:317-25. (6.2.1.3, 6.2.1.4, 6.2.3.1)

Youden, W. J. 1937. Use of incomplete block replications in estimating tobacco-mosaic virus. *Contr. Boyce Thompson Inst.* 9:41-48. (7.8.1)

———. 1940. Experimental designs to increase accuracy of greenhouse studies. *Contr. Boyce Thompson Inst.* 11:219-28. (7.8.1)

Youden, W. J., and W. S. Connor. 1953. The chain block design. *Biometrics* 9:127-40. (6.2.2.4)

Young, D. M., and R. G. Romans. 1948. Assay of insulin with one blood sample per rabbit per test day. *Biometrics* 4:122-31. (Ex. 8.5)

Zelen, M. 1957. The analysis of incomplete block designs. *J. Am. Stat. Assoc.* 52:204-17. (6.2.1.3)

INDEX